Grassland Dynamics
An Ecosystem Simulation Model

‘Science is built of facts the way a house is built of bricks; but an accumulation of facts is no more science than a pile of bricks is a house’
Henri Poincaré (1902), *La Science et l’Hypothèse*

‘Except in so far as they contribute to theories and generalisations the scientific mind is not interested in facts’
Vincent Wigglesworth (1957), *Parasitology* 47, 1–15

‘The most important effects of anything are unintended and unforeseen’
After *Dipak Nandy* (1975), *Six Laws of Social Policy*

GRASSLAND DYNAMICS

An Ecosystem Simulation Model

John H.M. Thornley

CEH Senior Research Fellow
NERC Institute of Terrestrial Ecology, Edinburgh

Honorary Research Associate
Institute of Grassland and Environmental Research, Aberystwyth

Honorary Principal Research Fellow
University of Reading

Formerly of the Grassland Research Institute, Hurley, UK

CAB INTERNATIONAL

CAB INTERNATIONAL
Wallingford
Oxon OX10 8DE
UK

Tel: +44 (0)1491 832111
Fax: +44 (0)1491 833508
E-mail: cabi@cabi.org

CAB INTERNATIONAL
198 Madison Avenue
New York, NY 10016–4314
USA

Tel: +1 212 726 6490
Fax: +1 212 686 7993
E-mail: cabi@cabi.org

A catalogue record for this book is available from the British Library, London, UK.

Library of Congress Cataloging-in-Publication Data
Thornley, J.H.M.
Grassland dynamics: an ecosystem simulation model / John H.M. Thornley
p. cm.
Includes index.
ISBN 0–85199–227–7 (alk. paper)
1. Grassland ecology—Computer simulation. I. Title.
QH87.7.T48 1998
577.4′01′13—dc21 97–34918
CIP

ISBN 0 85199 227 7

Typeset in Times by Columns Design Ltd, Reading
Printed and bound in the UK at the University Press, Cambridge

Contents

Preface

This book gives an account of a grassland ecosystem simulator which has been developed over some 20 years. The simulator is known as the 'Hurley Pasture Model', as its initial development took place at the Grassland Research Institute at Hurley in southern Britain. Recently, further development of the model has taken place at the NERC Institute of Terrestrial Ecology in Edinburgh, in conjunction with the BBSRC Institute of Grassland & Environmental Research in North Wyke, Devon, and some European Community grassland research programmes directed by Professor M.B. Jones of Trinity College, Dublin.

Most ecosystem modellers are too busy developing their models to take the time to write them up fully. Thus, except for some excellent work from the Dutch modelling groups in Wageningen, it is virtually impossible to find a full account of a plant ecosystem model in one place. This puts anybody starting out in this type of research, or seeking a broad view of what is involved, in a difficult position. I aim to provide at least a partial remedy to this impasse, by giving a complete account of how one plant ecosystem model works.

Arguably the most important aspect of a scientific model is that it should be as easy as possible to find out and understand the assumptions and mechanisms that are represented in the simulator. Only then is it possible to evaluate the predictions of the model in relation to its content, and in relation to observation. A transparent model allows the reader to decide if its content is appropriate to his objectives. Also, its transparency should permit easy modification or extension.

The task therefore is to explain how each and every process represented in the model works in detail, but also emphasizing the structure of the model, so that it can be seen how any process or the structure itself can be changed to meet particular needs. We also wish to present the material so that the reader who wishes to know how, say, maintenance respiration, or mineralization, is calculated, can, virtually, go straight there. When he gets there, he will find a mathematical statement and a biological explanation; he can also inspect the corresponding item in the computer program which may be downloaded.

The preferred language for describing quantitative physiology and ecology is mathematics, which, compared with current computer languages, is succinct, relatively timeless, and sufficient. The mathematical formalism required for deterministic dynamic models has been around for a long time, and for the present purpose, requires only basic algebra and calculus. I have observed how various models of my own and of former colleagues, written in mathematics, have been coded by others in the language of their choice and, indeed, some of our models have been used as typical examples of dynamic simulation by commercial software producers. However, some researchers prefer to read computer code, which also provides a means of verifying the mathematical statements. For these reasons the program is freely available. The program is written in ACSL (Chapter 10), a Fortran-based simulation language the use of which allows quite readable code to be written.

The model is of the type that is called mechanistic, or sometimes, process-based. It is my experience that the more mechanistic a model is, the more transparent it becomes, so long as de Wit's (1970) dictum of joining only two levels in the organizational hierarchy is approximately followed. Models have the reputation of being opaque; this may partly have arisen because many biological modellers are in unfamiliar territory when dealing with the concepts of models and mathematics. Ecosystem models are 'big', but are only complex in that they consist of a large number of simple elements. It is a worthwhile challenge to make these models more accessible than they are at present.

Another reason why 'big' models are found difficult is that they cover a wide range of disciplines, and few scientists are secure in all the relevant topics. The chapter headings of this book describe its contents best: plant, animal, soil, water, and environment. We attempt to follow Einstein's suggestion, that 'Things should be made as simple as possible, but no simpler'. This can of course be a matter of endless dispute. It seems to be difficult for a specialist, in, say, some area of crop physiology or soil science, to construct a balanced ecoystem model in which his own specialist area is simplified to an extent he would not accept when in his specialist role.

The content of a model determines the questions that can be asked of the model and the uses to which it can be put. Our primary objective is to understand the main features of grassland dynamics. To do this the model needs to be able to predict the principal consensus results of experiments and observations on grassland with moderate realism. Whether or not this is achieved can be difficult to decide in an area where experiment and observation give results that are often highly variable, and sometimes conflicting. Since a mechanistic model provides considerable scope for parameter twiddling, we aim only at satisfactory qualitative behaviour. 'Tuning', 'calibrating', or less kindly, 'fudging' the model, is left to others with more sharply defined requirements, and to those who feel it is reasonable to treat a mechanistic model as a gigantic regression equation, which I do not. De Wit pioneered crop modelling, and his intellect, experience and insight give his contributions and views unique weight and authority. In de Wit (1970), when discussing the possible use of a mechanistic model to fit data, he stated that this is 'a disastrous way of working ... Since there are many parameters and many equations involved this is not difficult. ... the technique reduces into the most cumbersome and subjective technique of curve fitting that can be imagined.' Also, see Hopkins and Leipold (1996) on the dangers of adjusting parameter values.

'Validation' is another problem area where confusion abounds. It needs to be said first that the validity of a model is a quantitative concept, ranging say between zero and unity, and second, that validity is not an attribute of a model itself. Validity depends on a relationship between a model and a set of objectives. A model may be valid for managing grassland in southern England (say), but fail in the wetter west of Britain, and also fail to contribute to our understanding of how process is related to outcome. It is self-evident that a model is valid as seen by the creator of the model, who set the modelling objectives. A model is generally less valid to other scientists, who may legitimately evaluate the model with their own objectives in mind. An important question is: is it reasonable for referees and editors to evaluate a model from the standpoint of their own private objectives, rather than the author's objectives? I think not, but the opposite is all too common.

An ecosystem model can be an absorbing and satisfying creation. A scientist who is applying such a model to a particular problem will find that his model evolves continually, and any account is out of date before it can be completed. Attempting to bolt on the latest modelling bells and whistles inevitably gives a model a short shelf-life. In trying to avoid this fate, I have taken a view as to what is essential in a plant ecosystem model, what is undesirable, and what is a dispensable bell or whistle. Thus this book is to be viewed as an example of how to construct a plant ecosystem model, tie the parts together, and describe the whole so that it is understandable, usable, and can be modified by others. It is not, and neither is it intended to be, the last word in grassland simulators. A plant ecosystem model is, inevitably, always an unfinished story.

There are no long lists of references, graphs or tables of data supporting the assumptions made in the model, or detailed accounts of other models. In a plant ecosystem model, filling all these *lacunae* authoritatively would require writing several more books. With modern bibliographic tools the reader can easily find up-to-date material in particular areas. However, I refer to the work that I have found useful, knowing well that there are many equally relevant references out there which have not been cited.

This book will be of use to students of ecology and agriculture as a 'casebook' example of a plant ecosystem model. It will also be of value to grassland agronomists and modellers, crop physiologists, and plant ecologists. There is no doubt in my mind that computer simulation, now providing undreamt-of possibilities for integration of knowledge, understanding of complexity and quantitative prediction, will be a growth area for many years. I hope that this book contributes to this process.

A bird's eye view of the Hurley Pasture Model is quickly obtained by looking at Fig. 2.1 (overview), and Figs 3.1, 4.1, 5.1, 6.1 for the plant, animal, soil and litter, and water submodels, and Tables 7.2, 7.3 (environment) and Table 7.4 (management). The four submodel figures can be used to locate the mathematical equations used for any process.

Chapters 3 to 7 can be read in any order. An ecosystem is truly an assembly of simultaneous processes even though our computers still cannot easily treat it as such. Chapters 1 and 2 contain introductory modelling material which may be omitted (but not Fig. 2.1). Readers more interested in what the model can and does predict rather than what is in it (an attitude which I do not wholely commend) may browse through the figures in Chapters 8 and 9.

I am much indebted to numerous plant and crop physiologists, soil and animal scientists, modellers, computing staff and visitors at the Institute of Terrestrial Ecology and at the late Grassland Research Institute, and also to researchers elsewhere whose time and hospitality I have enjoyed. The difficulties of listing the names of all who have helped me are so great that I do not attempt the task. Assistance with particular points is recognized in the text. Melvin Cannell has encouraged and supported me in many ways, and we have had numerous helpful discussions on various topics. Tony Parsons has been indispensable as my principal mentor in grassland physiology, a role he has filled with patience, tolerance and great kindness, although he does not agree with everything that I have said or done. Jim France, Deena Mobbs, Tony Parsons, Marcel Riedo and David Scholefield have read the whole or parts of an early draft and I am most grateful for their labour and their suggestions.

The program can be obtained by ftp to budbase.nbu.ac.uk/pub/tree/Book where a read.me file, the program source file (pasture.csl) and the ACSL command files and other files (Chapter 10) used to generate the figures in the book can be found. The most recent version of the Hurley Pasture Model is pasture.csl at budbase.nbu.ac.uk/pub/tree/Pasture.

Any comments on the material in the book will be welcome.

John Thornley
6 Makins Road, Henley-on-Thames, Oxon RG9 1PP, UK
e-mail: john.thornley@unixmail.nerc-bush.ac.uk
June 1997

1 Dynamic Models

1.1 Introduction

In this chapter the rationale for constructing biological models and the relevant elements of dynamic modelling theory are outlined.

D'Arcy Wentworth Thompson stated in his classic text *On Growth and Form* that: 'numerical precision is the very soul of science, and its attainment affords the best, perhaps the only criterion of the truth of theories and the correctness of experiments' (Thompson, 1942). To match experiment, observation and theory quantitatively requires that mathematics is used to formulate the theory. Mathematics does not itself provide the hypotheses or scientific laws that make up our theories, but it does provide a succinct and unambiguous language for expressing our scientific ideas, and it allows numerical solutions or predictions to be derived which are a consequence of those scientific ideas.

One reason for building a model is that we want to make predictions. A model of some sort, not necessarily mathematical, is always needed in order to make predictions, and this is often a goal in scientific work.

A second reason for building a model is that we want to understand how something works. 'Understanding' is taken to mean defining the relationships between the responses of a system and the mechanisms that are assumed to operate within the system. Although it may be possible to make accurate predictions about a system without an understanding as to how the system works, having an understanding cannot lessen our predictive capabilities, and will certainly extend the scope and possibly the accuracy of any predictions. Indeed, having an understanding, even if imperfect, will suggest ways in which we might intervene in the operations of the system and perhaps manipulate the outcomes.

A third reason for building models is that the model provides a method of studying complexity, and this is particularly relevant to plant ecosystems. A great deal is known separately about plants, grazing animals, soils and the environment. What does this knowledge add up to? Models provide a means of integrating many subsystem components and investigating how the subsystems interact. Without a model, this task seems to be beyond our unaided mental powers.

The plant ecosystem simulator described here is a *dynamic, deterministic, mechanistic* mathematical model. *Dynamic* refers to the model describing the time-course of various variables, such as soil organic matter, or plant water potential. *Deterministic* means that the model makes predictions of variables (e.g. dry matter per unit area) without any associated probability distribution; there are no random number generators in the model. *Mechanistic* implies that the model is based on assumptions about the mechanisms of processes represented in the model which are thought to be important in the system. Any predictions that the model makes can be traced back to what these processes are doing.

Use of the term *mechanistic* depends upon where you are standing. Our model is mechanistic at the highest level, the whole-system level. The plant submodel (Fig. 3.1) is also mechanistic, although the component of the plant submodel describing leaf photosynthesis is empirical or phenomenological (Fig. 3.3). A mechanistic model of leaf photosynthesis would be based on biochemical and other assumptions about within-leaf processes. De Wit (1970) has argued cogently against putting too many different levels of description into a single model. Too much complexity makes the model unduly large and cumbersome to work with, and it becomes difficult to tease out the cause–effect connections that are an essential part of understanding. The decision not to include a mechanistic leaf photosynthesis model is deliberate; it is feasible because there are phenomenological leaf-level models of photosynthesis which work very well as descriptors of leaf-level data. It is possible that the complexity of leaf photosynthetic response might be such that one is forced to a more mechanistic representation of leaf photosynthesis in order to capture that complexity, or that a specific modelling objective is to represent certain biochemical photosynthetic parameters explicitly within the ecosystem model, which then requires a more mechanistic representation of photosynthesis.

Our objectives in building a grassland ecosystem simulator are twofold: first, to gain a better understanding of grassland ecosystems, which means being able to successfully predict the consensus results of observation and experiment to date at all levels within the model; second, to provide a model that can be used to answer interesting 'what-if' questions. These what-if questions may involve scenarios that have not so far been experienced – possibly concerned with management, modified biological organisms, environment or climate regimes, which could be applicable in the future. Both these objectives require a mechanistic model: first, because mechanism is an essential part of understanding; second, because what-if questions may require changes within the system (e.g. modifying the intake parameters of grazing animals).

Also important is that the grassland ecosystem model can be run for hundreds or even thousands of years, providing indications of possible long-term consequences for such ecosystems. Experiments are inevitably short term, and the information which is thereby gleaned may be of limited value, or even positively misleading, when considering long-term results (e.g. Section 8.10). There seems to be little alternative to using models for such studies.

In many recent ecological studies, chaos, and its existence or non-existence, is much discussed (Berryman and Milstein, 1989; see May, 1976, and also Baker and

Gollub, 1996, for a general account of chaos theory). In some work, the mathematical treatment proposed may challenge credibility (e.g. Tilman and Wedlin, 1991), and elsewhere the difficult problem of distinguishing chaos from noise is discussed (e.g. Pool, 1989; Cazelles and Ferriere, 1992). In population models many different outputs, from stability to oscillations to chaos, can be obtained just by changing parameters (e.g. May, 1976). The reader may wonder if chaos plays a role in the grassland ecosystem model presented here. First we emphasize that there are no random processes within the model, nor is there noise in the driving variables (management and environment). Stochastic processes are avoided so that the process of pinning down cause and effect, prediction and mechanism, which is difficult enough in a large model, is as simple as possible. However, we found that the model, although pleasingly stable and robust under a wide range of conditions, is capable of sustained biennial oscillations with single parameter changes (Section 8.14). Oscillatory behaviour can lead to chaos, and simplifications and extensions of the present model have been used to study the potentially complex dynamics of grassland systems (Thornley *et al.*, 1995; Schwinning and Parsons, 1996a, b). To summarize, chaos is not important in the simulations presented here, although it could become relevant if the model is extended or modified.

Further discussion of the role of modelling in biology and of different modelling approaches can be found France and Thornley (1984) and Thornley and Johnson (1990).

1.2 State Variables

These are the variables that define *completely* the state of the system. The state variables usually denote quantities of substance or pool sizes, such as dry mass of a plant, or mass of water in the soil. The more state variables there are in a simulator, the more are the possible points of contact between the simulator and the part of the real world which is being simulated. The first thing to ask about a model is: what are the state variables? This tells a lot of about the scope of the model.

The state variables are independent of each other. For example, a simple plant model may have shoot dry mass, M_{sh} (kg), and root dry mass, M_{rt} (kg), as state variables, and the model may work in such a way that the shoot:root ratio (M_{sh}/M_{rt}) can vary. M_{sh} and M_{rt} change independently with the passage of time so that shoot:root ratio may alter. However, there could be some degree of correlation between growth in shoot and root dry masses as both depend for instance on photosynthetic supply. Total plant mass, M, is not a state variable, but is a dependent variable which can be calculated from the state variables by $M = M_{sh} + M_{rt}$. In a model, it is helpful to distinguish clearly between state variables and other dependent variables. Each state variable x will have a differential equation $dx/dt = \dots$, where t is the time variable. A dependent variable y is not represented by a differential equation, but is calculated from the state variables, usually algebraically by means of a statement y = function of the state variables x_i (i is an integer index). The time variable t is called the independent variable. When considering the output of a model it is often useful to calculate the values of dependent variables such as total dry mass and shoot:root ratio (M_{sh}/M_{rt}).

Variables such as air temperature (T_{air}) and relative humidity ($r_{el\text{-}hum}$) are called driving variables. Such environmental variables may depend on time *t*; they are supplied from outside the system and they 'drive' the system. At the current instant of time, the state of the system does not depend on T_{air} or $r_{el\text{-}hum}$. However, how the state variables change over the next time interval does depend on T_{air} and $r_{el\text{-}hum}$, so the future values of the state variables and the state of the system are affected by the driving variables. The plant ecosystem modeller assumes that T_{air} and $r_{el\text{-}hum}$ are particular functions of time *t*, or take certain values every day or every few minutes. The values ascribed to T_{air} and $r_{el\text{-}hum}$ do not depend on the plant ecosystem model. Driving variables such as T_{air}, $r_{el\text{-}hum}$, fertilizer application rates or animal stocking densities can be categorized as environmental or management variables (Chapter 7).

1.3 Parameters

Ideally, the values of all parameters in a mechanistic model would be determined by experiments at the level of the mechanism, or sometimes, by well-established theory. In practice, this is only possible for some parameters [e.g. animal intake parameters, Eqn (4.2a); leaf photosynthesis parameters, Eqns (3.2s), (3.2t); growth efficiency, Eqn (3.3e)]. Many parameters are only approximately known from experiment (e.g. many water relations parameters; Chapter 6), and there are always parameters which are not directly known from experiment, and must in some way be estimated [e.g. within-plant transport parameters and growth rate constants, Eqns (3.3b), (3.4b), (3.3c)]. Thus there is a great deal of scope for legitimately adjusting parameters.

Some biologists are much concerned with the particular values assigned to parameters, and may have difficulty accepting the more cavalier approach favoured by many modellers with backgrounds in applied mathematics or physics, who will pluck approximate order-of-magnitude parameter values seemingly 'out of the air', and be content to use those values, at least initially. The biologist, often with a traditional training in statistics which can give a particular perception of how the world operates, may be unsympathetic to the view that the precise numbers do not matter much. The modeller, perhaps well-schooled in the attitudes of the theoretical physicist, may take the position that the important aspects of the problem are model structure, qualitative behaviour, obtaining a qualitative understanding and a moderate degree of quantitative agreement. More precisely 'fixing-up' the numbers can easily be done later. It is often, but not always, a trivial matter.

De Wit (1970, p. 21) remarked:

> It will often be found that the results obtained with experimenting with the model and the actual system do not agree. In that case, the model may be adjusted such that a better agreement is obtained. Since there are many parameters and many equations involved this is not difficult. However, it is a disastrous way of working because the model degenerates from an explanatory [*mechanistic*] model in [*into*] a demonstrative [*empirical*] model which cannot be used anymore for extrapolation, and the technique reduces into the most cumbersome and subjective technique of curve fitting that can be imagined.

The italicized words in this quotation are mine. Hopkins and Leipold (1996) discuss some of the problems which can arise when adjusting the parameters of mechanistic models to obtain better agreement with particular data sets. They investigate the problem using a quite complex biochemical model with 19 rate constants, but their findings are equally applicable to ecosystem models. We list some of their conclusions.

1. When one rate constant is seriously in error, adjustment of a different rate constant gave the greatest improvement in model fit.
2. When a contaminant was present in the experiments, the effects could be hidden by adjusting the rate constants.
3. If an incorrect mechanism is assumed, the effects can sometimes be hidden by rate constant adjustment.
4. Parameters adjusted for one set of experimental conditions may give a poorer performance for other conditions than unadjusted parameters.

Realistically, some parameter adjustment or model tuning is generally required and is defensible. However, this must not be carried out too readily, or before the causes of model failure are clearly understood. It is very easy to degrade a model with *ad hoc* fixes, so that further rational development of the model becames impossible. This pragmatic procedure has been applied to the present model.

1.4 Differential Equations

The crux of a dynamic simulator is how the state variables change with time, t. This information is contained in a differential equation. For a model with a single state variable this can be written

$$\frac{dx}{dt} = f(x, P, E), \tag{1.1a}$$

where x is the state variable, f stands for some function of the quantities within the brackets, P denotes a set of parameters (or constants), and E stands for environmental and management variables. The environment E may be constant, or be a function of time t. The rate at which a state variable is changing depends only on the state of the system (i.e. the current values of the state variables), together with the values of parameters and environmental quantities.

Imagine that we are pouring water into a bucket of cross-sectional area A (m^2) at a rate of I_W (m^3 water day^{-1}), but the water is leaking out of the bucket at a rate proportional to the height of the water in the bucket, h (m). The volume leakage rate is $k_L h$ (m^3 day^{-1}) where k_L (m^2 day^{-1}) is a rate constant. The volume V (m^3) of water in the bucket is related to height h by $V = Ah$. The differential equation for volume V is

$$\begin{aligned} &\frac{dV}{dt} = I_W - k_L h = I_W - \left(\frac{k_L}{A}\right)V; \\ &I_W = 1\ \text{m}^3\ \text{water day}^{-1}, \qquad k_L = 2\ \text{m}^2\ \text{day}^{-1}, \qquad A = 1\ \text{m}^2, \\ &V(t=0) = 0\ \text{m}^3\ \text{water}. \end{aligned} \tag{1.1b}$$

The second and third lines of Eqns (1.1b) give all the numerical information needed so that the problem can be solved. $V(t = 0)$ is the volume of water present at time $t = 0$ day, assumed to be 0 m^3; I_W is the input rate of water; k_L (m^3 day^{-1}) is the leakage rate, and is a parameter of the type often called a rate constant with a 'per unit of time' in its dimensions; A (m^2) is the area of the cylindrical bucket. The ACSL source program for Eqns (1.1b) is given in Fig. 10.1.

The solution to Eqns (1.1b) is drawn in Fig. 1.1; it is similar to a growth equation known as the monomolecular equation (see Fig. 3.2, p. 77, of Thornley and Johnson, 1990). The ACSL command file to generate Fig. 1.1 is given in Fig. 10.2.

In the differential equation of Eqns (1.1b) there is an input term (I_W) and an output term ($k_L h$). In a biological model, any substance, such as plant soluble carbohydrate, animal protein, soil mineral nitrogen, is both produced and lost, often in a variety of ways. Eqns (1.1a) and (1.1b) can be written as

$$\frac{dx}{dt} = \text{inputs} - \text{outputs}, \tag{1.1c}$$

where the inputs are the processes that produce or transfer in the substance x (for soil mineral nitrogen, perhaps mineralization, atmospheric nitrogen deposition), and the outputs are the processes that utilize or transfer out x (e.g. nitrogen uptake by the plant, leaching). All the significant processes in the biology presented here are either chemical conversions, or transport.

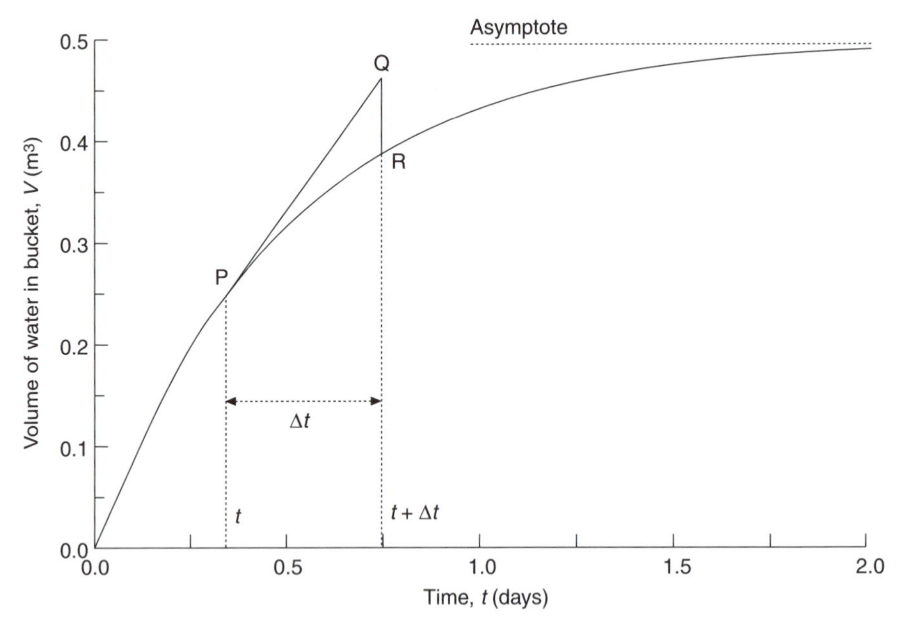

Fig. 1.1. Water flowing into a leaky bucket. The volume of water in the bucket V is shown as a function of time t. The curve is the solution of Eqn (1.1b) with the given parameters, and with an integration interval $\Delta t = 0.02$ day. Numerical integration by Euler's method as used throughout this book is also illustrated here. PQ is the tangent to the true solution at point P. Q is the point predicted by Euler's method [Eqn (1.2b)] with the Δt shown. The error is RQ.

In an ecosystem model there will be a differential equation of the type of Eqn (1.1c) for each state variable. In practice the outputs from one state variable will often be inputs to other state variables. For example, for carbon (C) substrate, the output flux of C substrate from the shoot pool to translocation is an input to the root C substrate pool (Fig. 3.1); the C output from the soil litter lignin pool is split between respiration to the atmosphere and inputs to two soil organic matter (SOM) pools (Fig. 5.1).

1.5 Numerical Integration

We rewrite Eqn (1.1a) in a simpler form, omitting explicit mention of parameters P and environment E:

$$\frac{dx}{dt} = f(x,t), \tag{1.2a}$$

where the rate of change of x depends on the current value of x and the current value of the time variable t. Eqn (1.2a) is a differential equation with t varying continuously.

To obtain a numerical solution we write it as a difference equation which is only approximately true:

$$x(t + \Delta t) \approx x(t) + \Delta x \qquad \text{where } \Delta x = f(x, t)\Delta t, \tag{1.2b}$$

where Δt is a finite time increment. Verbally, we can say that the value of x at time $t + \Delta t$ is approximately equal to its value at time t plus the time increment (Δt) $\times$ the rate at which x is changing at time t [$f(x, t)$].

The approximation is shown in Fig. 1.1. P is the point (x, t) writing $x \equiv V$, the volume of water in the bucket. The slope of PQ is $f(x, t)$. Q is the approximation given by Eqn (1.2b) and is the point ($x + \Delta x$, $t + \Delta t$). R is the true solution and is the point [$x(t + \Delta t)$, $t + \Delta t$]. It is not a bad approximation provided that the slope at P is not too different from the slope at R. The error with the Δt shown in Fig. 1.1 is QR, and is called the truncation error. This method of numerical integration is used throughout this book, and is known as Euler's method.

If the integration interval Δt is too large, then errors may be unacceptably big, or the model may become unstable and go off the rails. A good rule of thumb is that halving the interval Δt should not change the results over the total time period simulated by more than about 2%. The integration interval Δt should be small enough so that the proportional change in x in time Δt, $\Delta x/x$, is less than 0.5; note that doubling Δt doubles Δx [Eqn (1.2b)]. Ecosystem models are large and execution time is important. A practical procedure is to increase Δt until the model becomes unstable, and then go back to the last stable value of Δt. Halving this stable value gives a check as to whether the integration is giving acceptable accuracy.

If the integration interval Δt is too small, then Δx may be so small compared with x in Eqn (1.2b) that it is lost in the rounding errors which depend on how many bytes are used to store numbers in the computer. We use double precision (16 significant decimal digits) for the grassland model, so that this is not a problem. It is a problem using single precision (7 significant decimal digits).

In the grassland ecosystem model, the basic unit of the time variable (t) is 1 day. An integration interval of Δt = 1/64 = 0.015625 day (or 22.5 min) has been found to work satisfactorily for most situations. Occasionally the program fails because of numerical instability, e.g. at high temperatures where rate constants are higher, or at high atmospheric CO_2 concentrations when some small pools are changing very rapidly. It is then necessary to reduce Δt to 1/128 day. It is advantageous to use an integration interval which can be stored exactly in binary, e.g. $\Delta t = 2^{-6}$ day.

1.6 Kinetics

A mechanistic model represents processes that occur at some rate. The rate is calculated by an assumed *kinetic* expression. Most of the kinetic expressions used in the grassland model are one of three types described below. Suppose there is a pool of substance x, that the amount of substance in the pool is x (kg), and that the rate at which material leaves the pool is O_x (kg day^{-1}).

1.6.1 Linear or mass-action kinetics

The output O_x is proportional to the pool size x, with

$$O_x = kx. \tag{1.3a}$$

Here k (day^{-1}) is a constant. Many simple chemical reactions are of this type.

1.6.2 Michaelis–Menten kinetics

This gives a diminishing-returns response with an asymptote, and has the form

$$O_x = v_m \frac{x}{K + x}. \tag{1.3b}$$

v_m (kg day^{-1}) is the maximum rate of the reaction, and K (kg) gives the size of the pool when the rate is half-maximal: when $x = K$, $O_x = \frac{1}{2}v_m$. This equation occurs in biochemistry when the conversion of a substrate x to some product is catalysed by an enzyme in a simple manner. The curve with $\xi = 0$ in Fig. 3.3 is of this type.

1.6.3 Sigmoidal kinetics

For low values of x the output rate O_x increases faster than x; then diminishing returns occur and an asymptote is reached. The equation is

$$O_x = v_s \frac{x^q}{K_s^q + x^q}. \tag{1.3c}$$

v_s (kg day^{-1}) is the maximum rate, K_s (kg) gives the size of the pool when the rate is half-maximal: when $x = K_s$, $O_x = \frac{1}{2}v_s$, and q determines the sharpness of the response. This equation occurs in some more complex enzyme reactions. It is sigmoidal only for $q > 1$. If $q = 1$, then Eqn (1.3c) is the same as Eqn (1.3b). The curve used for animal intake [Fig. 4.2, Eqn (4.2a)] is a sigmoidal response curve with $q = 3$.

Further discussion of kinetic equations is given by Thornley and Johnson (1990, pp. 51–63).

1.7 Summary of the Modelling Process

The procedure used for describing the parts of the grassland ecosystem model is as follows:

1. Define the state variables, x_i and give the units of the x_i.
2. For each state variable write down a differential equation, in the form dx_i/dt = inputs − outputs, and give the starting or initial value of each x_i at time $t = 0$.
3. For each input or output, give an expression saying how that input or output is calculated from the current value of the state variables, with any modifying factors (temperature, water potential), environmental quantitites, and give values for all required parameters and constants, with units.

Some variables are passed between the submodels. These may be flux variables, which correspond to flows of material (C or N) between the submodels, or they may be so-called property variables. Property variables are variables that exist instantaneously, such as a C substrate concentration, or root water potential. For example, when N uptake by the root is calculated in the plant submodel, the concentrations of the soil mineral N pools (N_{amm}, N_{nit}) are required to do this, so that the values of the soil state variables N_{amm} and N_{nit} are supplied to the plant submodel. The N uptake rate by the root is a flux with dimensions of amount of N time^{-1}, which is an output from the soil submodel and an input to the plant submodel.

The influence of the soil submodel on the plant submodel can be by transfer of material such as C or N; or by transfer of information represented in a property variable which affects the rate of some process. All state variables are property variables. Variables such as total leaf area index (which is the sum of the leaf area indices in the different age categories) or N substrate concentration in the root (derived by dividing the amount of N substrate by the size of the root), are also property variables but are not state variables with a differential equation. Property variables including state variables exist at a point in time. Flux variables have dimensions of time^{-1}; their instantaneous existence is notional because a flux cannot be measured by an observation at a single point in time. The measurement must extend over a time interval. This distinction between states and rates, or properties and fluxes, is helpful when constructing dynamic models.

Overview of the Pasture Model

2

2.1 Introduction

The Hurley Pasture Model has a history going back some 20 years and, like most such models, it has evolved by small increments and by large steps. While the early work was essential for the development of the model, it is not helpful to give a historical account. The paper by Thornley and Verberne (1989) gives references to earlier work.

A plant ecosystem model is a substantial construction, and such models can only remain relevant and reflect current understanding if there is a continuing long-term effort to document, develop and apply the models. Such a commitment is commonly made for meteorological models used for weather forecasting, but is not made for plant ecosystem models. Perhaps this is because the science content of meteorological models is far less contentious than that of plant ecosystem models. Whether or not this is efficient is arguable. Fragmented short-term programmes as are currently in vogue in many countries may not provide value, at least, in the plant ecosystem modelling area.

A model is only as good as its parts. Progress is nearly always made at the submodel level. The research programme which, for many years, was associated with the Hurley Pasture Model, has investigated many submodels in depth, for instance: respiration, allocation, leaf and canopy photosynthesis, temperature responses in biology, intake and diet preference, mixed swards, spatial processes, and so on. An in-depth modelling exercise at the submodel level is invaluable for providing a simplified approach, more appropriate to an ecosystem-level model, whose limits of accuracy and biological realism are well-understood. However, not all components of the model have been evaluated with equal depth and rigour, and therefore some unevenness in treatment is inevitably present.

We do not attempt a general comparison of the Hurley Pasture Model with other models. Satisfactory and reasonably up-to-date descriptions of plant ecosystem models are hard to find. Objectives of modellers inevitably and rightly differ. Specific comparisons are best made at the process or submodel level, and such focused comparison and evaluation is included in the text.

Some recently published grassland models are due to Hunt *et al.* (1991), Verberne (1992), Parton *et al.* (1993), and Chen and Coughenour (1994). The work of Parton, Hunt, Coughenour and colleagues in North America probably represents the most comprehensive grassland modelling effort anywhere at the time of writing. Their publications contain much of value. The Netherlands model of Verberne (1992) is carefully and clearly described but few simulations are given.

The emphasis in the grassland model presented here is mostly concerned with the systematic definition of model content and demonstrating the value of relating that content mechanistically to prediction. It can be especially valuable to relate interesting or surprising model predictions to assumed mechanisms. Often we find that there is no single simple grassland response to, say, CO_2, or temperature. The predicted response depends, qualitatively and quantitatively, on other factors. This is one reason why a model is needed: to integrate the variety of sometimes conflicting observed responses within a common framework.

Some recent books on grassland are by Jones and Lazenby (1988), Holmes (1989) and Whitehead (1995). Loomis and Connor (1992) give a valuable account of most aspects of crop ecology. Frequent reference is made to these four books in the description of the model.

2.2 The Model

An overview of the model is given in Fig. 2.1, which shows the flows of C and N. It comprises three components: plant (*pl*, Chapter 3), animal (*an*, Chapter 4) and soil

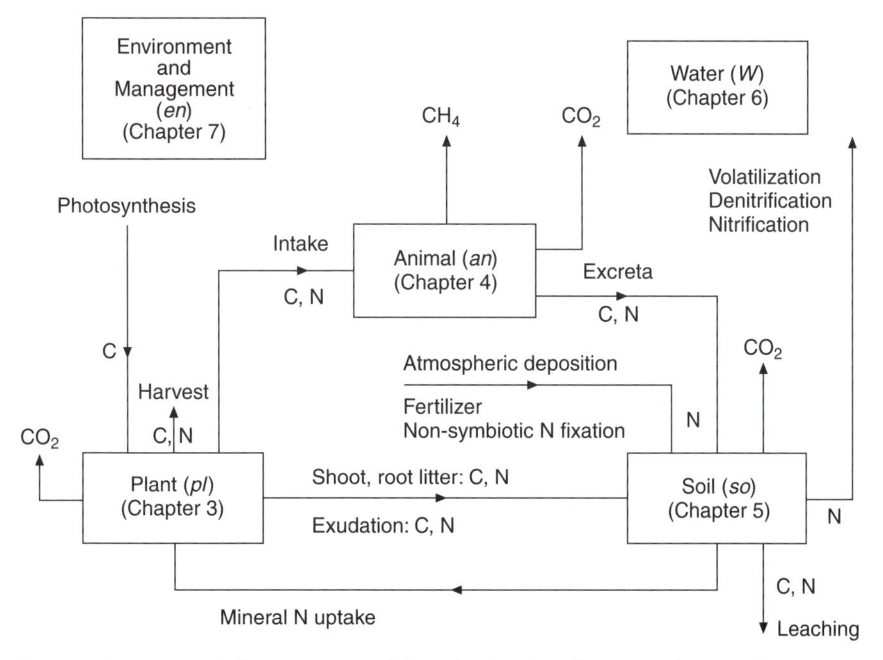

Fig. 2.1. Overview of the pasture model. Only the C, N flows are drawn. The abbreviations (*en, W, pl, an, so*) used to denote the different submodels or modules are given.

(*so*, Chapter 5), each identified by a two-letter abbreviation, and each represented by a submodel. The presence of the water submodel (*W*, Chapter 6), and the environment and management module (*en*, Chapter 7) are indicated in Fig. 2.1, but these do not correspond to discrete entities, as do the plant, animal and soil submodels. Water fluxes occur between the environment, soil and plant. The animal is assumed to make a negligible contribution to the water budget. The environment and management module has no state variables. The environment provides driving variables such as radiation, rainfall, relative humidity, temperature and wind; management defines fertilizer, harvesting, and animal stocking regimes. We assume that the environment also provides sinks/sources of infinite capacity for exchanges of carbon (C), nitrogen (N) and water (W). Thus, the CO_2 concentration of the atmosphere affects the grassland ecosystem, but the ecosystem does not affect the global atmospheric CO_2 concentration.

The model attempts to give a unified view of the role of C and N in the plant–animal–soil system, and how the influence of these major substrates is modulated by water and the environment.

2.3 Notation

The crux of a good notation is that it can be read and understood without searching the text for a definition or looking up symbols in tables. However, it is no easy task to construct a notation that is logical, not too cumbersome, and is readily grasped. Some readers may object that the traditional notation in their corner of the grassland ecosystem has not been used. A consistent mathematical notation which is carried over into the computer program makes the ACSL computer program much easier to work with, owing to the powerful interactive possibilities available with ACSL (Section 2.5; Chapter 10).

Every state variable is described by an equation such as Eqn (1.1c), with inputs and outputs. The inputs and outputs are denoted by I or O, with subscripts. All flows are of carbon (C), nitrogen (N) or water (W), so the first subscript describes the substance of the input/output flux, e.g. I_C, O_W. An output flux leaves pool i (say) in the direction of pool j, so we write $O_{W,i\rightarrow j}$. The C input flux $I_{C,k\rightarrow l}$ enters the pool l from the direction of pool k. Sometimes an output flux from one pool (i) may equal the input flux to another pool (j), so $I_{C,i\rightarrow j} = O_{C,i\rightarrow j}$, but this is not always true. Many C output fluxes are partially lost to the environment by respiration, and some N output fluxes similarly have more than one destination.

2.4 Units

Units are a particular problem and stumbling-block with plant ecosystem models because of the range of scales which is covered. Some constants and conversions that have been found useful are listed in the Appendix. The plant physiologist is often accustomed to moles and seconds; the crop physiologist may prefer grams, minutes or hours; the agronomist likes to speak of kilograms, hectares and the year or growing season; and the ecologist may choose any of these. It is helpful to adhere to a single set of units within the main part of a model. Any conversions to what might be considered more appropriate units are made outside the main part of the model. By following this procedure, the model is independent of conversion factors, and the modeller is never in doubt as to the units of quantities within the model. This facilitates checking the dimensional consistency of equations, which is essential when modelling. Thus we use 'absurd' numbers like 0.0015 sheep m^{-2} for animal stocking density, rather than 15 sheep ha^{-1} with factors of 10^4 in certain equations.

The base SI units for mass and length are kilogram (kg) and metre (m) (Royal Society, 1975), and these are used in the model. The base SI unit for time, the second (s), is only used for some quantities and initial calculations of photosynthesis and transpiration, which are converted to day time units. The day (day) is the base time unit in the model. Daily sums, yearly sums and averages are calculated outside the main part of the model.

Regrettably the SI committees have defined the quantity of substance, the mole (mol), so that 1 mole of ^{12}C has a mass of 0.012 kg, rather than a mass of 12 kg, which would be more consistent with the rest of SI and also more useful (Thornley and Johnson, 1991). Other researchers have also encountered this problem: e.g. Nobel, in his excellent textbook (Nobel, 1991), writes on p. 151 (and elsewhere) when referring to concentrations, that 1 mM = 1 mol m^{-3}. We depart occasionally from the standard SI mole so that the basic equations of the model are not cluttered with factors of 10^3, but this is explained where it occurs and should not cause difficulties.

Photosynthesis is often discussed in terms of moles CO_2 m^{-2} s^{-1}. However, quantities such as net primary production are preferred in kg C m^{-2} $year^{-1}$, with a convenient magnitude in the range 0.5–2 kg C m^{-2} $year^{-1}$. The model uses mass units for CO_2 fluxes (kg CO_2 etc), with conversions made to other units outside the kernel of the model (1 kg CO_2 = 1000 g kg^{-1} / 44.01 g mol^{-1} = 22.7 moles CO_2; see Appendix).

Similarly the radiation units used are J, both for total radiation and for PAR (photosynthetically active radiation). There is no exact conversion between radiation energy (J) and photons (mol) as this conversion is wavelength-dependent: 1 J of radiation of a given wavelength = 1 J × [wavelength (m) / 0.120 (J m $mole^{-1}$)] moles of photons. 0.120 (J m $mole^{-1}$) = Planck's constant (6.6262×10^{-34} J s) × velocity of light (2.998×10^{8} m s^{-1}) × Avogadro constant (6.022×10^{23} $mole^{-1}$). 1 J of radiation of wavelength 0.55 μm converts to 0.55×10^{-6} / 0.120 = 4.6 μmol. A wavelength of 0.55 μm is in the yellow part of the spectrum. Robson and Sheehy (1981) and Ludlow (1983) use conversion factors of 4.6 μmol $(J\ PAR)^{-1}$.

CO_2 'concentrations' in vpm (volume per million volumes) or μmol mol^{-1} are widely used, although these are more akin to relative concentrations or partial pressures. Their use could be misleading because a CO_2 concentration in vpm is independent of temperature and pressure, whereas the number or mass of CO_2 molecules impinging on a surface per unit area per unit time depends on both temperature and pressure. The model mostly uses a 'real' concentration of kg CO_2 m^{-3} but also occasionally vpm.

Water relations and in particular water potential are another problem area. The most common unit for water potential is the SI unit for pressure, the pascal (Pa). This unit has nothing to recommend it other than familiarity. Pa is equivalent to an energy per unit volume (J m^{-3}). However, the density of water varies with temperature and pressure, and it is preferable to define water potential in terms of energy per unit mass of water (J kg^{-1}). J kg^{-1} is a unit of more appropriate magnitude for use than Pa, since 100 J kg^{-1} ≡ 1 bar ≡ 10^5 Pa ≈ 1 atmosphere. It is also an 'intuitive' unit since it is easy to imagine lifting 1 kg of water to a height of 10 m (a tall house) against an acceleration due to gravity of 10 m s^{-2}, requiring an energy of 100 J.

Alternative units are provided in the text, tables and figures to ease the path of the reader. Some useful conversion factors are summarized in the Appendix.

2.5 The Computer Program

The source program file, pasture.csl, can be downloaded by ftp from budbase.nbu.ac.uk/pub/tree/Book (Section 10.3). The program statements are kept as close to the mathematics as possible – this is facilitated by using a purpose-built Continuous System Simulation Language (CSSL), in our case, ACSL (Advanced Continuous Simulation Language; Mitchell and Gauthier, 1993). However, the chapters here are differently ordered from the computer program. An introduction to ACSL is given in Chapter 10.

A 'front-end' such as ACSL can be used because the problem has a well-defined and standard structure. This greatly reduces the amount of 'house-keeping' code which would be needed if programming the problem directly in Basic or Fortran. Writing house-keeping code takes much time and effort. It can greatly obscure the biological structure of the problem. And it is easy to get wrong. The penalty is the cost of a commercial modelling package, and the lack of portability and compatibility that ensues. The other positive features of ACSL (and similar CSSL packages) are: easy coding and non-procedural statements which can go in any order, e.g. that which best describes the biology; easy interaction with the program at execution time facilitating debugging, program evaluation and development; good input/output statements including generating graphs and data files. However, the programming enthusiast will find it straightforward to translate an ACSL program into Fortran or whatever. This is asserted with confidence, as various ACSL programs similar to the pasture simulator have been translated into Fortran, Pascal and C.

3 Plant Submodel

3.1 Introduction

The plant submodel represents the growth of a vegetative grass crop and its response to light, temperature, nitrogen, water, harvesting and grazing. The scheme is illustrated in Fig. 3.1.

The submodel has 21 state variables. Of these, 16 are mass state variables, and five are morphological state variables (leaf area with four age categories, root density). The shoot is described by: lamina structural dry mass (DM) ($M_{lam,i}$, i = 1 to 4), sheath + stem structural DM ($M_{ss,i}$, i = 1 to 4), lamina area ($L_{AI,i}$, i = 1 to 4), and C and N substrate pools ($M_{CS,sh}$, $M_{NS,sh}$). Lamina area is partially decoupled from lamina structural DM, providing for a variable incremental specific leaf area. The root is described by structural DM ($M_{rt,i}$, i = 1 to 4), root density (ρ_{rt}), and C and N substrate pools ($M_{CS,rt}$, $M_{NS,rt}$).

Each of the structural DM variables and the lamina area variable is subscripted with i and has four age categories. The representation of tissue age makes a major difference to the dynamic responses of the model and the realism of its predictions.

Often modellers attempt to simplify their models by ignoring small and relatively labile pools, which are usually substrate pools. However, small pools may support large fluxes and can be highly influential in determining activities, of, say, shoot and root. It can be preferable to represent such small pools directly and accept the accompanying difficulties (small integration intervals, possible stability problems), rather than use empirical devices which are often clumsy, too rigid, and block off routes for further model development.

The use of C and N substrate pools allows the allocation of growth between shoot and root to be modelled using the transport-resistance approach (Thornley, 1972). This was chosen after recent experience that there are substantial differences between the dynamic responses predicted by a teleonomic (goal-seeking) allocation model (as used by Thornley and Verberne, 1989) and a transport-resistance allocation model (Thornley, 1995). Elsewhere it has been argued that the transport-resistance model of allocation should be the method of first choice for modelling

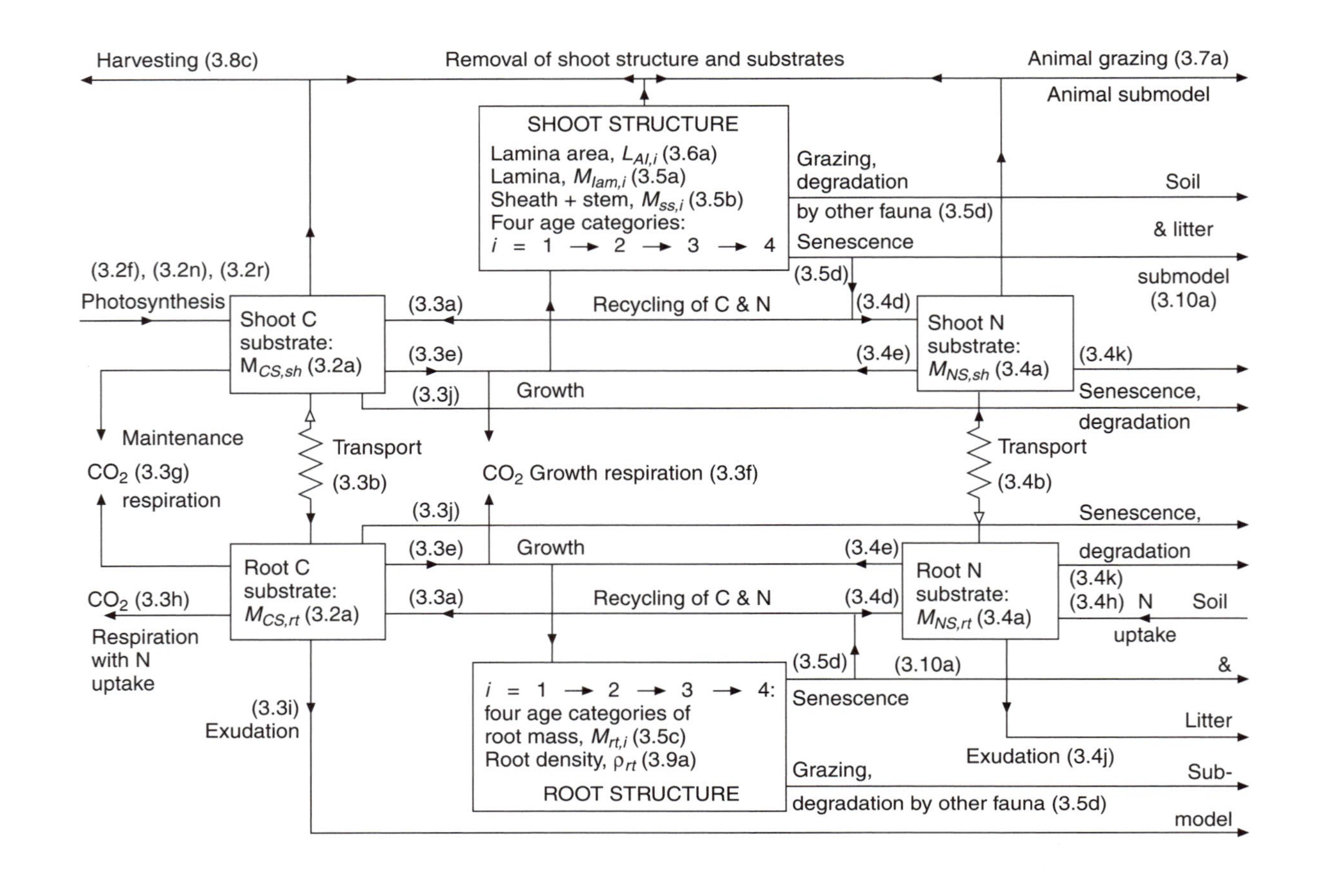

Fig. 3.1. Plant submodel. The state variables are shown in the boxes. Equation numbers point to fluxes or to differential equations.

allocation, and only if it fails should it be modified as indicated by the mode of failure (Thornley, 1997). Modification of the approach rather than its abandonment is required because it seems beyond doubt that allocation is the result of transport and chemical conversion processes, even if these processes are subject to evolved control mechanisms.

The plant submodel is driven by the C input from photosynthesis and the N input from N uptake from the soil mineral N pools. The environmental parameters that affect the plant submodel are radiation, CO_2 concentration, day length, air temperature, soil temperature, and rainfall through the water submodel. There are fluxes of both shoot and root litter, and root substrate C and N by exudation to the soil and litter submodel. The soil submodel only influences plant growth through the soil mineral N pools of ammonium N and nitrate N, apart from water. Management may profoundly affect plant growth through removal of C and N in cutting and grazing regimes, giving fluxes of shoot dry mass to harvesting and to grazing animals. Symbiotic nitrogen fixation is not explicitly represented, although, possibly, it could be introduced as a flux into the root N substrate pool (but see Section 7.4). Non-symbiotic N fixation in the soil produces a flux into the soil N ammonium pool (Fig. 5.1).

A major simplification in the model which some will find unacceptable is that there is no representation of reproductive development. There are several reasons for this. First, in well-managed grassland there is usually little reproductive growth; grazing and cutting are applied in such a way that this does not occur, because of the low digestibility and palatability associated with the reproductive crop. Second, even in an unmanaged grassland ecosystem, the populations and activities of grazers may be such as effectively to suppress reproductive growth. Last, it is difficult to model reproductive growth with other than an *ad hoc* treatment (e.g. chapter 7 of Thornley and Johnson, 1990), which is, arguably, less acceptable in a mechanistic model than omitting the process entirely.

First, we define some variables in terms of the state variables for later use. Then, for each state variable, a differential equation is given, and the equations defining the input/output processes are presented. State variables, their initial values, parameters and all other variables are listed in Table 3.1.

Table 3.1. Symbols of plant submodel. DM denotes dry mass.

State variables	Description	Initial value
$L_{AI,i}$, $i = 1–4$	Area of leaf lamina compartments (3.6a)	0.2 m^2 (m^2 ground)$^{-1}$
$M_{CS,rt}$, $M_{CS,sh}$	Mass of root and shoot substrate C (3.2a)	0.006, 0.01 kg substrate C m^{-2}
$M_{lam,i}$, $i = 1–4$	Structural mass of leaf lamina compartments (3.5a)	0.01 kg structural DM m^{-2}
$M_{NS,rt}$, $M_{NS,sh}$	Mass of root and shoot substrate N (3.4a)	0.003, 0.001 kg substrate N m^{-2}
$M_{rt,i}$, $i = 1–4$	Structural mass of root compartments (3.5c)	0.05 kg structural DM m^{-2}
$M_{ss,i}$, $i = 1–4$	Structural mass of sheath + stem compartments (3.5b)	0.04 kg structural DM m^{-2}
ρ_{rt}	Root density (3.9a)	1 kg structural DM (m^3 soil)$^{-1}$

Continued over

Table 3.1. *Continued*

Parameters	Description	Value
c_{gs,CO_2}	Effect of ambient CO_2 on stomatal conductance (3.2u)	2
$c_{hcan,LAI}$	Proportionality between leaf area index and canopy height (3.8a)	0.026 m [(m^2 leaf)/(m^2 ground)]$^{-1}$
$c_{LAI,i \to an}$	Tissue age weighting factors for animal intake from shoot lamina (3.7e)	1, 0.75, 0.5, 0.25
$c_{SLA,C}$, $c_{SLA,W}$	Effect of substrate C and shoot relative water content on leaf area expansion (3.6c)	2.5 [kg substrate C (kg structural DM)$^{-1}$]$^{-1}$, 1
$c_{SLA,max}$	Maximum value of specific leaf area (3.6c)	25 m^2 (kg structural DM)$^{-1}$
$c_{T,\alpha}$	Effect of temperature on photosynthetic parameter α (3.2s)	0.015 (°C)$^{-1}$
$c_{ss \to an}$	Weighting constant for sheath + stem component of animal intake relative to lamina (3.7d)	0 (for sheep)
$c_{ss,i \to an}$	Tissue age weighting factors for animal intake from shoot sheath + stem (3.7f)	1, 1, 1, 1
c_{uNamm}, c_{uNnit}	Respiratory cost of N uptake from soil ammonium and nitrate pools (3.3h)	0.4, 0.5 kg substrate C (kg N)$^{-1}$
$c_{uN,2}$, $c_{uN,3}$, $c_{uN,4}$	N uptake root tissue weighting constants (3.4h)	0.5, 0.25, 0.1
$d_{rt,min}$	Minimum root depth (3.9c)	0.02 m
$f_{C,plX}$, $f_{N,plX}$	Fractions of structural C and N in plant structure (3.1i)	0.45, 0.02 kg C, N (kg structural DM)$^{-1}$
$f_{CS,sh \to li,min}$, $f_{CS,rt \to li,min}$	Minimum fractions of substrate C lost to litter from shoot, root (3.3j)	0.3, 0.3
f_{lam}	Fraction of growth allocated to lamina (3.5a)	0.7
$f_{N,rtli,re,max}$, $f_{N,shli,re,max}$	Maximum fractions of structural N recycled from structural litter fluxes (3.4c)	0.4, 0.4
$f_{NS,sh \to li,min}$, $f_{NS,rt \to li,min}$	Minimum fractions of substrate N lost to litter from shoot, root (3.4k)	0.3, 0.3
$g_{s,max}$, $g_{s,min}$	Maximum and minimum stomatal conductances at 350 vpm CO_2 (3.2u)	0.005, 0.00005 m s^{-1} (m^2 leaf)$^{-1}$
h_{harv}	Harvesting height (3.8b)	0.03 m
$J_{N,uN}$	Root uptake parameter (3.4h)	0.005 kg substrate N (kg structural DM)$^{-1}$
$K_{C,mai}$	Maintenance respiration parameter (3.3g)	0.03 kg substrate C (kg structural DM)$^{-1}$
$K_{CO_2,Pmax}$	Photosynthesis parameter (3.2t)	1.281×10^{-3} kg CO_2 m^{-3}
$K_{Csh \to li}$, $K_{Crt \to li}$	Michaelis–Menten constants for C substrate litter loss from shoot and root (3.3j)	0.05, 0.05 kg substrate C (kg structural DM)$^{-1}$
$K_{C,uN}$	Root uptake parameter (3.4h)	0.05 kg substrate C (kg structural DM)$^{-1}$
$K_{MXrt,uN}$	Root uptake parameter (3.4h)	1 kg structural DM m^{-2}
$K_{N_{eff}}$	Root uptake parameter (3.4h)	0.005 kg N m^{-2}
$K_{N,rec}$	N recycling parameter (3.4c)	0.02 kg substrate N (kg structural DM)$^{-1}$
$K_{Nsh \to li}$, $K_{Nrt \to li}$	Michaelis–Menten constants for N substrate litter loss from shoot and root (3.4k)	0.01, 0.01 kg substrate N (kg structural DM)$^{-1}$
$k_{CS,rt,ex20}$, $k_{NS,rt,ex20}$	Substrate C, N root exudation constants [(3.3i), (3.4j)]	0.02, 0.005 day^{-1}
k_{can}	Canopy extinction coefficient (3.2b)	0.5 m^2 ground (m^2 leaf)$^{-1}$
$k_{deg,rt,20}$, $k_{deg,sh,20}$	Root, shoot structure degradation rate constants [(3.5a)]	0.001, 0.001 day^{-1}
$k_{mai,rt1-4,20}$, $k_{mai,sh1-4,20}$	Root, shoot maintenance coefficients (3.3g)	0.02, 0.02, 0.015, 0.01 kg substrate C (kg structural DM)$^{-1}$ day^{-1}

Table 3.1. *Continued*

Parameters	Description	Value
$k_{turn,lam,20}$, $k_{turn,rt,20}$, $k_{turn,ss,20}$	Turnover rates for lamina, root and sheath + stem tissue [(3.5a), (3.5b), (3.5c)]	0.08, 0.08, 0.08 day^{-1}
$P_{max,20}$	Photosynthetic parameter (3.2t)	2×10^{-6} kg CO_2 m^{-2} day^{-1}
$q_{W,ph}$, $q_{W,pl}$, $q_{W,uN}$	Effect of water activity on photosynthesis (3.2s), plant biochemistry (3.11d), and N uptake (3.4h)	2, 20, 3
$q_{\rho C}$, $q_{\rho N}$	Allometric constants for scaling within-plant C, N transport [(3.3b), (3.4b)]	1, 1
$r_{C:N,rec-li}$	Ratio of C to N in the recyclable component of structural DM (3.3a)	2.7 kg C (kg N)$^{-1}$
$r_{mm,CS}$, $r_{mm,NS}$	Relative molecular masses of C and N substrates (3.1e)	28.5, 62
T_0, T_{max}, T_{ref}	Temperature function parameters (3.11a)	0, 30, 20°C
$Y_{G,pl}$	Conversion yield for plant growth (3.3e)	0.75
$\alpha_{m,15}$	Photosynthetic efficiency (3.2s)	1×10^{-8} kg CO_2 J^{-1}
β	Photorespiration parameter (3.2s)	0.3×10^{-6} kg CO_2 m^{-2} s^{-1}
$\theta_{gs,max}$, $\theta_{gs,min}$	Shoot relative water content when stomata are fully open, closed (3.2u)	0.85, 0.65
μ_{20}	Growth rate constant (3.3c)	400 [kg substrate C (kg structural DM)$^{-1}$]$^{-1}$ [kg substrate N (kg structural DM)$^{-1}$]$^{-1}$
ξ	Photosynthetic sharpness parameter (3.2c)	0.95
ρ_{C20}, ρ_{N20}	Resistivity coefficients for C, N transport between shoot and root [(3.3b), (3.4b)]	0.2, 2 m^2 day (kg structural DM)$^{q_{\rho,C,N}}$
$\rho_{rt,max}$, $\rho_{rt,min}$	Maximum and minimum values of root density (3.9b)	1, 0.1 kg structural DM (m^3 soil)$^{-1}$
$\sigma_{uN,20}$	Root uptake constant at 20°C (3.4h)	1 kg N (kg root structural DM)$^{-1}$ day^{-1}
τ	CO_2 conductance parameter (3.2s)	0.0015 m s^{-1}
χ_{leaf}	Leaf transmission coefficient (3.2b)	0.1

Other principal plant variables	Description	Units
C_{rt}, C_{sh}, $C_{S,pl}$	Substrate C concentration in root, shoot (3.1h), plant (3.1k)	kg substrate C (kg structural DM)$^{-1}$
$C_{tot,sh}$, $N_{tot,sh}$	Total C and N concentrations in shoot (3.1i)	kg total C (kg total DM)$^{-1}$
d_{rt}	Root depth (3.9c)	m
f_{harv}	Fraction of above ground DM harvested (3.8b)	
$f_{N,rtli,re}$, $f_{N,shli,re}$	Fractions of N in root, shoot litter recycled (3.4c)	
$f(T)$	Temperature function (3.11a)	
$f_{T,rt}$, $f_{T,sh}$	Effect of temperature on root, shoot processes (3.11b)	
$f_{T,Pmax}$, $f_{T,\alpha}$	Effect of temperature on photosynthetic parameters P_{max} (3.11c) and α (3.2s)	
$f_{W,rt}$, $f_{W,sh}$	Effect of water on root, shoot processes (3.11d)	
$f_{W,uN}$	Effect of water on N uptake [(3.4h), (6.6e)]	
$f_{X,rt}$, $f_{X,sh}$	Fractions of structural DM in root and shoot (3.11)	
G_{rt}, G_{sh}	Growth rates of root, shoot (3.3c)	kg structural DM m^{-2} day^{-1}
g_s	Leaf stomatal conductance (3.2u)	m s^{-1} (m^2 leaf)$^{-1}$
h_{can}	Canopy height (3.8a)	m
$I_{C,pl \to an}$, $I_{N,pl \to an}$	Inputs of total C, N to animals (3.7b)	kg C, N m^{-2} day^{-1}

Continued over

Table 3.1. *Continued*

Other principal plant variables	Description	Units
$I_{CS,en\rightarrow sh}$	Input of substrate C to shoot substrate C pool (3.2o)	kg substrate C m^{-2} day^{-1}
$I_{CS,re,rt}$, $I_{CS,re,sh}$, $I_{NS,re,rt}$, $I_{NS,re,sh}$	Input of substrate C, N to root, shoot substrate C, N pools from recycling [(3.3a), (3.4d)]	kg substrate C, N m^{-2} day^{-1}
$I_{DM,pl\rightarrow an,gnd}$	Input of shoot DM to animals (3.7a)	kg DM m^{-2} day^{-1}
$I_{LAI,1}$	Input of lamina area to 1st lamina pool (3.6b)	m^2 lamina (m^2 ground)$^{-1}$ day^{-1}
$I_{NS,so\rightarrow rt}$	N uptake from soil mineral pools (3.4h)	kg substrate N m^{-2} day^{-1}
j_{leaf}	Light incident on a leaf surface within the canopy (3.2b)	J m^{-2} s^{-1}
L_{AI}	Leaf area index (3.1b)	m^2 leaf (m^2 ground)$^{-1}$
$M_{CS,pl}$, $M_{NS,pl}$	Mass of substrate C, N in plant (3.1c)	kg substrate C, N m^{-2}
$M_{CS,lam}$, $M_{NS,lam}$	Masses of substrate C and N in shoot lamina (3.1d)	kg C, N m^{-2}
$M_{S,i}$, i = *lam, pl, rt, sh*	Total substrate mass in lamina, plant, root, shoot (3.1e)	kg storage m^{-2}
M_{lam}, M_{rt}, M_{sh}, M_{pl}	Total dry masses of lamina, root, shoot, plant (3.1f)	kg DM m^{-2}
$M_{X,i}$, i = *lam, pl, rt, sh, ss*	Mass of structural DM in lamina, plant, root, shoot, sheath + stem (3.1a)	kg structural DM m^{-2}
N_{eff}	Effective soil mineral N pool from which plant uptake occurs (3.4g)	kg N m^{-2}
N_{min}	Soil mineral N pool (3.4f)	kg N m^{-2}
N_{rt}, N_{sh}, $N_{S,pl}$	substrate N concentration in root, shoot (3.1h), plant (3.1k)	kg substrate C (kg structural DM)$^{-1}$
$N_{sh,lamA}$	N content per unit area in shoot lamina (3.1j)	kg N (m^2 lamina)$^{-1}$
$O_{CS,rt\rightarrow en,m}$, $O_{CS,sh\rightarrow en,m}$	Output of substrate C for root, shoot maintenance (3.3g)	kg substrate C m^{-2} day^{-1}
$O_{CS,rt,G}$, $O_{CS,sh,G}$, $O_{NS,rt,G}$, $O_{NS,sh,G}$	Output of substrate C, N for root, shoot growth [(3.3e), (3.4e)]	kg substrate C, N m^{-2} day^{-1}
$O_{CS,rt\rightarrow en,uN}$	Output of substrate C from root for N uptake (3.3h)	kg substrate C m^{-2} day^{-1}
$O_{CS,rt\rightarrow so,ex}$, $O_{NS,rt\rightarrow so,ex}$	Output of substrates C, N from root to soil by exudation [(3.3i), (3.4j)]	kg substrate C, N m^{-2} day^{-1}
$O_{CS,sh\rightarrow an}$, $O_{NS,sh\rightarrow an}$	Output of substrate C, N from shoot to grazing animals [3.7h), (3.7i)]	kg substrate C, N m^{-2} day^{-1}
$O_{CS,sh\rightarrow hv}$, $O_{NS,sh\rightarrow hv}$	Output of substrate C, N from shoot to harvesting (3.8c)	kg substrate C m^{-2} day^{-1}
$O_{CS,sh\rightarrow rt}$	Output of substrate C from shoot to root due to translocation (3.3b)	kg substrate C m^{-2} day^{-1}
$O_{C,sh\rightarrow so}$, $O_{N,sh\rightarrow so}$, $O_{C,rt\rightarrow so}$, $O_{N,rt\rightarrow so}$	Outputs of total C, N from shoot, root to litter (3.10b)	kg total C, N m^{-2} day^{-1}
$O_{CS,sh\rightarrow so}$, $O_{NS,sh\rightarrow so}$, $O_{CS,rt\rightarrow so}$, $O_{NS,rt\rightarrow so}$	Outputs of substrate C, N from shoot, root to litter [(3.3j), (3.4k)]	kg total C, N m^{-2} day^{-1}
$O_{CX,sh\rightarrow so}$, $O_{NX,sh\rightarrow so}$, $O_{CX,rt\rightarrow so}$, $O_{NX,rt\rightarrow so}$	Outputs of structural C, N from shoot, root to litter (3.10a)	kg structural C, N m^{-2} day^{-1}
$O_{LAI,i\rightarrow an}$, $O_{LAI,i\rightarrow hv}$	Outputs of lamina area (i = 1–4) to grazing animals (3.7g), to harvesting (3.8c)	m^2 leaf m^{-2} day^{-1}
$O_{NS,rt\rightarrow sh}$	Output of substrate N from root to shoot due to translocation (3.4b)	kg substrate N m^{-2} day^{-1}
$O_{X,lam\rightarrow an}$, $O_{X,lam,i\rightarrow an}$, $O_{X,lam,i\rightarrow hv}$	Outputs of structural DM from all laminar pools (3.7d), laminar pools (i = 1–4) to grazing animals (3.7e), to harvesting (3.8c)	kg structural DM m^{-2} day^{-1}
$O_{X,pl\rightarrow an}$	Output of structural DM from plant to animal (3.7c)	kg structural DM m^{-2} day^{-1}
$O_{X,rt\rightarrow li}$, $O_{X,sh\rightarrow li}$	Outputs of root, shoot structural DM to litter (3.5d)	kg structural DM m^{-2} day^{-1}
$O_{X,ss\rightarrow an}$, $O_{X,ss,i\rightarrow an}$, $O_{X,ss,i\rightarrow hv}$	Outputs of structural DM from all sheath + stem pools (3.7d), sheath + stem pools (i = 1–4) to grazing animals (3.7f) and to harvesting (3.8c)	kg structural DM m^{-2} day^{-1}

Table 3.1. *Continued*

Other principal plant variables	Description	Units
P_{can}	Instantaneous canopy gross photosynthetic rate [(3.2f), (3.2n), (3.2r)]	kg CO_2 (m^2 ground)$^{-1}$ day^{-1}
P_{leaf}	Leaf gross photosynthetic rate (3.2e)	kg CO_2 (m^2 leaf)$^{-1}$ s^{-1}
P_{max}	Leaf light-saturated photosynthetic rate (3.2t)	kg CO_2 m^{-2} s^{-1}
$R_{G,rt}$, $R_{G,sh}$	Respiration rates from root, shoot growth (3.3f)	kg C m^{-2} day^{-1}
$R_{m,rt}$, $R_{m,sh}$	Respiration rates from root, shoot maintenance (3.3g)	kg C m^{-2} day^{-1}
R_{pl}, R_{rt}, R_{sh}	Total respiration rates from plant, root, shoot [(3.3m), (3.3l), (3.3k)]	kg C m^{-2} day^{-1}
$r_{C:N,Spl}$	Whole-plant C:N substrate ratio (3.1k)	kg C (kg N)$^{-1}$
$r_{sh:rt}$, $r_{sh:rt,X}$	Total dry mass and structural shoot:root ratios (3.1l)	
$r_{sh:rt,G}$	Shoot:root growth ratio (3.3d)	
S_{LA}	Specific leaf area (3.1g)	m^2 (kg lamina total DM)$^{-1}$
u_N	Total N uptake rate by root (3.4h)	kg N m^{-2} day^{-1}
u_{Namm}, u_{Nnit}	N uptake rates from soil ammonium and nitrate pools (3.4i)	kg N m^{-2} day^{-1}
α	Leaf photosynthetic efficiency (3.2s)	kg CO_2 J^{-1}
ν_{SLA}	Incremental specific leaf area (3.6b)	m^2 leaf (kg structural DM)$^{-1}$
$\rho_{rt,new}$	Root density of newly synthesized root mass (3.9b)	kg structural DM (m^3 soil)$^{-1}$
Animal variables		
$I_{C,pl\rightarrow an}$	Total C input to grazing animals from plant (3.7b)	kg C m^{-2} day^{-1}
$I_{DM,pl\rightarrow an}$	Input of DM from plant to grazing animals [(3.7a), (4.2a)]	kg plant DM animal^{-1} day^{-1}
$I_{DM,pl\rightarrow an,gnd}$	Input of DM from plant to grazing animals (3.7a)	kg plant DM m^{-2} day^{-1}
$I_{N,pl\rightarrow an}$	Total N input to grazing animals from plant (3.7b)	kg N m^{-2} day^{-1}
Environmental variables		
$C_{O_2,air}$, $C_{O_2,vpm}$	Ambient CO_2 concentration [(3.2s), (3.2u), (7.3a); Table 7.2]	kg CO_2 m^{-3}, vpm
$j_{PAR,sc}$	Instantaneous photosynthetically active radiation [(3.2b), (7.5a)]	J m^{-2} s^{-1}
T_{air}, T_{soil}	Air and soil temperatures [(3.11b), (7.5e); Section 7.5, Table 7.3, Fig. 7.2]	°C
Management variables		
h_{harv}	Harvesting height [harvesting regime, (3.8b), (7.6e)]	m
$n_{animals}$	Number of grazing animals [grazing regime, (3.7a), (7.6f)]	animals m^{-2}
Soil variables		
N_{amm}, N_{nit}	Soil ammonium, nitrate mineral pools (3.4f) (See Chapter 5, Fig. 5.1)	kg N m^{-2}
Water variables		
$a_{W,rt}$, $a_{W,sh}$	Chemical activity of water in root, shoot [(3.11d), (6.7a)]	
θ_{sh}, θ_{so}	Shoot, soil relative water contents [3.2u), (3.4h), (6.5d) with $i = sh$; (6.2f)]	

3.2 Some Definitions

The structural (X) dry masses of the shoot lamina (lam), the shoot sheath and stem (ss), the whole shoot (sh), the root (rt), and the plant as a whole (pl) are obtained by summing the components in the four tissue age categories, with

$$\begin{aligned}
M_{X,lam} &= M_{lam,1} + M_{lam,2} + M_{lam,3} + M_{lam,4},\\
M_{X,ss} &= M_{ss,1} + M_{ss,2} + M_{ss,3} + M_{ss,4},\\
M_{X,sh} &= M_{X,lam} + M_{X,ss},\\
M_{X,rt} &= M_{rt,1} + M_{rt,2} + M_{rt,3} + M_{rt,4},\\
M_{X,pl} &= M_{X,sh} + M_{X,rt}.
\end{aligned} \tag{3.1a}$$

Units are kg structural dry mass (DM) (m^2 ground)$^{-1}$.

The leaf area index is given by [m^2 leaf (m^2 ground)$^{-1}$]

$$L_{AI} = L_{AI,1} + L_{AI,2} + L_{AI,3} + L_{AI,4}. \tag{3.1b}$$

The total substrate C and N (CS, NS) masses in the plant (pl) are the sums of the masses in the shoot (sh) and the root (rt) (kg C, N m^{-2}):

$$M_{CS,pl} = M_{CS,sh} + M_{CS,rt}, \qquad M_{NS,pl} = M_{NS,sh} + M_{NS,rt}. \tag{3.1c}$$

The substrate C and N masses in the lamina are [kg C, N (m^2 ground)$^{-1}$)]:

$$M_{CS,lam} = \frac{M_{X,lam}}{M_{X,sh}} M_{CS,sh}, \qquad M_{NS,lam} = \frac{M_{X,lam}}{M_{X,sh}} M_{NS,sh}. \tag{3.1d}$$

It is assumed here that substrate is uniformly distributed within the shoot.

The substrate or storage (S) dry masses in the lamina (lam), shoot (sh), root (rt), and whole plant (pl) are (kg substrate DM m^{-2})

$$\begin{aligned}
M_{S,lam} &= \frac{M_{CS,lam} r_{mm,CS}}{12} + \frac{M_{NS,lam} r_{mm,NS}}{14},\\
M_{S,sh} &= \frac{M_{CS,sh} r_{mm,CS}}{12} + \frac{M_{NS,sh} r_{mm,NS}}{14},\\
M_{S,rt} &= \frac{M_{CS,rt} r_{mm,CS}}{12} + \frac{M_{NS,rt} r_{mm,NS}}{14},\\
M_{S,pl} &= M_{S,sh} + M_{S,rt},\\
r_{mm,CS} &= 28.5, \qquad r_{mm,NS} = 62.
\end{aligned} \tag{3.1e}$$

In this equation the relative molecular mass of C substrate $r_{mm,CS}$ is 342 (sucrose) / 12 = 28.5, and that for N substrate $r_{mm,NS}$ is 62 (nitrate).

The total dry masses of lamina, shoot, root and plant are, summing the structural and the substrate components [Eqns (3.1a), (3.1e)] (kg DM m^{-2})

$$\begin{aligned}
M_{lam} &= M_{X,lam} + M_{S,lam},\\
M_{sh} &= M_{X,sh} + M_{S,sh}, \qquad M_{rt} = M_{X,rt} + M_{S,rt},\\
M_{pl} &= M_{sh} + M_{rt}.
\end{aligned} \tag{3.1f}$$

Specific leaf area S_{LA} [m^2 (kg lamina total DM)$^{-1}$] is

$$S_{LA} = \frac{L_{AI}}{M_{lam}}. \tag{3.1g}$$

The substrate C and N concentrations in the shoot (*sh*) and the root (*rt*) are [kg C, N (kg structural DM)$^{-1}$]

$$C_{sh} = \frac{M_{CS,sh}}{M_{X,sh}}, \quad C_{rt} = \frac{M_{CS,rt}}{M_{X,rt}},$$
$$N_{sh} = \frac{M_{NS,sh}}{M_{X,sh}}, \quad N_{rt} = \frac{M_{NS,rt}}{M_{X,rt}}. \tag{3.1h}$$

These substrate concentrations are defined relative to the structural component of dry mass. Concentrations are also required in terms of mass of total C and total N (storage + structure) per unit mass of total dry mass. For the shoot, these are

$$C_{tot,sh} = \frac{M_{CS,sh} + f_{C,plX} M_{X,sh}}{M_{sh}}, \quad N_{tot,sh} = \frac{M_{NS,sh} + f_{N,plX} M_{X,sh}}{M_{sh}},$$
$$f_{C,plX} = 0.45 \text{ kg C (kg structural DM)}^{-1},$$
$$f_{N,plX} = 0.02 \text{ kg N (kg structural DM)}^{-1}. \tag{3.1i}$$

The f parameters used here are the fractions of structural C and N per unit of plant structural dry mass. It is assumed here that these are constant and do not depend on plant nutritional status. Thus the total C and N concentrations of plant dry mass can only change by means of changes in the pools of substrate C and N. The assumption of constant C and N contents of plant structural DM is weak, although to relax this assumption in a coherent manner could require several changes to this section of the model: storage pools with different turnover times, structural pools with different degrees of degradability and recycling, and structural pool composition depending on the availability of C and N substrates. Whitehead (1995, p. 268) gives results showing that, for perennial ryegrass herbage, N content varies with time of year and fertilizer N application, varying from 1.85% to 4.2%. In Italian ryegrass the variation is even greater, whereas in white clover N content is much more constant at about 4% or slightly more (Whitehead, 1995, pp. 269–270).

In the leaf lamina, the total N content per unit lamina area $N_{sh,lamA}$ [kg N (m^2 lamina)$^{-1}$] is, dividing fractional N content by the specific leaf area, S_{LA} [Eqn (3.1g)],

$$N_{sh,lamA} = \frac{N_{tot,sh}}{S_{LA}}. \tag{3.1j}$$

The mean substrate C and N concentrations, and the C:N substrate ratio in the plant are [kg C, N (kg structural DM)$^{-1}$]

$$C_{S,pl} = \frac{M_{CS,sh} + M_{CS,rt}}{M_{X,pl}}, \quad N_{S,pl} = \frac{M_{NS,sh} + M_{NS,rt}}{M_{X,pl}},$$
$$r_{C:N,Spl} = \frac{C_{S,pl}}{N_{S,pl}}, \tag{3.1k}$$

using Eqn (3.1a) for $M_{X,pl}$.

The fractions of structure in root (rt) and shoot (sh), and the structural and total dry mass shoot:root ratios are variable quantities given by [with Eqns (3.1a)]

$$f_{X,rt} = \frac{M_{X,rt}}{M_{X,pl}}, \quad f_{X,sh} = \frac{M_{X,sh}}{M_{X,pl}}, \quad r_{sh:rt,X} = \frac{M_{X,sh}}{M_{X,rt}}, \quad r_{sh:rt} = \frac{M_{sh}}{M_{rt}}. \tag{3.1l}$$

These are dimensionless.

3.3 Shoot and Root Substrate C Pools, $M_{CS,sh}$, $M_{CS,rt}$

The differential equations for these two pools are

$$\frac{dM_{CS,sh}}{dt} = I_{CS,en\to sh} + I_{CS,re,sh} - O_{CS,sh\to rt} - O_{CS,sh,G}$$
$$- O_{CS,sh\to en,m} - O_{CS,sh\to so} - O_{CS,sh\to an} - O_{CS,sh\to hv},$$
$$\frac{dM_{CS,rt}}{dt} = O_{CS,sh\to rt} + I_{CS,re,rt} - O_{CS,rt,G} \tag{3.2a}$$
$$- O_{CS,rt\to en,m} - O_{CS,rt\to en,uN} - O_{CS,rt\to so,ex} - O_{CS,rt\to so},$$
$$M_{CS,sh}(t=0) = 0.01, \quad M_{CS,rt}(t=0) = 0.006 \text{ kg substrate C m}^{-2}.$$

The initial values are set so that the initial substrate C concentrations are $C_{sh} = 0.05$ and $C_{rt} = 0.03$ kg substrate C (kg structural DM)$^{-1}$, using Eqns (3.1h), and calculating $M_{X,sh} = M_{X,rt} = 0.2$ kg structural DM m^{-2} at $t = 0$ day with Eqns (3.1a) and Table 3.1.

The two input terms to the shoot substrate C pool arise from canopy photosynthesis [$I_{CS,en\to sh}$, Eqn (3.2o)] and the recycling of C from senescing tissues, i.e. from the litter flux [$I_{CS,re,sh}$, Eqn (3.3a)]. The output terms are transport to the root [$O_{CS,sh\to rt}$, Eqn (3.3b)], utilization for growth including the growth respiration component [$O_{CS,sh,G}$, Eqn (3.3e)], maintenance [$O_{CS,sh\to en,m}$, Eqn (3.3g)], flux of substrate C with senescence to the soil [$O_{CS,sh\to so}$, Eqn (3.3j)], the grazing by the animals [$O_{CS,sh\to an}$, Eqn (3.7h)], and harvesting [$O_{CS,sh\to hv}$, Eqn (3.8c)].

The root substrate C equation has similar terms except that the photosynthesis input term is replaced by the transport term from the shoot [$O_{CS,sh\to rt}$, Eqn (3.3b)]. There are no grazing or harvesting outputs for roots. There is an output term for respiration associated with N uptake [$O_{CS,rt\to en,uN}$, Eqn (3.3h)], and an output term due to root exudation [$O_{CS,rt\to so,ex}$, Eqn (3.3i)] which connects to the soil submodel. The last output term $O_{CS,rt\to so}$ [Eqn (3.3j)] is the flux of substrate C to senescence.

3.3.1 Canopy photosynthesis, $I_{CS,en\to sh}$ (kg substrate C m^{-2} day^{-1})

Canopy photosynthesis is represented by the input term, $I_{CS,en\to sh}$ (kg C m^{-2} day^{-1}), in the first of Eqns (3.2a). There is no wholly satisfactory treatment of canopy photosynthesis (Boote and Loomis, 1991; Sands, 1995). There is, however, general agreement that leaf photosynthesis is well-described by a non-rectangular hyperbola [see Eqn (3.2e) below]. The non-rectangular hyperbola was introduced into photosynthesis research by Rabinowitch (1951), developed by Chartier (1966), and has been used by numerous authors (e.g. Hirose and Werger, 1987) as a phenomenological equation which gives a succinct and accurate summary of photosynthetic response. The problem is that some of the parameters of the non-rectangular hyperbola, notably the light-saturated rate P_{max} and perhaps less importantly the 'sharpness' parameter ξ [Eqn (3.2e)], are affected by leaf N content, the average light and temperature over the past 7 days or so and by leaf age (Berry and Björkman, 1980; Robson *et al.*, 1988, pp. 46–50; Battaglia *et al.*, 1996). Not only do these parameters acclimate to current light and temperature levels with a time constant of about 7 days, but light and temperature also affect leaf N levels, which are of course influenced also by soil N. Most models of photosynthesis are static (e.g. Chartier, 1966; Farquhar and von Caemmerer, 1982), whereas a simplified dynamic model, for example of the type proposed by Hahn (1991), seems to be required: in such a model the parameters of the non-rectangular hyperbola would appear as time-dependent variables influenced by environment (see also Evans *et al.*, 1993). Our treatment of canopy photosynthesis is also static. Three options for integrating from leaf to canopy level are provided.

Light profile in the canopy

This is calculated by assuming that the canopy is closed, and that light decreases on descending into the canopy according to the Monsi–Saeki equation (Monsi and Saeki, 1953). Below, $j_{PAR,sc}$ (J m^{-2} s^{-1}) is the photosynthetically active radiation (PAR) above the canopy (Fig. 7.4a), and according to the Monsi–Saeki formula, the light incident on a leaf surface at depth L'_{AI}, j_{leaf} (J m^{-2} s^{-1}), is

$$\begin{aligned} j_{leaf} &= \frac{k_{can}\, j_{PAR,sc}}{1 - \chi_{leaf}} \exp(-k_{can} L'_{AI}), \\ k_{can} &= 0.5 \text{ m}^2 \text{ ground (m}^2 \text{ leaf)}^{-1}, \qquad \chi_{leaf} = 0.1. \end{aligned} \tag{3.2b}$$

k_{can} is the canopy extinction coefficient; χ_{leaf} is the leaf transmission coefficient. This equation is plotted in Fig. 3.2 for two values of the extinction coefficient, k_{can}. The upright leaves of grass have a low extinction coefficient and light penetrates well into the canopy. This is seen for the graph with $k_{can} = 0.5$. More light is intercepted if the canopy extinction coefficient is higher (planophile leaves), as in the $k_{can} = 0.9$ plot. For $k_{can} = 0.5$, the canopy intercepts less light as shown by the higher value of j_{leaf} at the bottom of the canopy ($L_{AI} = 5$). However, the light is distributed more uniformly over the leaves. As the single leaf light response curve [Eqn (3.2e)] is non-linear (Fig. 3.3), it is possible for either of the values $k_{can} = 0.5$ or 0.9 to lead to a higher canopy photosynthesis (see Fig. 3.4), depending on the canopy leaf area index L_{AI} and on the light level above the canopy $j_{PAR,sc}$.

Leaf photosynthesis

The dependence of leaf gross photosynthesis P_{leaf} (kg CO_2 m^{-2} s^{-1}) on the light incident on the leaf j_{leaf} [Eqn (3.2b)] is described by a non-rectangular hyperbola:

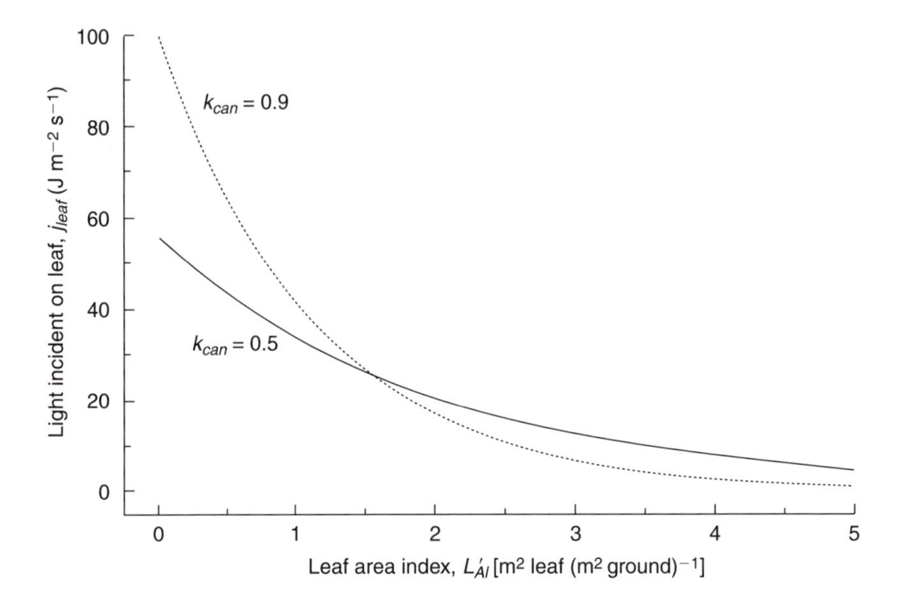

Fig. 3.2. Light incident on an average leaf at 'depth' L'_{AI} in the canopy calculated with Eqn (3.2b) assuming that the light incident on the top of the canopy is $j_{PAR,sc}$ = 100 J m^{-2} s^{-1}. Two different canopy extinction coefficients k_{can} are shown.

$$0 = \xi P_{leaf}^2 - P_{leaf}\left(\alpha j_{leaf} + P_{max}\right) + \alpha j_{leaf} P_{max}, \quad \xi = 0.95. \tag{3.2c}$$

This equation has three parameters: ξ, α and P_{max}, and is drawn in Fig. 3.3 for four values of ξ. ξ determines the sharpness of the knee of the curve. ξ = 0.95 is the value used in the grass simulator. There is evidence that ξ varies with growth light environment and with tissue N content (Hirose and Werger, 1987) although it is assumed to be constant through the canopy. α [kg CO_2 (J PAR)$^{-1}$] is the initial slope of the response curve and is sometimes called the photosynthetic efficiency. It is equivalent to the inverse of a quantum requirement: with 1 J of photosynthetically active radiation (PAR) = 4.26×10^{-6} mol of photons and 1 kg CO_2 = 1000 / 44 = 22.7 mol CO_2, an $\alpha = 10^{-8}$ kg CO_2 (J PAR)$^{-1}$ is equivalent to a quantum requirement of 4.26×10^{-6} / (22.7 × α) = 19 quanta per molecule of CO_2. This is a typical value for C_3 or C_4 leaves (Jones, 1992, pp. 204–205). α is a well-conserved quantity, varying greatly with ambient CO_2 concentration in C_3 species, a small dependence on temperature [Eqn (3.2s)], and little dependence on tissue N. P_{max} (kg CO_2 m^{-2} s^{-1}) is the asymptote of the light response curve. It is highly variable, being affected by tissue N, ambient CO_2, current temperature, growth temperature, growth light, and acclimating to CO_2 [Eqn (3.2t)]. A $P_{max} = 1 \times 10^{-6}$ kg CO_2 m^{-2} s^{-1} is equivalent to 22.7 μmol CO_2 m^{-2} s^{-1}.

The values for α and P_{max} used in plotting Fig. 3.3 are illustrative: in the model α and P_{max} are variables depending upon temperature, ambient CO_2 and other factors; they are calculated dynamically by Eqns (3.2s) and (3.2t) below. Note that if

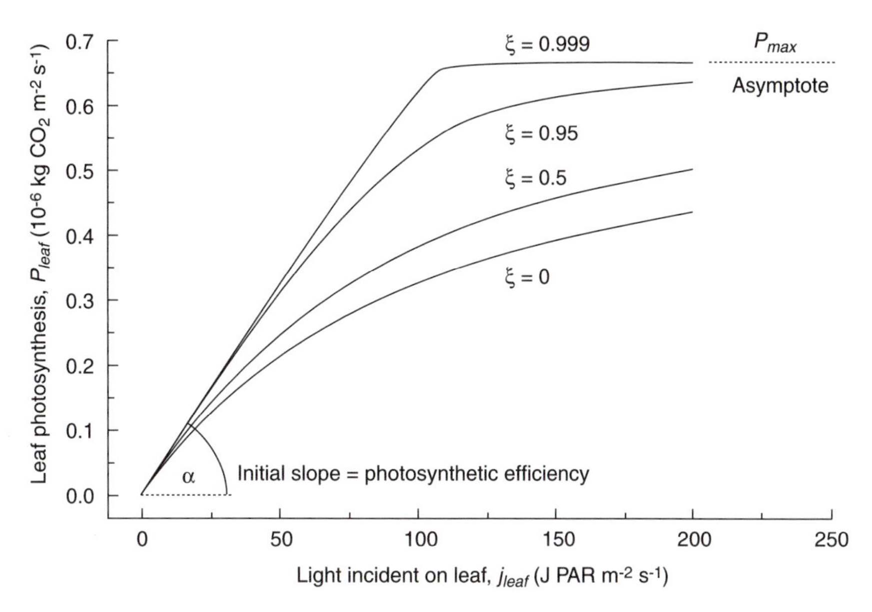

Fig. 3.3. Leaf photosynthesis:light response curve. Eqn (3.2e) is plotted for $\alpha = 0.636 \times 10^{-8}$ kg CO_2 J^{-1} [0.031 μmol CO_2 (mol PAR)$^{-1}$], $P_{max} = 0.667 \times 10^{-6}$ kg CO_2 m^{-2} s^{-1} (14 μmol CO_2 m^{-2} s^{-1}). These values are for 20°C at 350 vpm CO_2 without any nitrogen or water limitations. Four different values of the 'sharpness' parameter ξ are shown. 1 J PAR = 4.6 μmol PAR.

P_{max} is very large in Eqn (3.2c), so that the curves in Fig. 3.3 do not bend over, then the leaf photosynthetic rate is given by [retain just the terms with P_{max} in Eqn (3.2c)]

$$P_{leaf}(P_{max} \to \infty) = \alpha j_{leaf} . \tag{3.2d}$$

Eqn (3.2c) is a flexible equation which is capable of describing a wide range of light response curves. For us it has a valuable advantage that it can be integrated analytically through the canopy for a light regime distributed according to Eqn (3.2b) and Fig. 3.2, if it is assumed that the parameters of Eqn (3.2c) are the same throughout the canopy. This gives a computationally efficient way of calculating canopy photosynthesis, although arguably it may be biologically inaccurate.

Although Eqn (3.2c) is highly convenient, it is not possible to relate its parameters unambiguously to particular parts of the underlying biochemical machinery. As stated in an excellent paper by Stitt (1991),

> Although it has sometimes been assumed or asserted that photosynthesis is controlled by single 'limiting' factors … , this assertion is not easy to reconcile with the highly interactive and complex regulation of photosynthesis … , nor with the experimental evidence that control is shared between several enzymes and redistributes in a very flexible manner in other pathways … .

As discussed by Pachepsky *et al.* (1996), Eqn (3.2c) is equivalent to at least one of the versions of Farquhar's simplified biochemical model (Evans and Farquhar, 1991). However, for the reasons discussed by Stitt (1991), this is less of a justification for its use at the leaf level than the fact that, empirically, it works exceedingly well.

The non-rectangular hyperbola of Eqn (3.2c) is a quadratic in P_{leaf}. Solving for P_{leaf} gives

$$P_{leaf} = \frac{\alpha j_{leaf} + P_{max} - \sqrt{\left[\left(\alpha j_{leaf} + P_{max}\right)^2 - 4\xi \alpha j_{leaf} P_{max}\right]}}{2\xi}. \quad (3.2e)$$

Canopy photosynthesis

There are three options for calculating canopy photosynthesis in the model, obtained by using a program switch (s_Pcan).

NUMERICAL RECIPE FOR INSTANTANEOUS CANOPY PHOTOSYNTHESIS, P_{can}. A numerical recipe for instantaneous gross canopy photosynthesis P_{can} [kg C (m^2 ground)$^{-1}$ day^{-1}], can be written down which uses Eqn (3.2b) for j_{leaf}, the light incident on leaf surfaces at depth L'_{AI} in the canopy, and Eqn (3.2e) for the resulting gross photosynthesis of those leaves P_{leaf}. Adding up the contributions of layers of leaf area index $\Delta L'_{AI}$, therefore

$$\text{Initialize: } L'_{AI} = 0; \quad P_{can} = 0; \quad \Delta L'_{AI} = 0.2 \text{(say)};$$

$$\text{do} \left\{ j_{leaf} = \frac{k_{can} j_{PAR,sc}}{1 - \chi_{leaf}} \mathrm{e}^{-k_{can} L'_{AI}}; \right.$$

$$P_{leaf} = \frac{\alpha j_{leaf} + P_{max} - \sqrt{\left[\left(\alpha j_{leaf} + P_{max}\right)^2 - 4\xi \alpha j_{leaf} P_{max}\right]}}{2\xi}; \quad (3.2f)$$

$$\left. P_{can} := P_{can} + P_{leaf} \Delta L_{AI}; \quad L'_{AI} := L'_{AI} + \Delta L'_{AI}; \right\}$$

$$\text{while } L'_{AI} < L_{AI};$$

$$P_{can} := P_{can} - P_{leaf}\left(L'_{AI} - L_{AI}\right);$$

$$P_{can} := \frac{86{,}400 \times 12}{44} P_{can}.$$

The *do* statement is evaluated before the *while* statement, so the last evaluation of the *do* statement is for $L'_{AI} > L_{AI}$. The last but one statement corrects for this overshoot. The last line changes the units from s^{-1} to day^{-1}, and kg CO_2 to kg C. ': =' denotes 'is assigned the value of'.

This equation gives the instantaneous canopy photosynthetic rate, P_{can} (kg C m^{-2} day^{-1}) for given values of the light flux density incident on the canopy $j_{PAR,sc}$ (J m^{-2} s^{-1}; Fig. 7.4a) and the total L_{AI}. P_{can} depends on two canopy parameters [k_{can}, χ_{leaf}; Eqn (3.2b)] and three leaf parameters [α, P_{max}, ξ; Eqn (3.2c), Fig. 3.3]. Leaf area index increments of 0.2 are used here to step through the canopy, and the contributions to canopy photosynthesis are summed. Using this numerical approach, it is possible to make the leaf photosynthesis parameters depend, for instance, on depth within the canopy (L'_{AI}), or an average of the recent light level at that canopy depth, or tissue age, which may vary with position in the canopy, or tissue N content, which may be assumed to vary with depth in the canopy or be predicted by a submodel that calculates N distribution. One method of allowing for the decline in P_{max} with age is to assign different P_{max} values to different leaf age categories, and then to assume that the ratios of age categories is the same in all layers down the canopy (Parsons, personal communication). [Note: for high leaf area index canopies, Eqn (3.2f) can be far slower to compute than the analytical Eqn (3.2n) given below.]

ANALYTICAL EQUATION FOR CANOPY PHOTOSYNTHESIS, P_{can}. Alternatively, canopy gross photosynthesis, P_{can}, can be written as a single analytical expression where the contributions of all the leaves in the canopy are added by using an algebraic expression for the integral through the canopy, rather than doing the numerical summation as in Eqn (3.2f). This assumes that the leaf parameters in Eqn (3.2e) are the same down the canopy. Although this is untrue, especially for P_{max} (e.g. Sands, 1995), so long as the correct value for P_{max} at the top of the canopy is used, this method gives results of satisfactory accuracy rather efficiently.

First, write P_{leaf} as a function of leaf area index instead of j_{leaf} as given by Eqn (3.2e). Substitute for j_{leaf} in Eqn (3.2e) with Eqn (3.2b) to give

$$P_{leaf}\left(L'_{AI}\right) = \frac{p_0 e^{-k_{can}L'_{AI}} + P_{max} - \sqrt{\left[\left(p_0 e^{-k_{can}L'_{AI}} + P_{max}\right)^2 - 4\xi p_0 e^{-k_{can}L'_{AI}} P_{max}\right]}}{2\xi}. \quad (3.2g)$$

This equation has been simplified by defining a constant, p_0, with

$$p_0 = \frac{a k_{can} j_{PAR,sc}}{1 - \chi_{leaf}}. \quad (3.2h)$$

The integral for canopy photosynthesis is then

$$P_{can} = 86{,}400 \times \frac{12}{44} \times \int_0^{L_{AI}} P_{leaf}(L'_{AI}) \mathrm{d}L'_{AI}. \quad (3.2i)$$

The factors of 86,400 s day^{-1} and 12/44 kg C (kg CO_2)$^{-1}$ are as above. $P_{leaf}(L'_{AI})$ is given by Eqn (3.2g). L'_{AI} is a 'dummy' variable which varies from 0 at the top of the canopy to L_{AI} at the bottom of the canopy as the integration is performed. Substituting Eqn (3.2g) into Eqn (3.i) gives

$$P_{can} = \frac{86{,}400 \times 12}{2 \times 44\xi} \times \int_0^{L_{AI}} \left\{ p_0 e^{-k_{can}L'_{AI}} + P_{max} - \sqrt{\left[\left(p_0 e^{-k_{can}L'_{AI}} + P_{max}\right)^2 - 4\xi p_0 e^{-k_{can}L'_{AI}} P_{max}\right]} \right\} dL'_{AI}. \quad (3.2j)$$

This integral becomes simpler if the variables are changed. With

$$x = p_0 e^{-k_{can}L'_{AI}}, \quad \mathrm{d}x = -k_{can} p_0 e^{-k_{can}L'_{AI}} \mathrm{d}L'_{AI} = -k_{can} x \mathrm{d}L'_{AI}, \quad (3.2k)$$

Eqn (3.2j) becomes

$$P_{can} = \frac{86{,}400 \times 12}{2 \times 44 \xi k_{can}} \int_{p_0 e^{-k_{can} L_{AI}}}^{p_0} \frac{x + P_{\max} - \sqrt{\left[(x + P_{\max})^2 - 4\xi x P_{\max}\right]}}{x} \, dx. \tag{3.2l}$$

This integral is a standard form (Gradshteyn and Ryzhik, 1994, p. 102, 2.267, 1). Apart from the factor outside the integral sign, the integral is the function of x:

$$\begin{aligned} F(x) = {} & x - \sqrt{\left[(x + P_{max})^2 - 4\xi P_{max} x\right]} \\ & + P_{max} \ln\left\{\sqrt{\left[(x + P_{max})^2 - 4\xi P_{max} x\right]} - (2\xi - 1)x + P_{max}\right\} \\ & + (2\xi - 1) P_{max} \ln\left\{\sqrt{\left[(x + P_{max})^2 - 4\xi P_{max} x\right]} + x - (2\xi - 1) P_{max}\right\}. \end{aligned} \tag{3.2m}$$

Some constant terms given by Gradshteyn and Ryzhik's formulae have been omitted as these will cancel out when the next step is taken. Putting in the upper and lower limits of the integral, Eqn (3.2l) for canopy photosynthesis becomes

$$P_{can} = \frac{86{,}400 \times 12}{2 \times 44 \xi k_{can}} \left[F(p_0) - F\left(p_0 e^{-k_{can} L_{AI}}\right)\right]. \tag{3.2n}$$

Here p_0 [Eqn (3.2h)] is proportional to the light falling on the canopy, $j_{PAR,sc}$.

The input of substrate C to the shoot from the environment due to photosynthesis, $I_{CS,en \to sh}$, is given by

$$I_{CS,en \to sh} = P_{can}, \tag{3.2o}$$

where canopy photosynthesis can be obtained by Eqn (3.2f) or Eqn (3.2n). In either case it can be shown that the initial slope [units: kg C (J PAR)$^{-1}$] and the asymptote [units: kg C (m^2 ground)$^{-1}$ day^{-1}] of the canopy light response equation are (Thornley and Johnson, 1990, p. 249):

$$\begin{aligned} \frac{1}{86{,}400} \frac{dP_{can}}{dj_{PAR,sc}}\left(j_{PAR,sc} = 0\right) &= \frac{12}{44} \frac{\alpha}{1 - \chi_{leaf}} \left[1 - \exp(-k_{can} L_{AI})\right], \\ P_{can}\left(j_{PAR,sc} \to \infty\right) &= \frac{86{,}400 \times 12}{44} P_{max} L_{AI}. \end{aligned} \tag{3.2p}$$

The effects of extinction coefficient (k_{can}) and leaf area index (L_{AI}) on canopy photoynthesis at different light levels are illustrated in Fig. 3.4. The calculations are made with either Eqn (3.2f) or the equivalent Eqn (3.2n). It can be seen that at low leaf area index, the most efficient canopy is that with a high extinction coefficient ('prostrate' or 'planophile' leaves) which gives high interception of light. However, at higher L_{AI}, and at moderate to high light levels, the most efficient canopy is that with a low extinction coefficient (erect leaves; here the extinction coefficient k_{can} = 0.4) which allows light to penetrate into the canopy. At low light levels, all intercepted light is used efficiently – the leaf is operating on the steeper initial portion of the leaf response curves shown in Fig. 3.3. With a high L_{AI} and low light the extinction coefficient does not make much difference because most of the light is

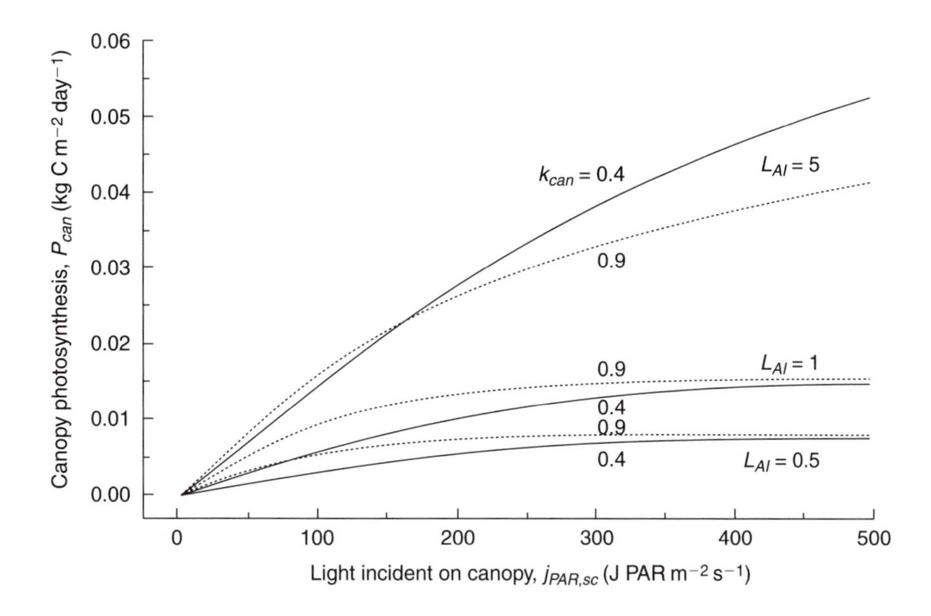

Fig. 3.4. Canopy photosynthesis:light response curve. Eqn (3.2n) is plotted for $\alpha = 0.636 \times 10^{-8}$ kg CO_2 J^{-1} [0.031 μmol CO_2 (mol PAR)$^{-1}$], $P_{max} = 0.667 \times 10^{-6}$ kg CO_2 m^{-2} s^{-1} (14 μmol CO_2 m^{-2} s^{-1}), $\xi = 0.95$, $\chi_{leaf} = 0.1$. These values are for 20°C at 350 vpm CO_2 without any nitrogen or water limitations. Two different values of canopy extinction coefficient are used, k_{can} = 0.4 and 0.9 m^2 ground (m^2 leaf)$^{-1}$, with three different values of the leaf area index, L_{AI}. 1 J PAR = 4.6 μmol PAR. 0.05 kg C m^{-2} day^{-1} = 48 μmol C m^{-2} s^{-1}.

intercepted and used with a similar efficiency. With a low L_{AI}, a smaller fraction of the light is intercepted so the efficiency of interception is important. The large difference between the k_{can} = 0.4 and k_{can} = 0.9 curves at low L_{AI} and low light arises from interception efficiency. The large difference at high L_{AI} and high light is due to the intercepted light being more evenly distributed over the canopy (caused by the low k_{can}) so that the leaves are operating on a more efficient portion of their photosynthetic response curve (Fig. 3.3).

CANOPY PHOTOSYNTHESIS ASSUMING P_{max} DECREASES DOWN THE CANOPY. The above derivation of canopy photosynthesis is based on the assumption that the leaf photosynthesis parameters [ξ, α and P_{max} of Eqn (3.2c)] do not change with depth in the canopy. There is much evidence that P_{max} depends strongly on the average light level experienced by a leaf, so that leaves lower down the canopy have lower P_{max} values (e.g. Fig. 2.7, p. 43, of Robson *et al.*, 1988; see also discussion above). The third of the canopy photosynthesis options provided in the model uses Charles-Edwards' assumption that P_{max} decreases down the canopy in the same way as does the light level, i.e. according to Eqn (3.2b) (Charles-Edwards *et al.* 1986, p. 58; also Sands, 1995).

Assume that the light incident on a leaf surface at depth L'_{AI} in the canopy, j_{leaf}

(J PAR m^{-2} s^{-1}) is given by the Monsi–Saeki Eqn (3.2b), and that the P_{max} parameter of the leaf response curve [Eqn (3.2e)] is described by

$$P_{max,LAI} = P_{max,0}\, e^{-k_{can} L'_{AI}}. \tag{3.2q}$$

$P_{max,0}$ (kg CO_2 m^{-2} s^{-1}) is the value of P_{max} at the top of the canopy. Substitute this equation for P_{max} and Eqn (3.2b) for j_{leaf} in Eqn (3.2e), and use the resulting equation in the integral in eqn (3.2i), to obtain an expression for canopy photosynthesis:

$$P_{can} = \frac{86{,}400 \times 12}{2 \times 44\xi} \times \frac{1 - e^{-k_{can} L_{AI}}}{k_{can}} \times \left\{ \frac{\alpha k_{can} j_{PAR,sc}}{1 - \chi_{leaf}} + P_{max,0} - \sqrt{\left[\left(\frac{\alpha k_{can} j_{PAR,sc}}{1 - \chi_{leaf}} P_{max,0}\right)^2 - \frac{4\xi\alpha k_{can} j_{leaf} P_{max,0}}{1 - \chi_{leaf}}\right]} \right\}. \tag{3.2r}$$

For the same P_{max} at the upper canopy surface, canopy photosynthesis is always decreased by the assumption that P_{max} decreases down the canopy. However, at low light levels the effect is small because P_{max} is less important than α. Also at low leaf area indices the effect is small because essentially the whole canopy is in the upper layer. These effects are illustrated in Table 3.2.

Table 3.2 compares canopy photosynthesis calculated assuming (row 4) that P_{max} is constant throughout the canopy, and in row 6, assuming P_{max} decreases according to Eqn (3.2q). The calculation has been performed for three light levels ($j_{PAR,sc}$) and two leaf area indices (L_{AI}). The second row $\alpha j_{PAR,sc}$ gives the canopy photosynthetic rate that would occur if the leaf photosynthetic response to light were linear with slope α (i.e. with $P_{max} \rightarrow \infty$), and all the incident light is converted into CO_2 with conversion yield α. It is only when this value is comparable or

Table 3.2. Canopy photosynthesis is calculated assuming first that the light-saturated leaf photosynthesis parameter P_{max} is constant through the canopy [row 4, Eqns (3.2n) or (3.2f)], or second that it decreases with light level through the canopy as in Eqns (3.2q) and (3.2r) (row 5). Parameters are: $\xi = 0.95$, $\chi_{leaf} = 0$, $k_{can} = 0.5$, $\alpha = 6.36 \times 10^{-9}$ kg CO_2 (J PAR)$^{-1}$ [0.031 mol CO_2 (mol PAR)$^{-1}$], $P_{max} = P_{max,0} = 0.667 \times 10^{-6}$ kg CO_2 m^{-2} s^{-1} (15 µmol CO_2 m^{-2} s^{-1}). The calculation is for three values of the light flux density ($j_{PAR,sc}$) and two values of leaf area index (L_{AI}).

$j_{PAR,sc}$ (J PAR m^{-2} s^{-1}) [µmol m^{-2} s^{-1}]	20 [92]		100 [460]		300 [1380]	
$\alpha j_{PAR,sc}$ (10^{-6} kg CO_2 m^{-2} s^{-1}) [µmol CO_2 m^{-2} s^{-1}]	0.127 [2.9]		0.636 [14]		1.91 [43]	
Leaf area index, L_{AI}	1	5	1	5	1	5
(1) Canopy photosynthesis with constant P_{max} (10^{-6} kg CO_2 m^{-2} s^{-1}) [µmol CO_2 m^{-2} s^{-1}]	0.050 [1.14]	0.116 [2.64]	0.243 [5.52]	0.573 [13]	0.567 [12.9]	1.52 [35]
(2) Canopy photosynthesis with decreasing P_{max} (10^{-6} kg CO_2 m^{-2} s^{-1}) [µmol CO_2 m^{-2} s^{-1}]	0.050 [1.14]	0.116 [2.64]	0.240 [5.45]	0.560 [12.7]	0.482 [11]	1.12 [25]

greater than P_{max} that P_{max} can limit canopy photosynthesis: that is, for above-canopy irradiances of at least 100 J m^{-2} s^{-1}. For southern Britain, the average light level during the day (annual mean) is *c.* 110 J PAR m^{-2} s^{-1}. Even if P_{max} is limiting photosynthesis, assuming a variable P_{max} within the canopy will not affect canopy photosynthesis in a low L_{AI} canopy. At an irradiance of 300 J m^{-2} s^{-1}, at L_{AI} of 1, assuming decreasing P_{max} in the canopy decreases canopy photosynthesis by 15%; at L_{AI} of 5, the decrease is 26%. If the parameter ξ is decreased, this increases the influence of P_{max} on photosynthetic rate at lower values of irradiance (Fig. 3.3); this increases the differences between rows 4 and 6 of Table 3.2.

The results in Table 3.2 suggest that the effect of a decreasing P_{max} going down the canopy has less impact on canopy photosynthesis than having the correct value of P_{max} at the top of the canopy. This is especially true for temperate grassland where light levels are moderate and the leaf area index is limited by grazing or cutting.

Leaf photosynthesis parameters

The single leaf photosynthesis curve [Eqn (3.2c)] has three parameters, ξ α and P_{max}. It is assumed that ξ, the parameter determining the sharpness of the knee of the leaf response curve (Fig. 3.3), is unaffected by plant or environmental variables, although there is evidence that increasing tissue N decreases ξ (θ of Fig. 3 of Hirose and Werger, 1987), and also that increasing growth light decreases ξ (Fig. 5 of Prioul *et al.*, 1980).

Photosynthetic efficiency α depends on ambient CO_2 concentration, leaf temperature and leaf water status. It is calculated with

$$\begin{aligned}
&\alpha = \alpha_{m,15} f_{CO_2,\alpha} f_{T,\alpha} f_{W,ph}, \\
&\alpha_{m,15} = 1\times10^{-8}\ \text{kg CO}_2\ \text{J}^{-1}[0.05\ \text{mol CO}_2\ (\text{mol PAR})^{-1}], \\
&f_{CO_2,\alpha} = 1 - \frac{\beta}{\tau C_{O_2,air}}, \qquad \beta = 0.3\times10^{-6}\ \text{kg CO}_2\ \text{m}^{-2}\text{s}^{-1}, \\
&\qquad \tau = 0.0015\ \text{m s}^{-1}, \\
&f_{T,\alpha} = 1 \qquad \text{if } T_{air} \le 15°\text{C}, \\
&\qquad 1 - c_{T,\alpha}(T_{air} - 15) \qquad \text{if } T_{air} > 15°C, \qquad c_{T,\alpha} = 0.015(°\text{C})^{-1}, \\
&f_{W,ph} = (a_{W,sh})^{q_{W,ph}}, \quad q_{W,ph} = 2.
\end{aligned} \tag{3.2s}$$

The first factor $f_{CO_2,\alpha}$ allows empirically for the effect of ambient CO_2 in suppressing photorespiration in C_3 plants; this increases photosynthetic efficiency [cf. equation 2.37 of Charles-Edwards (1981)]. $C_{O_2,air}$ (kg CO_2 m^{-3}) is the ambient CO_2 concentration [Eqn (7.3a)]. The expression used for the CO_2 dependence of α is not valid below about 300 vpm CO_2. The second factor $f_{T,\alpha}$ represents that increasing temperature above 15°C decreases α by 1.5% per °C (Ehleringer and Björkman, 1977; Loomis and Connor, 1992, p. 263). Leaf temperature is assumed equal to air temperature. The third factor, $f_{W,ph}$, allows empirically for the effect of shoot water status on α; shoot water activity ($a_{W,sh}$) is calculated from shoot water potential (ψ_{sh}) in Eqn (6.7a); the exponent $q_{W,ph}$ is smaller than that used for general plant biochemistry [Eqn (3.11d), Section 3.12], giving the result that water stress has a smaller effect on α than on other plant rate constants (Section 6.8). There is no evidence that

α responds to N nutrition (Hirose and Werger, 1987; Evans and Terashima, 1988; Gastal and Bélanger, 1993).

The light-saturated photosynthetic rate parameter P_{max} depends here on ambient CO_2 concentration, temperature, water status, relative stomatal opening and tissue N concentration. Leaf age effects are ignored. P_{max} is computed with

$$
\begin{aligned}
P_{max} &= P_{max,20} f_{CO_2,P_{max}} f_{T,P_{max}} f_{W,ph} f_{gs,P_{max}} f_{N,P_{max}}, \\
P_{max,20} &= 2\times10^{-6}\,\text{kg CO}_2\left(\frac{2000}{44} = 45.5\,\mu\text{mol CO}_2\right)\text{m}^{-2}\text{s}^{-1}, \\
f_{CO_2,P_{max}} &= \frac{1}{1+K_{CO_2,P_{max}}/C_{O_{2,air}}}, \\
K_{CO_2,P_{max}} &= 1.281\times10^{-3}\,\text{kg CO}_2\ \text{m}^{-3} \qquad (3.2t) \\
& \quad (700\ \text{vpm CO}_2,\ 20^\circ\text{C and atmospheric pressure}), \\
f_{gs,P_{max}} &= \frac{g_s}{g_{s,max}}, \\
f_{N,P_{max}} &= 0.5 \quad \text{if N}_{sh,lamA} \le 0.001, \\
f_{N,P_{max}} &= 0.5\left(1+\frac{N_{sh,lamA}-0.001}{0.001}\right) \quad \text{if } 0.001 < N_{sh,lamA} < 0.002, \\
f_{N,P_{max}} &= 1 \quad \text{if } 0.002 \le N_{sh,lamA}.
\end{aligned}
$$

$P_{max,20}$ is a notional maximum value when all the f modifiers are unity, including that for CO_2. $f_{CO_2,P_{max}}$ allows empirically for the effect of CO_2 on P_{max}. For an ambient CO_2 concentration of 350 vpm at 20°C [see Eqn (7.3a) and after], $f_{CO_2,Pmax}$ = 1/3, and in the absence of limitation due to temperature, water, stomatal closure or nitrogen, $P_{max,20}$ at 350 vpm CO_2 is 0.667 × 10^{-6} kg CO_2 m^{-2} s^{-1} (× 1000/44 = 15.2 μmol CO_2 m^{-2} s^{-1}). Robson *et al.* (1989, Fig. 2.7, p. 43) report a range of P_{max} values, from 0.2 to 1.2 × 10^{-6} kg CO_2 m^{-2} s^{-1}, depending on growth light environment. We use a rounded value in the centre of this range, remembering that it is notoriously difficult to estimate the position of an asymptote from hyperbolic light response data. The temperature function for P_{max} ($f_{T,Pmax}$) is given in Eqn (3.11c), in which the optimum temperature depends on ambient CO_2 (Long, 1991), although the effect of this variation of optimum temperature with CO_2 is negligible in temperate environments. The water status function ($f_{W,ph}$) is as in Eqn (3.2s) (Section 6.8). The stomatal factor $f_{gs,P_{max}}$ decreases P_{max} as stomatal conductance g_s falls below its maximum value [Eqn (3.2u)]. Note that water stress can decrease P_{max} both through $f_{W,ph}$ and also through possible stomatal closure. The effect of leaf lamina nitrogen content [Eqn (3.1j)] on P_{max}, $f_{N,P_{max}}$, is a ramp function, after Kristensen and Lantinga (personal communication) and Hirose and Werger (1987), with a minimum value of 0.5 when the laminar N concentration, $N_{sh,lamA}$, is below 0.001 kg N m^{-2} (this corresponds to a leaf N concentration of *c.* 2.2% of dry mass), a maximum value of 1 when $N_{sh,lamA}$ is above 0.002 kg N m^{-2}, and varying linearly in between (Woledge and Pearse, 1985, Fig. 2). Because the intercept of the sloping part of the ramp function passes through the origin, and leaves with N concentrations below 0.001 kg N m^{-2} are very uncommon, this is equivalent to assuming a linear dependence of $f_{N,P_{max}}$ on $N_{sh,lamA}$ with a maximum value (Parsons, personal communication).

The relationship between N supply and single leaf photosynthesis is much investigated (e.g. Evans, 1989; Pettersson and McDonald, 1994), as are possible relationships between canopy N distribution, canopy photosynthesis and plant growth (e.g. Dewar, 1996; Medlyn, 1996). Both topics are controversial and arguably need addressing with a mechanistic model combining leaf photosynthesis and the allocation of C and N substrates within the canopy. Here variable N content in the canopy is ignored in the default calculations of canopy photosynthesis (but see Eqn (3.2r) and above).

Leaf stomatal conductance g_s [m s^{-1} (m^2 leaf)$^{-1}$] is given by

$$
\begin{aligned}
& f_{gs,CO_2} = \frac{1 + c_{gs,CO_2}}{1 + c_{gs,CO_2} \dfrac{C_{O_2,vpm}}{350}}, \qquad c_{gs,CO_2} = 2, \\
& g_{s,max} = f_{gs,CO_2} g_{s,max,350}, \qquad g_{s,min} = f_{gs,CO_2} g_{s,min,350}, \\
& g_{s,max,350}, g_{s,min,350} = 0.005, 0.00005 \,\mathrm{m\ s^{-1}} \left(\mathrm{m^2\ leaf}\right)^{-1}, \\
& g_s = g_{s,min} \qquad \text{if } \left(\theta_{sh} \le \theta_{gs,min} \text{ or } j_{leaf} = 0\right), \\
& g_s = g_{s,min} + \frac{\left(\theta_{sh} - \theta_{gs,min}\right)\left(g_{s,max} - g_{s,min}\right)}{\theta_{gs,max} - \theta_{gs,min}} \\
& \qquad \text{if } \left[\left(\theta_{gs,min} < \theta_{sh} \le \theta_{gs,max}\right) \text{ and } j_{leaf} > 0\right], \\
& g_s = g_{s,max} \qquad \text{if } \left(\theta_{sh} \ge \theta_{gs,max} \text{ and } j_{leaf} > 0\right), \\
& \theta_{gs,min}, \theta_{gs,max} = 0.65, 0.85.
\end{aligned}
\tag{3.2u}
$$

The acclimation response of stomatal conductance to atmospheric CO_2 concentration is defined by the factor f_{gs,CO_2} with parameter c_{gs,CO_2}. When the CO_2 concentration $C_{O_2,vpm}$ = 350 vpm, the factor f_{gs,CO_2} = 1. When the CO_2 concentration is doubled to 700 vpm, the factor $f_{gs,CO2}$ takes the value $(1 + c_{gs,CO_2})/(1 + 2c_{gs,CO_2})$ = 3/5 for c_{gs,CO_2} = 2, reducing leaf stomatal conductance by 40% (Morison, 1985; Polley *et al.*, 1993). θ_{sh} is the shoot relative water content [Eqn (6.5d), i = sh]. The fully open and closed leaf stomatal conductances at 350 vpm CO_2 are $g_{s,max,350}$ and $g_{s,min,350}$. The shoot relative water contents when the stomata are closed or fully open are $\theta_{gs,min}$ and $\theta_{gs,max}$. Leaf stomatal conductance varies linearly with shoot relative water content θ_{sh} between the fully open and the completely closed positions. Jones (1992, p. 145) describes a range of 0.002–0.008 m s^{-1} for the maximal stomatal leaf stomatal conductances for cultivated C_3 grasses. Kelliher *et al.* (1995) report values of 0.006 and 0.012 m s^{-1} for natural vegetation and crops. The stomata are closed in the dark [radiation flux j_{leaf} = 0, Eqn (3.2b)].

Zur and Jones (1981) used an exponential dependence of stomatal resistance on turgor pressure, remarking that research indicates that stomatal aperture is quite constant over a wide range of leaf water potential, but closure is rapid when a critical leaf water potential is reached. Ludlow (1980) reported a dependence of stomatal conductance on vapour density which depends on leaf water potential in

Panicum maximum. Ludlow (1987) suggested that there is no unique relationship between stomatal conductance and leaf water potential. Jones (1992) discussed stomatal response to light, intercellular CO_2, water status, humidity, temperature and leaf age; he also mentioned a dependence on nutrition. However, he did not describe a model of stomatal conductance suitable for the present application. Jensen *et al.* (1993) assumed that stomatal conductance depends on leaf water potential, PAR, leaf temperature and vapour density deficit. Ogink-Hendriks (1995, equation 7) used a model where stomatal conductance depends on global radiation, humidity deficit, air temperature and soil moisture deficit. Jones (1988, Fig. 6.9, table 6.3) gives data relating stomatal closure to water potential for six grasses: the threshold water potential for partial stomatal closure varies over the range -600 to -1500 J kg^{-1} (-0.6 to 1.5 MPa) and the water potential for complete closure is -1200 to > -3500 J kg^{-1} (-1.2 to > -3.5 MPa). Dewar (1995) proposed a mechanistic interpretation of an empirical model of stomatal response (Leuning, 1995). Dewar's method would require further development before it could be used here. Models such as that of Leuning which assume that a property variable (g_s) depends on a process variable (assimilation rate) can be misleading (Section 1.7) and may lead to awkward *implicit* relationships (an implicit relationship is one in which, for example, stomatal conductance depends on assimilation rate which in turn depends on stomatal conductance). A more fundamental (mechanistic) restatement of the problem avoids implicit relationships, and is usually more transparent.

3.3.2 Recycled C from litter fluxes, $I_{CS,re,sh}$, $I_{CS,re,rt}$ (kg substrate C m^{-2} day^{-1})

The shoot and root litter fluxes are assumed to be structural dry mass from age category 4 (Fig. 3.1) [Eqns (3.5d)]. It is assumed that variable fractions ($f_{N,shli,re}$, $f_{N,rtli,re}$) of the N in these structural litter fluxes can be recycled internally to the substrate N pools, the fractions depending on the N concentrations in shoot and root [Eqns (3.4c)] (Hunt, 1983). With default parameter and initial values, 80% and 58% of shoot and root litter structural N fluxes are recycled to the N substrate pools at $t = 0$ (1 January). First, the recycled fluxes of N ($I_{NS,re,sh}$, $I_{NS,re,rt}$) are calculated from the litter fluxes [Eqns (3.4d)]. Then, assuming there is a fixed C:N ratio, $r_{C:N,rec\text{-}li}$, in the recyclable litter, the recycled C fluxes are computed:

$$I_{CS,re,sh} = r_{C:N,rec-li} I_{NS,re,sh}, \qquad I_{CS,re,rt} = r_{C:N,rec-li} I_{NS,re,rt},$$
$$r_{C:N,rec-li} = 2.7 \text{ kg C (kg N)}^{-1}. \tag{3.3a}$$

It is assumed that the litter fluxes are only of structural dry mass, without any associated substrate C or N. Substrate C and N are retained in the plant.

3.3.3 Transport of substrate C from shoot to root, $O_{CS,sh\rightarrow rt}$ (kg substrate C m^{-2} day^{-1})

This is assumed to be proportional to the substrate C concentration difference, $C_{sh} - C_{rt}$ [Eqns (3.1h)], divided by a resistance, $r_{C,sh\rightarrow rt}$ [m^2 day (kg structural DM)$^{-1}$; this is the reciprocal of a flux density]. The resistance is calculated by adding resistances associated with shoot and root, $r_{C,sh}$ and $r_{C,rt}$. The resistances associated with shoot and root are obtained by scaling a resistivity coefficient ρ_{C20} according to structural dry mass with a dimensionless allometric constant $q_{\rho C}$ and adjusting for temperature and water status. Therefore

$$O_{CS,sh\to rt} = \frac{C_{sh} - C_{rt}}{r_{C,sh\to rt}},$$

$$r_{C,sh\to rt} = r_{C,sh} + r_{C,rt},$$

$$r_{C,sh} = \frac{r_{C,sh20}}{f_{T,sh}f_{W,sh}}, \quad r_{C,rt} = \frac{r_{C,rt20}}{f_{T,rt}f_{W,rt}}, \tag{3.3b}$$

$$r_{C,sh20} = \frac{\rho_{C20}}{M_{X,sh}^{q_{\rho C}}}, \quad r_{C,rt20} = \frac{\rho_{C20}}{M_{X,rt}^{q_{\rho C}}},$$

$$\rho_{C20} = 0.2 \text{ m}^2\text{day (kg structural dry mass)}^{q_{\rho C}-1}, \quad q_{\rho C} = 1.$$

The value of ρ_{C20} is chosen so that satisfactory values of C flux and C substrate concentration gradient coexist. The temperature and water functions, $f_{T,sh}$, $f_{T,rt}$, $f_{W,sh}$, $f_{W,rt}$, are obtained in Eqns (3.11b), (3.11d). Plant structural dry mass components $M_{X,sh}$, $M_{X,rt}$ are given by Eqns (3.1a).

3.3.4 Outputs of substrate C for shoot and root growth, $O_{CS,sh,G}$, $O_{CS,rt,G}$ (kg substrate C m^{-2} s^{-1})

The shoot and root growth rates, G_{sh}, G_{rt} (kg structural DM m^{-2} day^{-1}), are

$$G_{sh} = \mu_{20} f_{T,sh} f_{W,sh} C_{sh} N_{sh} M_{X,sh},$$

$$G_{rt} = \mu_{20} f_{T,rt} f_{W,rt} C_{rt} N_{rt} M_{X,rt},$$

$$\mu_{20} = 400\left[\text{kg C substrate (kg structural DM)}^{-1}\right]^{-1} \tag{3.3c}$$

$$\times\left[\text{kg N substrate (kg structural DM)}^{-1}\right]^{-1}\text{day}^{-1}.$$

μ_{20} is a 20°C growth rate constant. This is modified by temperature $f_{T,sh}$ [Eqn (3.11b)] and water $f_{W,sh}$ [Eqn (3.11d)]. DM denotes dry mass. Shoot growth depends on the total shoot structural dry mass, $M_{X,sh}$ [Eqn (3.1a)], multiplied by the shoot substrate C and N concentrations, C_{sh}, N_{sh} [Eqns (3.1h)]. Root growth rate G_{rt} is treated similarly. The shoot:root growth ratio is defined by

$$r_{sh:rt,G} = \frac{G_{sh}}{G_{rt}}. \tag{3.3d}$$

'C substrate' denotes mono- and di-saccharides and degradable polysaccharides but is not a sharply defined concept. Neither is 'N substrate'. With the temperature and water f factors equal to unity, C_{sh} = 0.05 (cf. Gill *et al.*, 1989, Fig. 3.1) and N_{sh} = 0.01 (Jeffrey, 1988, Fig. 5.5), whole-shoot specific growth rate is 400 × 0.05 × 0.01 = 0.2 day^{-1}. This value is compatible with Fig. 2.5 of Robson *et al.* (1988).

The output fluxes of substrate C from the shoot and root substrate C pools required to support growth rates of G_{sh} and G_{rt} are

$$O_{CS,sh,G} = f_{C,plX}\frac{G_{sh}}{Y_{G,pl}}, \quad O_{CS,rt,G} = f_{C,plX}\frac{G_{rt}}{Y_{G,pl}}, \quad Y_{G,pl} = 0.75. \tag{3.3e}$$

$f_{C,plX}$ is the fractional C content of plant structural dry mass, assumed the same for shoot and root [Eqn (3.1i)]. $Y_{G,pl}$ is the conversion yield or growth efficiency of the growth process: that is, the utilization of 1 kg of substrate C yields $Y_{G,pl}$ kg of C in the form of structural dry mass. The value of $Y_{G,pl}$ is quite precisely determined empirically (McCree, 1970; Thornley and Johnson, 1990, p. 270) and theoretically (Penning de Vries *et al.*, 1974; and e.g. Thornley and Johnson, 1990, equations 12.23f,g, p. 353).

The fraction utilized but not converted into plant structure is respired, giving shoot and root growth respiration rates of

$$R_{G,sh} = O_{CS,sh,G}(1 - Y_{G,pl}), \quad R_{G,rt} = O_{CS,rt,G}(1 - Y_{G,pl}). \tag{3.3f}$$

3.3.5 Outputs of substrate C for shoot and root maintenance, $O_{CS,sh \to en,m}$, $O_{CS,rt \to en,m}$ (kg substrate C m^{-2} day^{-1})

These are assumed to be proportional to the components of structural dry mass (DM), with different weighting for the different age categories, according to

$$\begin{aligned}
O_{CS,sh \to en,m} &= R_{m,sh} = f_{C,plX} \frac{C_{sh}}{C_{sh} + K_{C,mai}} f_{T,sh} f_{W,sh} \Big[k_{mai,sh1,20} \left(M_{lam,1} + M_{ss,1} \right) \\
&\quad + k_{mai,sh2,20} \left(M_{lam,2} + M_{ss,2} \right) + k_{mai,sh3,20} \left(M_{lam,3} + M_{ss,3} \right) \\
&\quad + k_{mai,sh4,20} \left(M_{lam,4} + M_{ss,4} \right) \Big], \\
O_{CS,rt \to en,m} &= R_{m,rt} = f_{C,plX} \frac{C_{rt}}{C_{rt} + K_{C,mai}} f_{T,rt} f_{W,rt} \\
&\quad \times \left(k_{mai,rt1,20} M_{rt,1} + k_{mai,rt2,20} M_{rt,2} + k_{mai,rt3,20} M_{rt,3} + k_{mai,rt4,20} M_{rt,4} \right), \\
K_{C,mai} &= 0.03 \text{ kg C substrate (kg structural DM)}^{-1}, \\
k_{mai,sh1-4,20} &= k_{mai,rt1-4,20} = \\
&\quad 0.02,\ 0.02,\ 0.015,\ 0.01 \text{ kg C substrate (kg C structural DM)}^{-1} \text{ day}^{-1}.
\end{aligned} \tag{3.3g}$$

$f_{C,plX}$ is the carbon content of plant structure, defined in Eqn (3.1i). The next Michaelis–Menten term reduces the rate of maintenance respiration at values of shoot or root substrate C C_{sh} (C_{rt}) below $K_{C,mai}$ = 0.03. Maintenance is modified for temperature [$f_{T,sh}$, $f_{T,rt}$, Eqns (3.11b)] and water status [$f_{W,sh}$, $f_{W,rt}$, Eqns (3.11d)]. The $k_{mai,shi,20}$, $k_{mai,rti,20}$, i = 1–4, are 20°C rate constants.

Some authors adjust the value of the maintenance rate constant according to tissue N content rather than tissue age as it is commonly observed that tissue maintenance rates correlate with tissue N content (e.g. Ryan, 1995). However, the two approaches are much the same as tissue N content generally decreases with increasing tissue age (Jeffrey, 1988, Fig. 5.6; Gill *et al.*, 1989, Fig. 3.1). Robson *et al.* (1988, p. 57) stress the sensitivity of maintenance to many factors: mineral nutrition, stage of plant development, light, temperature and growth conditions. For grasses, they give a range of 6–90 mg CO_2 (g DM)$^{-1}$ day^{-1}, which converts to 0.004–0.06 kg C substrate respired (kg C in DM)$^{-1}$ day^{-1}. This wide range highlights our lack of understanding of maintenance processes – the rate parameters in the above equation are regarded as being adjustable. Penning de Vries (1975) pioneered theoretical studies of maintenance respiration in plants. A recent in-depth study is by Bouma (1995) in his thesis, parts of which have been published (Bouma and De Visser, 1993; Bouma *et al.*, 1992, 1994). See also Thornley (1970), Amthor (1984) and Thornley and Johnson (1990, chapter 11).

3.3.6 Output of substrate C to the grazing animals $O_{CS,sh\to an}$ (kg substrate C m^{-2} day^{-1})

This flux is calculated in Eqn (3.7h). Outputs to the grazing animals are discussed in Section 3.7.

3.3.7 Output of substrate C to harvesting $O_{CS,sh\to hv}$ (kg substrate C m^{-2} day^{-1})

This flux is calculated in Eqn (3.8c). Fluxes of plant material to harvesting are discussed in Section 3.8.

3.3.8 Output of root substrate C for root N uptake, $O_{CS,rt\to en,uN}$ (kg substrate C m^{-2} day^{-1})

It is assumed that root uptake of mineral N has a respiratory cost, utilizing root substrate C and giving a respiratory flux of CO_2 of R_{uN}. Root uptake occurs from the soil N ammonium and nitrate pools, N_{amm}, N_{nit}, at rates of u_{Namm} and u_{Nnit} [kg N m^{-2} day^{-1}; Eqns (3.4i)]. The carbon cost associated with each of these fluxes is represented by parameters c_{uNamm} and c_{uNnit}. The respiratory flux is

$$\begin{aligned} &O_{CS,rt\to en,uN} = R_{uN} = R_{uNamm} + R_{uNnit}, \\ &R_{uNamm} = c_{uNamm}u_{Namm}, \qquad R_{uNnit} = c_{uNnit}u_{Nnit}, \\ &c_{uNamm} = 0.4, \qquad c_{uNnit} = 0.5 \text{ kg C substrate (kg N taken up)}^{-1}. \end{aligned} \tag{3.3h}$$

R_{uNamm} and R_{uNnit} are the components of respiration associated with the two uptake/reduction processes.

The cost of N from nitrate is assumed to be 25% higher than that of N from ammonium. Bouma (1995, chapter 7) calculates that the cost of N uptake from nitrate is 12% higher than from ammonium. He measured costs of N uptake from nitrate in the range 0.39–0.67 mol O_2 (mol NO_3^-)$^{-1}$, which converts to 0.33–0.57 kg C (kg N)$^{-1}$. Maintaining ion gradients can increase these values by up to 50%. Bloom *et al.* (1992) reported values of 0.49–1.08 mol O_2 (mol NO_3^-)$^{-1}$. There is reasonably good agreement between theory, experiment, and the rounded numbers used in Eqn (3.3h). Theory depends on assumptions concerning how many ATP are required per N transported (e.g. Penning de Vries *et al.*, 1974, assume 0.3), whether intra-cell transport costs are added, whether maintaining ion gradients is included, and where in the plant nitrate reduction takes place (see also Thornley and Johnson, 1990, pp. 320–322, 348–349).

3.3.9 Output of root substrate C to exudation, $O_{CS,rt\to so,ex}$ (kg substrate C m^{-2} day^{-1})

Root exudation of substrate C occurs at a rate of

$$O_{CS,rt\to so,ex} = k_{CS,rt,ex20} f_{T,rt} f_{W,rt} M_{X,rt} C_{rt}, \quad k_{CS,rt,ex20} = 0.02 \text{ day}^{-1}. \tag{3.3i}$$

$k_{CS,rt,ex20}$ is a rate constant. The flux is proportional to root structural dry mass [$M_{X,rt}$, Eqn (3.1a)] and root substrate C concentration [C_{rt}, Eqn (3.1h)]; it is modulated by root temperature [$f_{T,rt}$, Eqn (3.11b)] and root water status [$f_{W,rt}$, Eqn (3.11d)]. Under standard conditions (20°C, $f_{T,rt} = f_{W,rt} = 1$), 2% of the root substrate C is exuded per day. This exudation is to the soluble C pool C_{sol} in the soil model [Fig. 5.1; Eqn (5.3a)].

Running the model with defaults for one year, gross photosynthesis is 15,300 kg C ha^{-1} year^{-1}, C translocation to the root is 6200 kg C ha^{-1} year^{-1}, and C exudation by the root is 50 kg C ha^{-1} year^{-1}. In wheat, C deposition by roots can be up to 20% of assimilation (Barber and Martin, 1976; Martin, 1977; Keith *et al.*, 1986). Smucker (1984), using ^{14}C methods, reported that 20–40% of the C translocated to roots can be deposited. Vaadia (1960) reported diurnal variations in exudation rate and root pressure in decapitated sunflowers, which are also predicted by the model.

3.3.10 Fluxes of shoot and root substrate C to senescence, $O_{CS,sh\to so}$, $O_{CS,rt\to so}$ (kg substrate C m^{-2} day^{-1})

It is assumed that variable fractions $f_{CS,sh\to li}$, $f_{CS,rt\to li}$ of substrate C in the shoot and root are lost with the structural DM fluxes from the shoot and root which arise from senescence and from degradation [Eqns (3.5d)]. The fluxes are given by

$$O_{CS,sh\to so} = f_{CS,sh\to li} C_{sh} O_{X,sh\to li}, \qquad O_{CS,rt\to so} = f_{cs,rt\to li} C_{rt} O_{X,rt\to li},$$

$$f_{CS,sh\to li} = \frac{f_{CS,sh\to li,min} K_{Csh\to li} + C_{sh}}{K_{Csh\to li} + C_{sh}}, \quad f_{CS,rt\to li} = \frac{f_{CS,sh\to li,min} K_{Crt\to li} + C_{rt}}{K_{C_{rt}\to li} + C_{rt}}, \qquad (3.3j)$$

$$f_{CS,sh\to li,min} = f_{CS,rt\to li,min} = 0.3,$$

$$K_{Csh\to li} = K_{Crt\to li} = 0.05 \text{ kg substrate C (kg structural DM)}^{-1}.$$

A minimum of 30% of the substrate C is lost to senescence if the C substrate concentrations are low, rising to 100% if the substrate concentrations are high compared with, for example, $K_{Csh\to li}$.

3.3.11 Respiratory totals for shoot, root and plant

Total shoot respiration is, summing growth and maintenance components [Eqns (3.3f) and (3.3g)]

$$R_{sh} = R_{G,sh} + R_{m,sh}. \qquad (3.3k)$$

Total root respiration is, summing growth, maintenance and respiration associated with N uptake [Eqns (3.3f), (3.3g) and (3.3h)]

$$R_{rt} = R_{G,rt} + R_{m,rt} + R_{uN}. \qquad (3.3l)$$

Total plant respiration is

$$R_{pl} = R_{sh} + R_{rt}. \qquad (3.3m)$$

3.4 Shoot and Root Substrate N Pools, $M_{NS,sh}$, $M_{NS,rt}$

The differential equations for these two pools are

$$\frac{dM_{NS,sh}}{dt} = O_{NS,rt\to sh} + I_{NS,re,sh} - O_{NS,sh,G} - O_{NS,sh\to so} - O_{NS,sh\to an} - O_{NS,sh\to hv},$$
$$\frac{dM_{NS,rt}}{dt} = I_{NS,so\to rt} + I_{NS,re,rt} - O_{NS,rt\to sh} - O_{NS,rt,G} - O_{NS,rt\to so,ex} - O_{NS,rt\to so}, \quad (3.4a)$$
$$M_{NS,sh}(t=0) = 0.001, \quad M_{NS,rt}(t=0) = 0.003 \text{ kg N substrate m}^{-2}.$$

The initial values are set so that the initial substrate N concentrations are N_{sh} = 0.005 and N_{rt} = 0.015 kg substrate N (kg structural DM)$^{-1}$, using Eqns (3.1h), and calculating $M_{X,sh}$ and $M_{X,rt}$ at t = 0 with Eqns (3.1a) and Table 3.1.

The input and output terms to the shoot substrate N pool arise from transport of substrate N from the root [$O_{NS,rt\to sh}$, Eqn (3.4b)], recycling from litter [$I_{NS,re,sh}$, Eqn (3.4d)], utilization of substrate N for shoot growth [$O_{NS,sh,G}$, Eqn (3.4e)], flux of substrate N with senescence to the soil [$O_{NS,sh\to so}$, Eqn (3.4k)], grazing by the animals [$O_{NS,sh\to an}$, Eqn (3.7i)], and harvesting [$O_{NS,sh\to hv}$, Eqn (3.8c)].

The root substrate N equation has similar terms except that the principal input term is N uptake from the soil mineral N pools [$I_{NS,so\to rt}$, Eqn (3.4h)], the transport term to the shoot is an output rather than an input [$O_{NS,rt\to sh}$, Eqn (3.4b)], and a substrate N exudation term is included [$O_{NS,rt\to so,ex}$, Eqn (3.4j)].

3.4.1 Transport of substrate N from root to shoot, $O_{NS,rt\to sh}$ (kg substrate N m^{-2} day^{-1})

As with substrate C transport in Section 3.3.3, this is proportional to the substrate N concentration difference between root and shoot, $N_{rt} - N_{sh}$ [Eqns (3.1h)], divided by a resistance, $r_{N,rt\to sh}$ [m^2 day (kg structural DM)$^{-1}$]. This resistance is calculated by adding resistances associated with root and shoot, $r_{N,rt}$ and $r_{N,sh}$, which are obtained by scaling a resistivity coefficient ρ_{N20} with structural dry mass using a dimensionless allometric constant $q_{\rho N}$ and adjusting for temperature and water status. Thus

$$O_{NS,rt\to sh} = \frac{N_{rt} - N_{sh}}{r_{N,rt\to sh}},$$
$$r_{N,rt\to sh} = r_{N,rt} + r_{N,sh},$$
$$r_{N,rt} = \frac{r_{N,rt20}}{f_{T,rt} f_{W,rt}}, \quad r_{N,sh} = \frac{r_{N,sh20}}{f_{T,sh} f_{W,sh}}, \quad (3.4b)$$
$$r_{N,rt20} = \frac{\rho_{N20}}{M_{X,rt}^{q_{\rho N}}}, \quad r_{N,sh20} = \frac{\rho_{N20}}{M_{X,sh}^{q_{\rho N}}},$$
$$\rho_{N20} = 2 \text{ m}^2 \text{ day (kg structural dry mass)}^{q_{\rho N}-1}, \qquad q_{\rho N} = 1.$$

The value of ρ_{N20} is such that N fluxes and N substrate concentration gradients are acceptable. Temperature and water functions, $f_{T,sh}$, $f_{T,rt}$, $f_{W,sh}$, $f_{W,rt}$, are given in Eqns (3.11b), (3.11d). Plant structural dry mass components $M_{X,sh}$, $M_{X,rt}$ are defined in Eqns (3.1a).

3.4.2 Recycled N from litter fluxes, $I_{NS,re,sh}$, $I_{NS,re,rt}$ (kg substrate N m^{-2} day^{-1})

Tissue turnover gives rise to shoot and root fluxes of structural dry mass from tissue age category 4 (Fig. 3.1) [Eqns (3.5a), (3.5b), (3.5c)]. It is assumed that variable fractions ($f_{N,shli,re}$, $f_{N,rtli,re}$) of the N in these litter fluxes can be recycled internally to the plant substrate N pools; the remainder goes into the soil and litter submodel. These variable fractions depend on the substrate N concentrations N_{sh}, N_{rt} according to

$$f_{N,shli,re} = \frac{f_{N,shli,re,max} K_{N,rec}}{N_{sh} + K_{N,rec}}, \quad f_{N,rtli,re} = \frac{f_{N,rtli,re,max} K_{N,rec}}{N_{rt} + K_{N,rec}};$$
$$f_{N,shli,re,max} = f_{N,rtli,re,max} = 0.4, \tag{3.4c}$$
$$K_{N,rec} = 0.02 \text{ kg N substrate (kg structural DM)}^{-1}.$$

If the substrate N concentration in the shoot, N_{sh}, is zero, a maximum fraction $f_{N,shli,re,max}$ of the structural N flux in the litter is recycled. As N_{sh} increases, recycling decreases, with a half-maximum shoot N concentration of $K_{N,rec}$. Root litter N recycling is similar.

These two fractions are applied to the N in the structural dry mass fluxes from shoot and root to litter [Eqns (3.5d)] to give the fluxes of N recycled into the substrate N pools in the shoot and the root, with

$$I_{NS,re,sh} = f_{N,shli,re} f_{N,plX} O_{X,sh \to li},$$
$$I_{NS,re,rt} = f_{N,rtli,re} f_{N,plX} O_{X,rt \to li}. \tag{3.4d}$$

$f_{N,plX}$ is the fraction of N in plant structural dry mass [Eqn (3.1i)].

3.4.3 Outputs of substrate N for shoot and root growth, $O_{NS,sh,G}$, $O_{NS,rt,G}$ (kg substrate N m^{-2} s^{-1})

The requirements of N from the substrate N pools for growth are calculated directly from the shoot and root growth rates, G_{sh}, G_{rt} (kg structural DM m^{-2} day^{-1}), given in Eqns (3.3c). These are

$$O_{NS,sh,G} = f_{N,plX} G_{sh}, \quad O_{NS,rt,G} = f_{N,plX} G_{rt}. \tag{3.4e}$$

The fractional content of N in plant structural dry mass, $f_{N,plX}$, is given in Eqn (3.1i).

3.4.4 Output of substrate N to the grazing animals $O_{NS,sh \to an}$ (kg substrate N m^{-2} day^{-1})

This flux is calculated in Section 3.7 [Eqn (3.7i)], which discusses the outputs to the grazing animals.

3.4.5 Output of substrate N to harvesting $O_{NS,sh \to hv}$ (kg substrate N m^{-2} day^{-1})

This flux is calculated in Section 3.8 [Eqn (3.8c)], which discusses the outputs to harvesting.

3.4.6 N uptake by the root, $I_{NS,so \to rt}$ (kg substrate N m^{-2} day^{-1})

N uptake occurs from the soil ammonium and nitrate pools, N_{amm}, N_{nit} (Fig. 5.1). A single plant N uptake term is used, so this includes N uptake that might result from mycorrhizal activity, which is not represented explicitly in the model. Our approach does not distinguish between convective (mass flow) and diffusive transport, and ignores root geometry. A state-of-the-art mechanistic model of coupled ammonium and nitrate uptake, using two-dimensional finite element analysis, numerical methods and Michaelis–Menten kinetics, has recently been described by Abbès *et al.* (1996). Their analysis would not be suitable for our application.

First, a total soil mineral N concentration N_{min} is defined by

$$N_{min} = N_{amm} + N_{nit}. \tag{3.4f}$$

For the purpose of plant uptake, these two pools are combined differently to give an effective soil mineral N pool, N_{eff} (kg N m^{-2}), according to

$$\begin{aligned} N_{eff} &= N_{amm} + f_{T,uN,nit} N_{nit}, \\ x &= 1-(1-0.5)\frac{20-T_{soil}}{20-10}, \\ f_{T,uN,nit} &= \max(0.0, \min(x, 1.0)). \end{aligned} \tag{3.4g}$$

This equation allows for the fact that the *relative* rate of uptake from the ammonium and nitrate pools varies with temperature (Clarkson *et al.*, 1986, 1992; Wild *et al.*, 1987; Whitehead, 1995, p. 18; Macduff, personal communication; but see Hatch and Macduff, 1991) as shown in Fig. 3.5. $f_{T,uN,nit}$ describes the uptake rate of nitrate relative to ammonium. The last line in Eqn (3.4g) restricts $f_{T,uN,nit}$ to the range (0, 1). At low temperatures the root is less efficient at taking up nitrate. The N uptake rate is calculated by

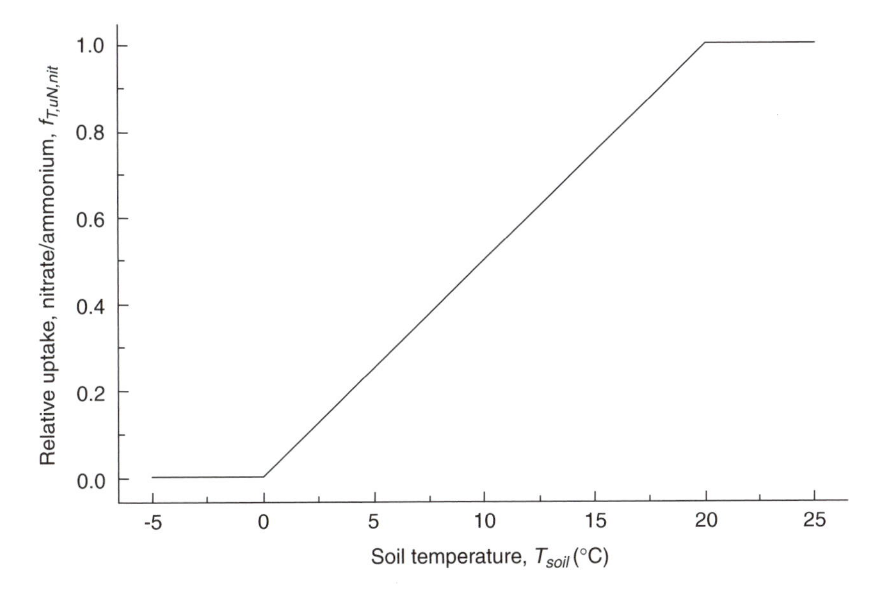

Fig. 3.5. Temperature dependence of plant N uptake from the soil nitrate pool relative to that from the soil ammonium pool. See Eqn (3.4g) *et seq.*

$$I_{NS,so\to rt} = u_N = \sigma_{uN} \frac{N_{eff}}{N_{eff} + K_{N_{eff}}} \frac{\left(M_{rt,1} + c_{uN,2}M_{rt,2} + c_{uN,3}M_{rt,3} + c_{uN,4}M_{rt,4}\right)}{\left[1 + \frac{K_{C,uN}}{C_{rt}}\left(1 + \frac{N_{rt}}{J_{N,uN}}\right)\right]\left(1 + \frac{M_{X,rt}}{K_{MXrt,uN}}\right)},$$

$$\sigma_{uN} = \sigma_{uN,20} f_{T,rt} f_{W,rt} f_{W,uN}, \quad f_{W,uN} = \left(\frac{\theta_{so}}{\theta_{so,\max}}\right)^{q_{W,uN}},$$

$$\sigma_{uN,20} = 1 \text{ kg N (kg root structural DM)}^{-1} \text{ day}^{-1}, \tag{3.4h}$$

$$K_{N_{eff}} = 0.005 \text{ kg N m}^{-2}, \quad c_{uN,i}, \quad i = 2, 3, 4 = 0.5, 0.25, 0.1,$$

$$K_{C,uN} = 0.05 \text{ kg C substrate (kg structural DM)}^{-1},$$

$$J_{N,uN} = 0.005 \text{ kg N substrate (kg structural DM)}^{-1}, \quad q_{W,uN} = 3,$$

$$K_{MXrt,uN} = 1 \text{ kg root structural DM m}^{-2}.$$

The uptake constant, $\sigma_{uN,20}$, is modified for root temperature [$f_{T,rt}$, Eqn (3.11b)], root water status [$f_{W,rt}$, eqn(3.11d)] and with $f_{W,uN}$ for soil relative water content θ_{so} [eqn (6.2f); see also Section 3.12]. The maximum value of soil relative water content $\theta_{so,max}$ is given in Table 6.1. There is a Michaelis–Menten dependence on the effective soil mineral pool N_{eff} with constant $K_{N_{eff}}$. The components of root structural dry mass [Eqn (3.1a)] are weighted according to tissue age with the $c_{uN,i}$ parameters. The denominator makes uptake dependent on root carbon substrate concentration (C_{rt}) with constant $K_{C,uN}$: N uptake is an active process requiring C substrate. Uptake is inhibited by high levels of nitrogen substrate in the root (N_{rt}), with inhibition constant $J_{N,uN}$: inhibition is predicted by models of membrane transport in which the release of the ion transported to the inside of the membrane is reversible. The last term in the denominator ensures that uptake does not simply scale with root mass, but diminishing returns are obtained as total root structural dry mass [Eqns (3.1a)] approaches the value of $K_{MXrt,uN}$ = 1 kg root structural DM m^{-2}. In nutrient culture it may be appropriate to set $K_{MXrt,uN} = 1 \times 10^{10}$ (a large number) so that the diminishing returns do not operate. Note that this seemingly minor assumption produces major effects on the shoot:root allocation patterns (Fig. 3.11 and associated discussion). Note also that C uptake by the shoot is already of the diminishing-returns type (Fig. 3.9c), so that this assumption ensures that N uptake by the root behaves similarly.

The values of the parameters in Eqn (3.4h) are adjusted so that the values of the concentrations of N in the soil and C and N substrates in the plant are acceptable, and the uptake rate is sufficient to give a satisfactory rate of plant growth. Our 20°C value of $\sigma_{uN,20}$ of 1.0 kg N (kg root structural DM)$^{-1}$ day^{-1} is about two orders of magnitude larger than the specific absorption rates from solution of about 1 mg N (g root fresh weight)$^{-1}$ day^{-1} reported by Clarkson *et al.* (1992, Fig. 8) at temperatures of 3–9°C. With default parameter and steady-state values, on 1 January the N uptake rate is about 150 kg N ha^{-1} year^{-1} (Fig. 8.2).

The N uptake rates from the soil ammonium and nitrate pools separately are

$$u_{Namm} = u_N \frac{N_{amm}}{N_{eff}}, \quad u_{Nnit} = u_N \frac{f_{T,uN,nit} N_{nit}}{N_{eff}}. \tag{3.4i}$$

u_N, N_{eff} and $f_{T,uN,nit}$ are defined in Eqns (3.4h) and (3.4g).

3.4.7 Output of root substrate N to exudation, $O_{NS,rt\to so,ex}$ (kg substrate N m^{-2} day^{-1})

Root exudation of substrate N occurs at a rate of

$$O_{NS,rt\to so,ex} = k_{NS,rt,ex20} f_{T,rt} f_{W,rt} M_{X,rt} N_{rt}; \; k_{NS,rt,ex20} = 0.005 \text{ day}^{-1}. \tag{3.4j}$$

$k_{NS,rt,ex20}$ is a rate constant. The flux is proportional to root structural dry mass [$M_{X,rt}$, Eqn (3.1a)] and root substrate N concentration [N_{rt}, Eqn (3.1h)]; it is modulated by root temperature [$f_{T,rt}$, Eqn (3.11b)] and root water status [$f_{W,rt}$, Eqn (3.11d)]. Under standard conditions (20°C, $f_{T,rt} = f_{W,rt} = 1$), 0.5% of the root substrate C is exuded per day. This exudation is to the soil ammonium pool N_{amm} in the soil model [Fig. 5.1; Eqn (5.4a)]. Running the model with default values for one year predicts that 20 kg N ha^{-1} year^{-1} is exuded. This may be compared with plant uptake of 500 kg N ha^{-1} year^{-1}. I have not found any measurements of N exudation in grassland, but in wheat, Janzen (1990) observed N deposition over a growing season of 26 and 76 mg N (3 plants)$^{-1}$ in low N and high N fertility treatments. Assuming a plant density of 200 plants m^{-2}, these values are equivalent to exudation rates of about 20 and 50 kg N ha^{-1} over the season.

3.4.8 Fluxes of shoot and root substrate N to senescence, $O_{NS,sh\to so}$, $O_{NS,rt\to so}$ (kg substrate N m^{-2} day^{-1})

It is assumed that variable fractions $f_{NS,sh\to li}$, $f_{NS,rt\to li}$ of substrate N in the shoot and root are lost with the structural DM fluxes arising from senescence and degradation from the shoot and root [Eqns (3.5d)]. The fluxes are given by

$$\begin{aligned}
&O_{NS,sh\to so} = f_{NS,sh\to li} N_{sh} O_{X,sh\to li}, \qquad O_{NS,rt\to so} = f_{NS,rt\to li} N_{rt} O_{X,rt\to li}, \\
&f_{NS,sh\to li} = \frac{f_{NS,sh\to li,min} K_{Nsh\to li} + N_{sh}}{K_{Nsh\to li} + N_{sh}}, \quad f_{NS,rt\to li} = \frac{f_{NS,rt\to li,min} K_{Nrt\to li} + N_{rt}}{K_{Nrt\to li} + N_{rt}}, \\
&f_{NS,sh\to li,min} = f_{NS,rt\to li,min} = 0.3, \\
&K_{Nsh\to li} = K_{Nrt\to li} = 0.01 \text{ kg substrate N (kg structural DM)}^{-1}.
\end{aligned} \tag{3.4k}$$

A minimum of 30% of the substrate N is lost to senescence if the N substrate concentrations are low, rising to 100% if the substrate concentrations are high compared with, for example, $K_{Nsh\to li}$.

3.5 Shoot and Root Structural Pools, $M_{lam,1-4}$, $M_{ss,1-4}$, $M_{rt,1-4}$

There are eight state variables for the four age categories of lamina (*lam*) and sheath + stem (*ss*) structural dry mass (Fig. 3.1). The differential equations for the four lamina structural dry mass pools are

$$\frac{dM_{lam,1}}{dt} = f_{lam}G_{sh} - k_{turn,lam}M_{lam,1} - k_{deg,sh}M_{lam,1}$$
$$-O_{X,lam,1\to an} - O_{X,lam,1\to hv},$$
$$\frac{dM_{lam,2}}{dt} = k_{turn,lam}M_{lam,1} - k_{turn,lam}M_{lam,2} - k_{deg,sh}M_{lam,2}$$
$$-O_{X,lam,2\to an} - O_{X,lam,2\to hv},$$
$$\frac{dM_{lam,3}}{dt} = k_{turn,lam}M_{lam,2} - k_{turn,lam}M_{lam,3} - k_{deg,sh}M_{lam,3}$$
$$-O_{X,lam,3\to an} - O_{X,lam,3\to hv}, \qquad (3.5a)$$
$$\frac{dM_{lam,4}}{dt} = k_{turn,lam}M_{lam,3} - k_{turn,lam}M_{lam,4} - k_{deg,sh}M_{lam,4}$$
$$-O_{X,lam,4\to an} - O_{X,lam,4\to hv},$$
$$M_{lam,1-4}(t=0) = 0.01 \text{ kg structural dry mass (m}^2\text{ ground)}^{-1},$$
$$f_{lam} = 0.7, \qquad k_{turn,lam} = k_{turn,lam20}f_{T,sh} / f_{W,sh}, \qquad k_{turn,lam20} = 0.08 \text{ day}^{-1},$$
$$k_{deg,sh} = f_{T,sh}k_{deg,sh,20}, \qquad k_{deg,sh,20} = 0.001 \text{ day}^{-1}.$$

The initial (t = 0) values of the four variables $M_{lam,1-4}$ are as given. f_{lam} is the fraction of current growth G_{sh} [Eqn (3.3c)] allocated to lamina rather than sheath + stem. Robson's results (1973, Fig. 3) suggest a value of 0.7, which is consistent with observed field data (Parsons, personal communication). f_{lam} is assumed constant. Each compartment turns over at a rate of $k_{turn,lam}$. This gives each pool an output which enters the next pool without loss of C or N. The output from the oldest pool, $M_{lam,4}$, provides a structural dry mass flux, some of which is recycled [Eqns (3.4d), (3.3a)] and the rest of which enters the soil and litter submodel as litter [Eqns (3.10b)]. At 20°C the value of the turnover rate $k_{turn,lam,20}$ corresponds to a half-life of about 8 days. With four compartments this gives a default value for total leaf half-life of about 1 month. This half-life may be compared with the observation that there are on average three live fully expanded leaves per tiller, and during the English summer, the rate of leaf appearance and loss is about one leaf per tiller per 10 days (Robson, 1973; Davies, 1977; Parsons, 1988, pp. 138, 148). The turnover rate is modified for temperature [$f_{T,sh}$, Eqn (3.11b)]. Increased water stress reduces $f_{W,sh}$ below unity [Eqn (3.11d)]; this increases leaf turnover rate, decreasing leaf longevity, leaf area index and transpiration. A non-specific 'degradation' term, with rate constant $k_{deg,sh}$ provides a small rate of loss from each compartment. This term is only modified with temperature ($f_{T,sh}$) and not according to water status ($f_{W,sh}$), and acts equally on lamina (*lam*) and on sheath + stem [*ss*, Eqn (3.5b)]. This term contributes substantially to the stability of leaf area index. It is envisaged that $k_{deg,sh}$ represents pest and disease processes. The rate constant of 0.001 day^{-1} gives each compartment a half-life of about 2 years due to this process. The output terms $O_{X,lam,i\to an}$ and $O_{X,lam,i\to hv}$, i = 1–4, are outputs to grazing animals and to harvesting [Eqns (3.7e), (3.8c)].

Similarly, the differential equations for the four sheath + stem structural dry mass pools are

$$\frac{dM_{ss,1}}{dt} = (1 - f_{lam}) G_{sh} - k_{turn,ss} M_{ss,1} - k_{deg,sh} M_{ss,1}$$
$$- O_{X,ss,1 \to an} - O_{X,ss,1 \to hv},$$
$$\frac{dM_{ss,2}}{dt} = k_{turn,ss} M_{ss,1} - k_{turn,ss} M_{ss,2} - k_{deg,sh} M_{ss,2}$$
$$- O_{X,ss,2 \to an} - O_{X,ss,2 \to hv},$$
$$\frac{dM_{ss,3}}{dt} = k_{turn,ss} M_{ss,2} - k_{turn,ss} M_{ss,3} - k_{deg,sh} M_{ss,3} \qquad (3.5b)$$
$$- O_{X,ss,3 \to an} - O_{X,ss,3 \to hv},$$
$$\frac{dM_{ss,4}}{dt} = k_{turn,ss} M_{ss,3} - k_{turn,ss} M_{ss,4} - k_{deg,sh} M_{ss,4}$$
$$- O_{X,ss,4 \to an} - O_{X,ss,4 \to hv},$$
$$M_{ss,1-4}(t = 0) = 0.04 \text{ kg structural dry mass } (\text{m}^2 \text{ ground})^{-1},$$
$$k_{turn,ss} = k_{turn,ss20}\, f_{T,sh} / f_{W,sh}, \qquad k_{turn,ss20} = 0.08 \text{ day}^{-1}.$$

The four root structural dry mass pools have the differential equations

$$\frac{dM_{rt,1}}{dt} = G_{rt} - k_{turn,rt} M_{rt,1} - k_{deg,rt} M_{rt,1},$$
$$\frac{dM_{rt,2}}{dt} = k_{turn,rt} M_{rt,1} - k_{turn,rt} M_{rt,2} - k_{deg,rt} M_{rt,2},$$
$$\frac{dM_{rt,3}}{dt} = k_{turn,rt} M_{rt,2} - k_{turn,rt} M_{rt,3} - k_{deg,rt} M_{rt,3}, \qquad (3.5c)$$
$$\frac{dM_{rt,4}}{dt} = k_{turn,rt} M_{rt,3} - k_{turn,rt} M_{rt,4} - k_{deg,rt} M_{rt,4},$$
$$M_{rt,1-4}(t = 0) = 0.05 \text{ kg structural dry mass } (\text{m}^2 \text{ ground})^{-1},$$
$$k_{turn,rt} = k_{turn,rt20}\, f_{T,rt} / f_{W,rt}, \qquad k_{turn,rt20} = 0.08 \text{ day}^{-1},$$
$$k_{deg,rt} = f_{T,rt} k_{deg,rt20}, \qquad k_{deg,rt20} = 0.001 \text{ day}^{-1}.$$

Root growth rate G_{rt} [Eqn (3.3c)] provides the first root compartment with new material. The turnover rate $k_{turn,rt}$ is modified for temperature [$f_{T,rt}$, Eqn (3.11b)] but inversely for root water status [$f_{W,rt}$, Eqn 3.11d], so that increased water stress increases root turnover. The turnover rate of 0.08 day^{-1} at 20°C corresponds to a half-life per compartment of about 8 days, or a root half-life of one month. There is little definite information on the length of life of grass roots, which is much affected by defoliation regime (Whitehead, 1995, p. 91). There is evidence that when swards are defoliated every 3–4 weeks, roots can live some 5–6 months (Garwood, 1967b; Troughton, 1981). The average length of life of prairie grassland roots has been estimated at 2–5 years (Dahlman and Kucera, 1965; Sims and Singh, 1978). Arguably, it may be important to distinguish between fine roots and coarse roots in respect of activity and turnover rates. The degradation term is as in Eqn (3.5a). There are no output terms for grazing or harvesting of roots.

The output fluxes of structural dry mass from the shoot and the root from senescence and degradation are, taking the $k_{turn,..}$ term in each of the fourth of Eqns (3.5a), (3.5b) and (3.5c), and adding the $k_{deg,sh}$, $k_{deg,rt}$ terms in all these equations, with Eqns (3.1a):

$$O_{X,sh4\to li} = k_{turn,lam}M_{lam,4} + k_{turn,ss}M_{ss,4}, \qquad O_{X,rt4\to li} = k_{turn,rt}M_{rt,4},$$
$$O_{X,sh\to deg} = k_{deg,sh}\left(M_{X,lam} + M_{X,ss}\right), \qquad O_{X,rt\to deg} = k_{deg,rt}M_{X,rt}, \tag{3.5d}$$
$$O_{X,sh\to li} = O_{X,sh4\to li} + O_{X,sh\to deg}, \qquad O_{X,rt\to li} = O_{X,rt4\to li} + O_{X,rt\to deg}.$$

Note that these output fluxes towards litter are not equal to the input fluxes to the soil and litter submodel owing to C and N recycling [Section 3.10; Eqns (3.4d), (3.3a)].

3.6 Leaf Lamina Area Pools, $L_{AI,1-4}$

Lamina area is partially decoupled from lamina structural dry mass (DM), providing for a variable specific leaf area for currently synthesized material [Eqn (3.6c)]. Again four age categores are used. The differential equations for the four pools are

$$\frac{dL_{AI,1}}{dt} = I_{L_{AI,1}} - k_{turn,lam}L_{AI,1} - k_{deg,sh}L_{AI,1} - O_{LAI,1\to an} - O_{LAI,1\to hv},$$
$$\frac{dL_{AI,2}}{dt} = k_{turn,lam}L_{AI,1} - k_{turn,lam}L_{AI,2} - k_{deg,sh}L_{AI,2} - O_{LAI,2\to an} - O_{LAI,2\to hv},$$
$$\frac{dL_{AI,3}}{dt} = k_{turn,lam}L_{AI,2} - k_{turn,lam}L_{AI,3} - k_{deg,sh}L_{AI,3} - O_{LAI,3\to an} - O_{LAI,3\to hv}, \tag{3.6a}$$
$$\frac{dL_{AI,4}}{dt} = k_{turn,lam}L_{AI,3} - k_{turn,lam}L_{AI,4} - k_{deg,sh}L_{AI,4} - O_{LAI,4\to an} - O_{LAI,4\to hv},$$
$$L_{AI,1-4}(t=0) = 0.2\ \text{m}^2\ \text{lamina area}\ (\text{m}^2\ \text{ground})^{-1}.$$

The initial values in Eqns (3.6a) with those in Eqns (3.5a) give an initial specific leaf area of 20 m^2 (kg structural DM)$^{-1}$ (e.g. see Parsons *et al.*, 1991a, p. 626). The production of new leaf lamina area $I_{LAI,1}$ is given below [Eqn (3.6b)]. The turnover and degradation terms are calculated as before [Eqns (3.5a)]. The rates of loss of lamina area to grazing animals $O_{LAI,i\to an}$ and harvesting $O_{LAI,i\to hv}$, i = 1–4, are calculated in Eqns (3.7g) and (3.8c).

The input to the first pool, $I_{LAI,1}$, is related to the structural dry mass input [f_{lam} G_{sh} of the first of Eqns (3.5a)] by a variable specific leaf area v_{SLA} [Eqn (3.6c)], to give

$$I_{LAI,1} = v_{SLA} f_{lam} G_{sh}. \tag{3.6b}$$

f_{lam} is a constant [Eqns (3.5a)]. G_{sh} is the shoot growth rate [Eqns (3.3c)]. The incremental specific leaf area v_{SLA} depends on shoot C substrate concentration C_{sh} [Eqn

(3.1h)] and on shoot relative water content, θ_{sh} [Eqn (6.5d) with $i = sh$], according to

$$v_{SLA} = c_{SLA,max}\left(1 - c_{SLA,C}\, C_{sh}\right)\left[1 - c_{SLA,W}\left(1 - \theta_{sh}\right)\right],$$
$$c_{SLA,max} = 25 \text{ m}^2 \text{ leaf (kg structural DM)}^{-1}, \qquad (3.6c)$$
$$c_{SLA,C} = 2.5\left[\text{kg C substrate (kg structural DM)}^{-1}\right]^{-1}, \qquad c_{SLA,W} = 1.$$

The maximum specific leaf area $c_{SLA,max}$ is attained in a leaf with low C substrate concentration [$C_{sh} \approx 0$, Eqn (3.1h)] and high relative water content [$\theta_{sh} \approx 1$, Eqn (6.5d)]. This corresponds to low light/CO_2 and high temperature, and non-limiting water. Increasing C substrate concentration C_{sh} decreases SLA linearly. Increasing water stress decreases shoot relative water content θ_{sh} and this decreases SLA. A temperature effect is not incorporated explicitly, but operates indirectly: increased temperature may reduce C substrate concentration C_{sh} giving thinner leaves, but increased temperature may increase water stress leading to thicker leaves. The direct effects on SLA in Eqn (3.6c) and indirect effects via growth rate [Eqn (3.3c)] and shoot:root allocation [Eqn (3.3d)] can be complex.

The factors affecting specific leaf area are not well understood (see e.g. Zur and Jones, 1981; Hsiao and Bradford, 1983; Ludlow, 1987). Quite variable results have been observed, e.g. Thomas and Norris reported that in S23 ryegrass leaf extension is insensitive to water stress under field conditions (Hughes, 1974, p. 32). Jones (1988, p. 220) presents data showing the rate of leaf extension in three grasses: in *Lolium perenne* leaf extension is depressed when water potential falls below -200 J (kg water)$^{-1}$ (-0.2 MPa) and approaches zero at -1600 J kg^{-1} (-1.6 MPa).

3.7 Outputs to the Grazing Animals

There are 14 separate outputs from plant pools to the animal. Eight of these are from the lamina and sheath + stem structural dry mass (DM) pools: $O_{X,lam,i\to an}$, $O_{X,ss,i\to an}$ (kg structural DM m^{-2} day^{-1}), i = 1–4. There are four fluxes of leaf lamina area $O_{LAI,i\to an}$ [m^2 (m^2 ground)$^{-1}$ day^{-1}] associated with the lamina structural DM fluxes. Finally, there are two fluxes from the shoot substrate C and N pools, $O_{CS,sh\to an,gr}$ and $O_{NS,sh\to an,gr}$ (kg substrate C, N m^{-2} day^{-1}).

The dry mass (DM) flux into the animal, $I_{DM,pl\to an}$ (kg DM animal^{-1} day^{-1}), is given in Eqn (4.2a). The DM flux into the animals per unit ground area is [kg DM (m^2 ground)$^{-1}$ day^{-1}]

$$I_{DM,pl\to an,gnd} = n_{animals} I_{DM,pl\to an}. \qquad (3.7a)$$

$n_{animals}$ is the number of animals per unit area [Eqn (7.6f)].

The total C and N fluxes from the plant into the animal per unit ground area are (kg C, N m^{-2} day^{-1})

$$I_{C,pl\to an} = C_{tot,sh} I_{DM,pl\to an,gnd}, \qquad I_{N,pl\to an} = N_{tot,sh} I_{DM,pl\to an,gnd}. \qquad (3.7b)$$

The total C and N concentrations, $C_{tot,sh}$, $N_{tot,sh}$ of shoot dry mass are given by Eqns (3.1i).

The total DM flux of $I_{DM,pl\rightarrow an}$ removes a flux of structural dry mass of

$$O_{X,pl\rightarrow an} = \frac{M_{X,sh}}{M_{sh,tot}} I_{DM,pl\rightarrow an,gnd} \tag{3.7c}$$

with units of kg structural DM m^{-2} day^{-1} and where $M_{X,sh}$ and M_{sh} are given by Eqns (3.1a) and (3.1f). This structural dry mass flux may be removed from lamina and/or sheath + stem, depending on the characteristics of the grazing animals and on what is on offer. The structural DM fluxes from lamina ($O_{X,lam\rightarrow an}$) and sheath + stem ($O_{X,ss\rightarrow an}$) to the animals are

$$O_{X,lam\rightarrow an} = \frac{M_{X,lam}}{M_{X,lam} + c_{ss\rightarrow an} M_{X,ss}} O_{X,pl\rightarrow an},$$

$$O_{X,ss\rightarrow an} = \frac{c_{ss\rightarrow an} M_{X,ss}}{M_{X,lam} + c_{ss\rightarrow an} M_{X,ss}} O_{X,pl\rightarrow an}, \tag{3.7d}$$

$$c_{ss\rightarrow an} = 0 \text{ for sheep.}$$

$c_{ss\rightarrow an}$ is a dimensionless weighting constant taken to be zero as default. This default approximates to continous grazing where animals ingest predominantly lamina (Morris, 1969; Hodgson *et al.*, 1981). The lamina and sheath + stem structural dry masses, $M_{X,lam}$ and $M_{X,ss}$, are given in Eqns (3.1a).

The total flux of laminar DM, $O_{X,lam\rightarrow an}$, is taken from the four age categories of lamina according to

$$O_{X,lam,i\rightarrow an} = \frac{c_{LAI,i\rightarrow an} L_{AI,i}}{z_{LAI,gr}} O_{X,lam\rightarrow an}, \qquad i = 1-4,$$

$$z_{LAI,gr} = \sum_{i=1}^{4} c_{LAI,i\rightarrow an} L_{AI,i}, \tag{3.7e}$$

$$c_{LAI,i\rightarrow an}(i = 1-4) = 1,\ 0.75,\ 0.5,\ 0.25.$$

The values of the constant $c_{LAI,i\rightarrow an}$, i = 1–4, allow us to weight intake according to leaf age category. Here we weight intake in favour of the younger shoot material. However, even without this weighting, when the model is run dynamically, intake will be predominantly from the younger age categories (Johnson and Parsons, 1985b). It has been argued (Johnson and Parsons, 1985b) that, in reality, these simple dynamics, rather than active selection, explain why animals are observed to eat predominantly the youngest leaves (Morris, 1969; Hodgson *et al.*, 1981), and also how increases in the intensity of grazing can actually *increase* the proportion of young leaves in the sward (Parsons *et al.*, 1988b, table 2, p. 9). The $L_{AI,i}$ i = 1–4, are the leaf area indices in the four age categories.

Similarly, the fluxes of sheath + stem material removed from the four age categories of sheath + stem are

$$O_{X,ss,i\to an} = \frac{c_{ss,i\to an} M_{ss,i}}{z_{ss,gr}} O_{X,ss\to an}, \qquad i = 1-4,$$

$$z_{ss,gr} = \sum_{i=1}^{4} c_{ss,i\to an} M_{ss,i}, \tag{3.7f}$$

$$c_{ss,i\to an}(i = 1-4) = 1,\ 1,\ 1,\ 1.$$

Equal weighting factors are used for the four age categories of sheath + stem.

The fluxes of lamina area, $O_{LAI,i\to an}$, accompanying the structural DM fluxes in Eqns (3.7e) are

$$O_{LAI,i\to an} = \frac{O_{X,lam,i\to an}}{M_{lam,i}} L_{AI,i}. \tag{3.7g}$$

$M_{lam,i}$ is the structural DM in the ith lamina compartment [Eqn (3.5a)].

The output of substrate C to the grazing animals, $O_{CS,sh\to an}$, is obtained from the total dry mass input to the grazing animals, $I_{DM,pl\to an,gnd}$ (kg DM m^{-2} day^{-1}) [Eqn (3.7a)], with

$$O_{CS,sh\to an} = I_{DM,pl\to an,gnd} \frac{M_{CS,sh}}{M_{sh}}. \tag{3.7h}$$

M_{sh} [Eqn (3.1f)] is the total shoot DM. Eqn (3.7h) assumes that the shoot substrate C [mass, $M_{CS,sh}$, Eqn (3.2a)] is uniformly distributed over the shoot.

Likewise, the the output of substrate N to the grazing animals, $O_{NS,sh\to an}$, is

$$O_{NS,sh\to an} = I_{DM,pl\to an,gnd} \frac{M_{NS,sh}}{M_{sh}}. \tag{3.7i}$$

3.8 Outputs to Harvesting

Management determines the pulse harvesting function, $p_{ulse,harv}$ of Eqn (7.6d), the value of which is 0 or 1. This function switches harvesting on or off. When harvesting is switched on, the function $p_{ulse,harv}$ has a value 1 for a period of 1 day (Fig. 7.6). At a single harvest, harvested material is removed at a uniform rate calculated at the beginning of the day over a period of 1 day. This avoids any double-valued functions which would occur if harvesting were assumed to be instantaneous. Whether or not material is harvested depends on whether the height of the canopy, h_{can} (m), is above the harvesting height, h_{harv} (m). Canopy height is assumed to be proportional to leaf area index, with

$$h_{can} = c_{hcan,LAI} L_{AI},$$

$$c_{hcan,LAI} = 0.026\,\mathrm{m}\left[(\mathrm{m}^2\ \mathrm{leaf})/(\mathrm{m}^2\ \mathrm{ground})\right]^{-1}. \tag{3.8a}$$

Thus a canopy of leaf area index L_{AI} equal to unity has a height of 2.6 cm (default) (e.g. Parsons *et al.*, 1994). The fraction of the above-ground dry mass harvested, f_{harv}, is assumed to be

$$\text{if } h_{can} < h_{harv}, f_{harv} = 0,$$
$$\text{if } h_{can} \geq h_{harv}, f_{harv} = \frac{h_{can} - h_{harv}}{h_{can}}. \tag{3.8b}$$

This assumes that shoot material is uniformly distributed with respect to height: e.g. if canopy height $h_{can} = 2 \times$ harvesting height h_{harv}, then at harvesting one half of the shoot material is removed with $f_{harv} = 0.5$. More elaborate and possibly realistic treatments, involving alternative distributions of shoot material with height could readily be incorporated (Parsons *et al.*, 1994, p. 190). The harvesting height h_{harv} is a management parameter [Eqn (7.6e)] with a default value of 0.03 m.

Because the harvested material is removed over one day at a rate computed at the beginning of the day when harvesting takes place, the output fluxes from the 14 shoot state variables are [in order, lamina structure (4), sheath + stem structure (4), lamina area (4), substrate C, N]

$$O_{X,lam,i\to hv} = P_{ulse,harv} f_{harv} M_{lam,i}, \qquad O_{X,ss,i\to hv} = P_{ulse,harv} f_{harv} M_{ss,i},$$
$$O_{LAI,i\to hv} = P_{ulse,harv} f_{harv} L_{AI,i}, \qquad i = 1-4, \tag{3.8c}$$
$$O_{CS,sh\to hv} = P_{ulse,harv} f_{harv} M_{CS,sh}, \qquad O_{NS,sh\to hv} = P_{ulse,harv} f_{harv} M_{NS,sh}.$$

3.9 Root Density, ρ_{rt}

The average root density ρ_{rt}, in units of kg structural root DM (m^3 soil)$^{-1}$, is a state variable. ρ_{rt} obeys the differential equation

$$\frac{d\rho_{rt}}{dt} = \left(\frac{G_{rt}}{M_{X,rt}}\right)\left(\rho_{rt,new} - \rho_{rt}\right),$$
$$\rho_{rt}(t=0) = 1 \text{ kg structural dry mass (m}^3\text{ soil)}^{-1}]. \tag{3.9a}$$

$\rho_{rt,new}$ is the root density of newly synthesized root mass, which is being synthesized at the root structural mass growth rate G_{rt} [Eqn (3.3c)]. The proportional rate of synthesis of root is G_{rt} divided by the root structural dry mass $M_{X,rt}$ [Eqn (3.1a)] with units of day^{-1}. This equation says that the average root density is moving from its current average value ρ_{rt} towards the value of the root density which is currently being synthesized, $\rho_{rt,new}$. The rate of movement depends on the proportional rate of synthesis of root $G_{rt}/M_{X,rt}$.

The root density of newly synthesized root mass $\rho_{rt,new}$ depends on the soil relative water content θ_{so} [Eqn (6.2f)] according to

$$\rho_{rt,new} = \rho_{rt,max} - \left(\rho_{rt,max} - \rho_{rt,min}\right)\left(1 - \frac{\theta_{so}}{\theta_{so,\max}}\right),$$
$$\rho_{rt,max} = 1, \rho_{rt,min} = 0.1 \text{ [kg structural dry mass (m}^3\text{ soil)}^{-1}\text{]}. \tag{3.9b}$$

If the soil is at field capacity with $\theta_{so} = \theta_{so,max}$ [Eqn (6.2a)], then $\rho_{rt,new}$ has its maximum value of $\rho_{rt,max}$. As the soil dries out, $\rho_{rt,new}$ becomes closer to its minimum value of $\rho_{rt,min}$.

Root depth, d_{rt} (m) depends on the structural mass of root $M_{X,rt}$ and the average root density ρ_{rt}, with a minimum value of $d_{rt,min}$ (m)

$$d_{rt} = \text{maximum}\left(d_{rt,min}, \frac{M_{X,rt}}{\rho_{rt}}\right), \quad d_{rt,min} = 0.02 \text{ m.} \tag{3.9c}$$

Without this minimum value of 2 cm for the root depth, under some extreme conditions (e.g. low light), very fast leaching [Eqns (5.5e), (6.2e)] and a numerical instability in the soil nitrate pool gave execution failure.

3.10 Litter Fluxes

There are several fluxes of C and N directly from the plants into the soil surface and below-ground litter pools (Figs 3.1 and 5.1). The plant gives rise to structural dry matter fluxes [Fig. 3.1, Eqns (3.5d)] from senescence [e.g. the $k_{turn,lam}$ term in the fourth of Eqns (3.5a)], and from degradation [e.g. the $k_{deg,sh}$ terms in Eqns (3.5a)]. Structural fluxes are partitioned between the metabolic, cellulose and lignin litter pools (Fig. 5.1). The substrate fluxes [Eqn (3.3j), (3.4k)] which accompany the structural fluxes are placed directly in the metabolic compartments.

The structural fluxes of C and N from shoot and root senescence and degradation which enter the surface and soil litter pools (Fig. 5.1) are obtained by substracting the recycled fluxes of C and N [e.g. $I_{CS,re,sh}$, $I_{NS,re,sh}$, Eqns (3.3a), (3.4d)] from the total structural fluxes of C and N [e.g. Eqns (3.5d)], giving (kg structural C, N m^{-2} day^{-1})

$$\begin{aligned} O_{CX,sh\to so} &= f_{C,plX} O_{X,sh\to li} - I_{CS,re,sh}, \quad O_{NX,sh\to so} = f_{N,plX}\, O_{X,sh\to li} - I_{NS,re,sh}, \\ O_{CX,rt\to so} &= f_{C,plX} O_{X,rt\to li} - I_{CS,re,rt}, \quad O_{NX,rt\to so} = f_{N,plX}\, O_{X,rt\to li} - I_{NS,re,rt}. \end{aligned} \tag{3.10a}$$

$f_{C,plX}$ and $f_{N,plX}$ are the C and N fractions in plant dry mass [Eqn (3.1i)].

It is assumed that the senescing and degrading structural dry matter fluxes contain plant substrate C or N, according to Eqns (3.3j), (3.4k). The total C and N fluxes from shoot and root into the soil and litter submodel are therefore

$$\begin{aligned} O_{C,sh\to so} &= O_{CX,sh\to so} + O_{CS,sh\to so}, \qquad O_{C,rt\to so} = O_{CX,rt\to so} + O_{CS,rt\to so}, \\ O_{N,sh\to so} &= O_{NX,sh\to so} + O_{NS,sh\to so}, \qquad O_{N,rt\to so} = O_{NX,rt\to so} + O_{NS,rt\to so}. \end{aligned} \tag{3.10b}$$

With default parameters at zero time, the C:N ratios of the shoot and root litter fluxes are 28 and 17. Note that these are variable quantities, depending upon the fractions of structural N internally recycled [Eqns (3.4c)], and on the fractions of substrate C and N retained [Eqns (3.3j), (3.4k)].

3.11 Effects of Temperature on Plant Processes

Responses to temperature are arguably the most important of the responses of a plant ecosystem. They are difficult to analyse because temperature affects all rate parameters, with many qualitative and quantitative differences. A general discussion of the effects of temperature on plant processes is in Thornley and Johnson (1990, chapter 5). Some workers choose to use temperature response functions based on the Arrhenius equation (e.g. Farquhar, 1988; Schuster and Monson, 1990).

However, the Arrhenius function does not generally provide a convenient or accurate representation of the temperature dependence of biological processes. It is optimistic to suppose that multistep processes catalysed by proteins can be modelled by Arrhenius functions when few inorganic reactions obey the Arrhenius model satisfactorily. The theoretical basis of the Arrhenius model is also narrow. Other workers use temperature response functions based on a numerical table (e.g. de Wit *et al.*, 1978, p. 56; Parton *et al.*, 1993, Figs 2, 6), or with a fitted polynomial function (Kirschbaum and Farquhar, 1984; McMurtrie and Wang, 1993). Here it is assumed as default that most plant rate parameters have a sigmoidal cubic temperature function $f(T)$; this is a mathematically transparent function of the correct general shape. However, by altering a parameter q_{fT} this function can be varied continuously and made, for instance, quadratic or quartic in temperature T. This enables easy examination of the nature of the temperature response on the model predictions. The equation used is

$$f(T) = m_{fT} \frac{(T - T_0)^{q_{fT}} (T_0' - T)}{(T_{ref} - T_0)^{q_{fT}} (T_0' - T_{ref})} \quad \text{for } T_0 < T < T_0', \quad \text{else } f(T) = 0,$$

$$m_{fT} = 1, \quad q_{fT} = 2, \quad T_0 = 0°\text{C}, \quad T_{ref} = 20°\text{C} \quad T_0' = 45°\text{C}, \tag{3.11a}$$

$$T_{max} = \frac{1}{1 + q_{fT}} (T_0 + q_{fT} T_0') = 30°\text{C}, \quad T_{inf} = \frac{2T_0 + (q_{fT} - 1)T_0'}{1 + q_{fT}} = 15°\text{C}.$$

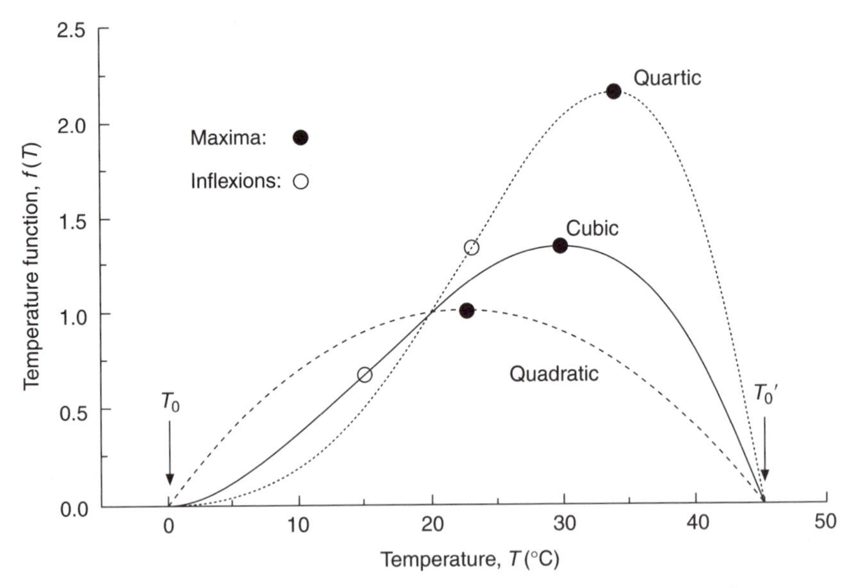

Fig. 3.6. Temperature function for plant and soil biochemical processes, $f(T)$. Eqn (3.11a) is drawn for $q_{fT} = 1$ (quadratic), 2 (cubic, default) and 3 (quartic). Respectively, the maxima of these curves are at 22.5°C, 30°C and 33.75°C. The cubic and quartic have inflections at 15°C and 22.5°C.

This equation is shown in Fig. 3.6 for three values of the parameter q_{fT} which give a quadratic (1), cubic (2, default) and quartic (3). m_{fT} is a multiplier which gives an easy way of changing all the rate constants which use $f(T)$ simultaneously. The temperature function is zero at $T = T_0$ and $T = T_0'$, unity at the reference temperature of T_{ref}, and has its maximum value at $T = T_{max}$. There may be a point of inflection at $T = T_{inf}$ for $q_{fT} > 1$.

Note that if q_{fT} is increased, then for a temperate climate with temperatures mostly between 0 and 20°C, the average value of $f(T)$ decreases (Fig. 3.6). If q_{fT} is increased to increase sigmoidicity, it may be necessary to increase the multiplier m_{fT} to obtain comparable results. For example, averaging $f(T)$ with $m_{fT} = 1$ over the year for a southern Britain soil temperature wave (Fig. 7.1), the average values are: quadratic ($q_{fT} = 1$), 0.66; cubic ($q_{fT} = 2$), 0.38; quartic ($q_{fT} = 3$), 0.24. The multiplier m_{fT} could be adjusted to give the same average value of $f(T)$ in a given climate, so that the effect of changing only the *shape* of the temperature function could be investigated.

Note also that in Eqn (3.11a) the high temperature limit T_0' is fixed at 45°C, and if q_{fT} is changed to change the sigmoidicity, then the temperature optimum T_{max} varies with q_{fT}, as shown in Fig. 3.6. An alternative to Eqns (3.11a) is to fix the temperature optimum T_{max} and allow the high temperature limit T_0' to vary with q_{fT} [see Eqn (3.11c) below].

The shoot is assumed to be at air temperature, T_{air}, and the root at soil temperature, T_{soil}. The temperature functions for shoot and root are

$$f_{T,rt} = f(T_{soil}), \qquad f_{T,sh} = f(T_{air}). \tag{3.11b}$$

The temperature dependence of the initial slope of the leaf photosynthesis:light response curve [Eqns (3.2c), (3.2s), Fig. 3.3] is calculated differently [Eqn (3.2s)]. This is a light-dependent mechanism which does not respond to temperature in the same way as most enzymatic processes.

The temperature dependence of the light-saturated photosynthetic rate parameter P_{max} [Eqn (3.2t)], denoted by $f_{T,Pmax}$, is computed with Eqn (3.11a) but fixing the temperature optimum at a value determined by the CO_2 concentration (Long, 1991). $f_{T,Pmax}$ is given by

$$f_{T,Pmax} = m_{fT} \frac{(T - T_0)^{q_{fT}} (T'_{0Pmax} - T)}{(T_{ref} - T_0)^{q_{fT}} (T'_{0Pmax} - T_{ref})} \qquad \text{for } T_0 < T < T'_{0,Pmax},$$

$$\text{else } f_{T,Pmax} = 0,$$

$$m_{fT} = 1, \qquad q_{fT} = 2, \qquad T_0 = 0°C, \qquad T_{ref} = 20°C,$$

$$T'_{0,Pmax} = \frac{(1 + q_{fT}) T_{max,Pmax} - T_0'}{q_{fT}}; \tag{3.11c}$$

$$T_{max,Pmax} = T_{max,Pmax,350} + (T_{max,Pmax,700} - T_{max,Pmax,350}) \frac{C_{O_2,vpm} - 350}{700 - 350},$$

$$T_{max,Pmax,700} = 35°C, \qquad T_{max,Pmax,350} = 30°C.$$

The temperature optimum $T_{max,Pmax}$ increases with CO_2 concentration $C_{O_2,vpm}$ (Table

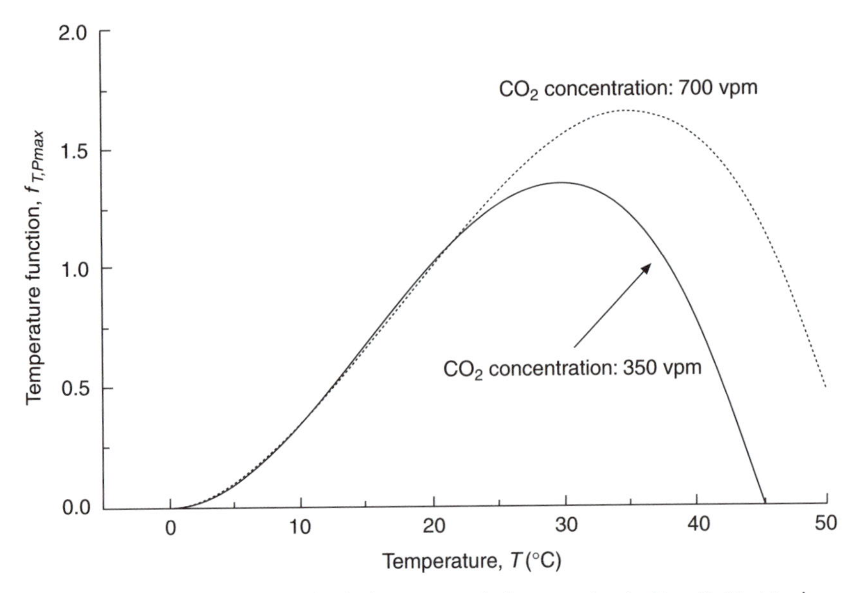

Fig. 3.7. Temperature function for light-saturated photosynthesis. Eqn (3.11c) is drawn for CO_2 concentrations of 350 vpm and 700 vpm.

7.2). As CO_2 concentration increases above 350 vpm, $T_{max,Pmax}$ increases and so does the high temperature limit of P_{max}, $T'_{0,Pmax}$ The temperature function $f_{T,Pmax}$ is shown in Fig. 3.7 for two values of CO_2 concentration. Changing the CO_2 concentration from 350 vpm to 700 vpm, the upper temperature limit for photosynthesis shifts from 45°C to 52.5°C; the temperature optimum shifts from 30°C to 35°C and $f_{T,Pmax}$ at the optimum increases from 1.35 to 1.65. As formulated here, the effect is insignificant in a temperate environment (Section 9.2).

3.12 Effects of Water on Plant Processes

The water submodel (Chapter 6) calculates the chemical activity of water in shoot and root, $a_{W,sh}$, $a_{W,rt}$ [Eqns (6.7a)]. These are obtained from the water potentials in shoot and root [ψ_{sh}, ψ_{rt}, Eqn (6.5c), $i = sh, rt$]. The effects of water activity on biochemical activity are discussed in Chapter 6 (Section 6.8). The basic rationale is that a biochemical rate constant ∝ enzyme activity ∝ enzyme conformation ∝ degree of hydration ∝ water activity. Simple empirical functions for the general influence of water stress on biochemical rates in the plant are assumed. For the shoot and root, the water functions are defined as

$$f_{W,sh} = a_{W,sh}^{q_{w,pl}}, \qquad f_{W,rt} = a_{W,rt}^{q_{w,pl}}; \qquad q_{W,pl} = 20. \tag{3.11d}$$

Increasing water stress causes water potential to decrease and the chemical activity of water decreases below unity. The high exponent $q_{W,pl}$ in Eqn (3.11d) causes biochemical rate constants to decrease sharply with drying (Fig. 6.5).

The effect of water on plant photosynthesis ($f_{W,ph}$) is calculated in Eqns (3.2s) where it is used to modify the parameters α and P_{max} in Eqns (3.2s) and (3.2t). The exponent is smaller, giving a smaller effect of water stress on α and P_{max} than on general plant biochemistry. I am not aware of work which defines the effects of water stress on the light reactions (α) and the dark reactions (P_{max}) in the absence of stomatal closure [Eqn (3.2u)].

The effect of soil water content (θ_{so}) on modifying N uptake by the root is calculated and applied in Eqns (3.4h). This is inevitably empirical and approximate for a model with a single spatial compartment for the root and a single soil horizon.

The simulation is halted when the pressure component of the shoot water potential, $\psi_{sh,pr}$ [Eqn (6.5b), $i = sh$], falls to -500 J (kg water)$^{-1}$ (= -5×10^5 Pa).

3.13 Responses of the Decoupled Plant Submodel

In this section, responses of the plant submodel, unaffected by diurnal or seasonal changes in environment, management (fertilizer, harvesting, grazing), water relations, or the soil submodel, are examined. The plant submodel displays its innate properties, without modification by driving variables which are time-dependent. These results might be compared with simulated swards in growth cabinets where above- and below-ground conditions are kept constant, although this idealized situation is seldom achieved. In this situation the significant environmental variables are: radiation, soil mineral N, and temperature (assuming air and soil temperatures are equal).

The responses are of three types: first, the time course of the system growing from small initial values to a steady state; second, the exponential-growth phase responses; and last, the steady-state responses when net growth is zero. In the first case, the system starts with arbitrary initial values, and after a transient period, settles down on to a stable growth trajectory which depends on environment but not on the initial values. For part of this stable growth trajectory, the system is growing exponentially, and intensive quantities such as shoot:root ratio, various concentrations, specific rates of growth and specific activities are fairly constant and well characterized. The exponential growth period is important because experiments are often directed at plants in this condition, because most plants and crops do undergo such a period of growth, and because, for managed grassland with intensive grazing or frequent cutting, exponential growth may be a better approximation to what is occurring than the steady state. The steady state is also well-characterized, usually independent of initial values, and can give valuable insights into the responses of the system. However, the steady-state is less applicable to field environments, and the assumption here of no reproductive phase may also be unrealistic for steady-state conditions.

It is self-evident that the parameters determining variables such as shoot:root allocation during an exponential growth phase may not be the same as those determining shoot:root allocation in a steady state. Such differences may apply to many physiological observables.

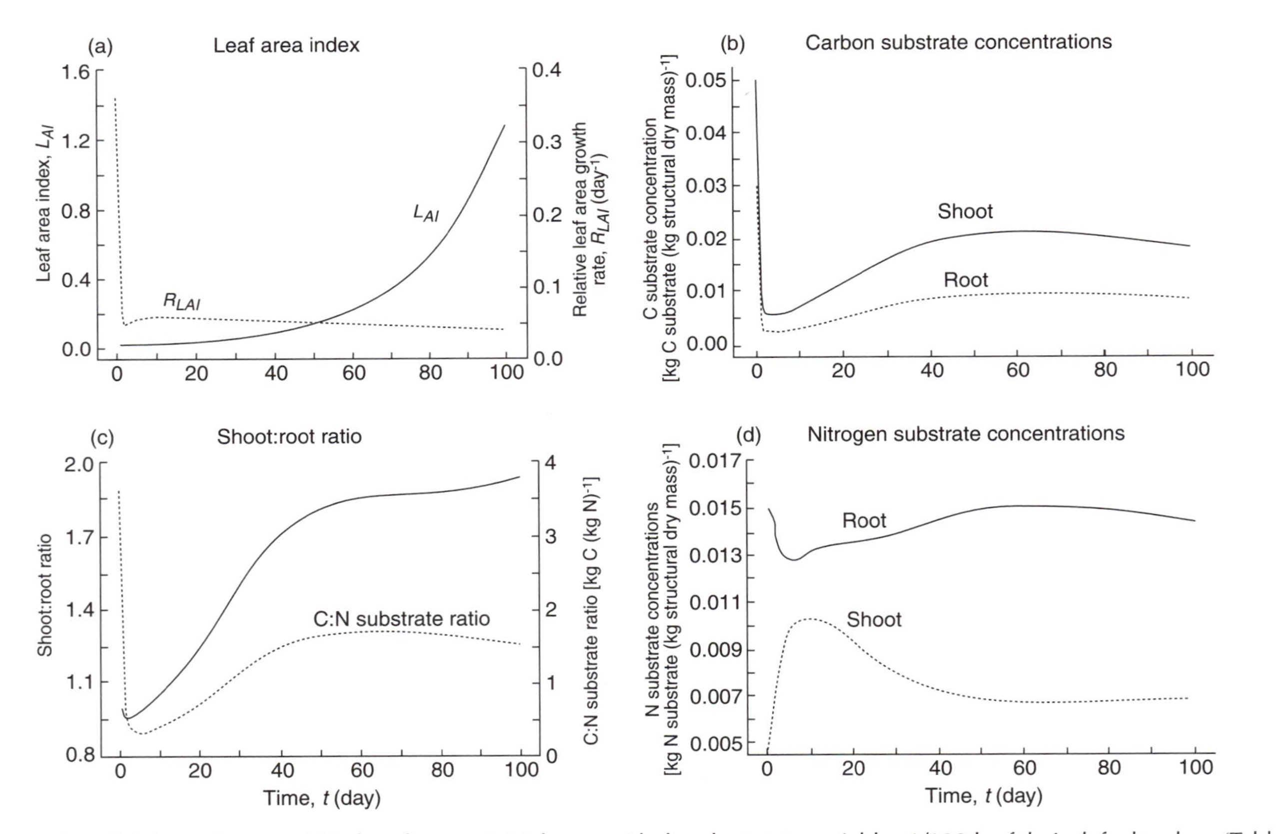

Fig. 3.8. Plant submodel dynamics over 100 days from an initial state with the plant state variables 1/100th of their default values (Table 3.1). Soil and litter, and water submodels disabled. No grazing. Constant environment of 20°C, 4.8 MJ PAR m^{-2} day^{-1} (511 μmol PAR m^{-2} s^{-1} during a 12 h day, Appendix), soil mineral concentration constant at 0.001 kg N m^{-2} [Eqns (5.5g) with (5.4a), (5.5a)], no diurnal or seasonal variation. (a) Leaf area index, L_{AI} [Eqn (3.1b)], and the relative net leaf area growth rate. (b) Shoot and root carbon substrate concentrations [Eqns (3.1h)]. (c) Shoot:root dry mass ratio, $f_{sh:rt}$ [Eqns (3.1l)] and whole-plant C:N substrate ratio [Eqns (3.1c)]. (d) Shoot and root nitrogen substrate concentrations [Eqns (3.1h)].

3.13.1 Dynamic responses

Starting from arbitrary initial values with very small plants, the time-course over a 100-day period is given in Fig. 3.8. It can be seen that about 10 days are required for the plant subsystem to settle down on to a smooth growth trajectory; this is a measure of the intrinsic response time of the plant submodel at 20°C. Leaf area index grows at a high initial relative growth rate as the initial high substrate pools are utilized (Fig. 3.8a). The substrate pools themselves exhibit oscillatory behaviour (Fig. 3.8b, d). This is reflected in the shoot:root ratio and the plant substrate C:N ratio (Fig. 3.8c).

In Fig. 3.9, the 100-day simulation of Fig. 3.8 has been continued for a further 900 days. Leaf area index (Fig. 3.9a) exhibits a logistic response (e.g. Thornley and Johnson, 1990, p. 79), with the net shoot growth rate being proportional to the slope of the L_{AI} curve. The shoot:root ratio (Fig. 3.9b) is quite different in the steady state (t = 1000 day) than when the plants are growing at a constant exponential rate (t = *c.* 60 day; Fig. 3.8a); also, the C:N substrate ratio in the plant as a whole changes considerably, and this is reflected in larger changes in the substrates in the shoot and root compartments (not shown). In Fig. 3.9c, gross canopy photosynthetic rate and net shoot growth rate are plotted against leaf area index, taking the data from the same 1000-day simulation. It can be seen that canopy photosynthetic rate follows a typical diminishing-returns response, whereas net shoot growth rate is parabolic $[x(x_m - x)]$, equalling zero at leaf area indices of 0 and (x_m =) 8.6. When analysing the stability of a pasture under grazing, it can be useful to assume that the response of net shoot growth to leaf area index is parabolic (Section 4.6.1).

3.13.2 Exponential-growth responses

These are illustrated in Fig. 3.10. Specific growth rate and shoot:root ratio are two important physiological variables. In Figs 3.10a, b these are plotted against specific shoot and root activity rather than radiation and soil mineral N to avoid plotting a canopy photosynthesis or an N uptake response, and to facilitate comparison with allocation models (e.g. Mäkelä and Sievänen, 1987; Thornley, 1997). Specific growth rate is only weakly asymptotic against specific shoot activity but is more strongly asymptotic against specific root activity (Figs 3.10a, b). Shoot:root ratio decreases with increasing shoot activity and increases with increasing root activity. The C:N substrate ratio also behaves predictably. These results are as expected with a transport-resistance approach to allocation and are in general agreement with experiment (Wilson, 1988). The temperature response in Fig. 3.10c is interesting, particularly as most published allocation submodels do not consider the role of temperature. The shoot:root ratio increases slightly with increasing temperature although this is accompanied by a large decrease in C:N substrate ratio. All substrate concentrations decrease with increasing temperature, but the C and N substrates do not decrease at the same rate. This is a consequence of the different relative temperature dependencies of photosynthesis and N uptake [Eqns (3.2s), (3.2t), (3.11c), (3.4h), (3.11a)], which increase the relative efficiency of N uptake as temperature increases. In detail, this result depends on the light level (in this case, 4.8 MJ PAR m^{-2} day^{-1}), because this determines whether the canopy (of low L_{AI}) is closer to the initial slope [Eqn (3.2s)] or the asymptote [Eqn (3.2t)] of the leaf

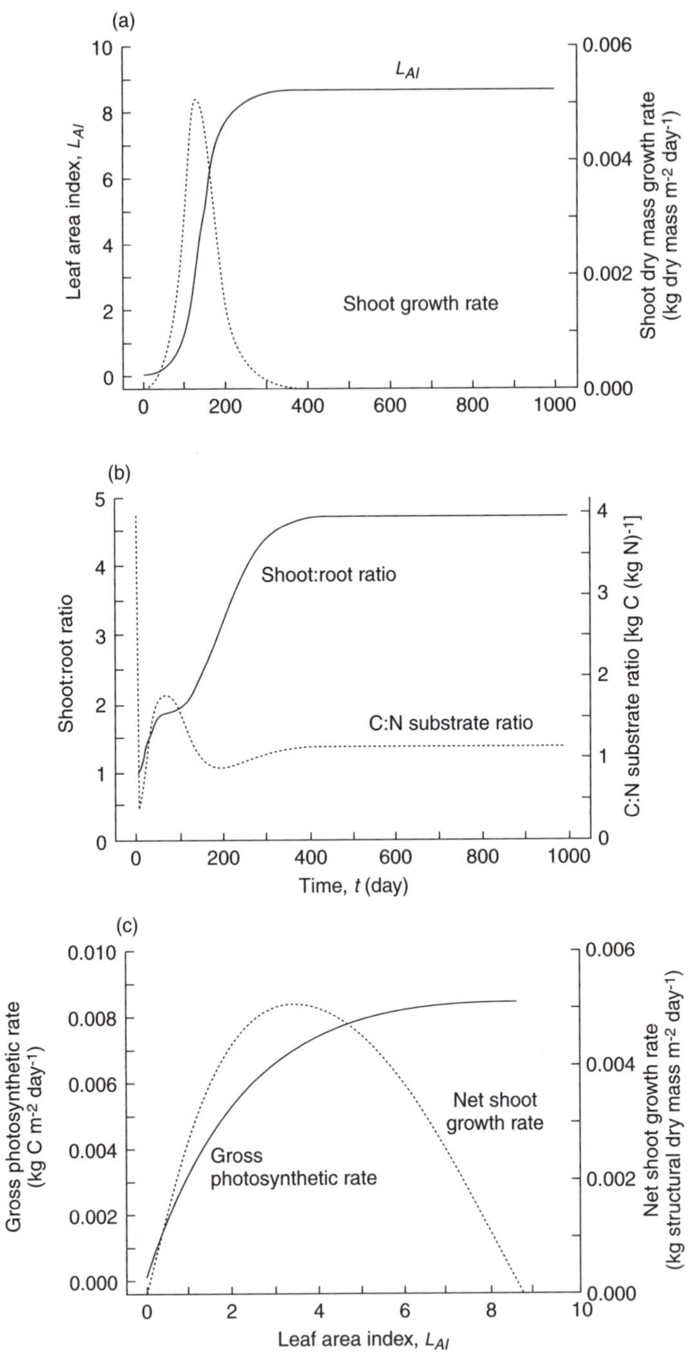

Fig. 3.9. Plant submodel dynamics over 1000 days. This is a continuation of the simulation of Fig. 3.8. (a) Leaf area index, L_{AI} [Eqn (3.1b)], and net shoot growth rate. (b) Shoot:root dry mass ratio, $f_{sh:rt}$ [Eqns (3.1l)], and whole-plant C:N substrate ratio [Eqns (3.1c)]. (c) Canopy gross photosynthetic rate [Eqn (3.2n)] and net shoot growth rate [differentiate the second of Eqns (3.1f)].

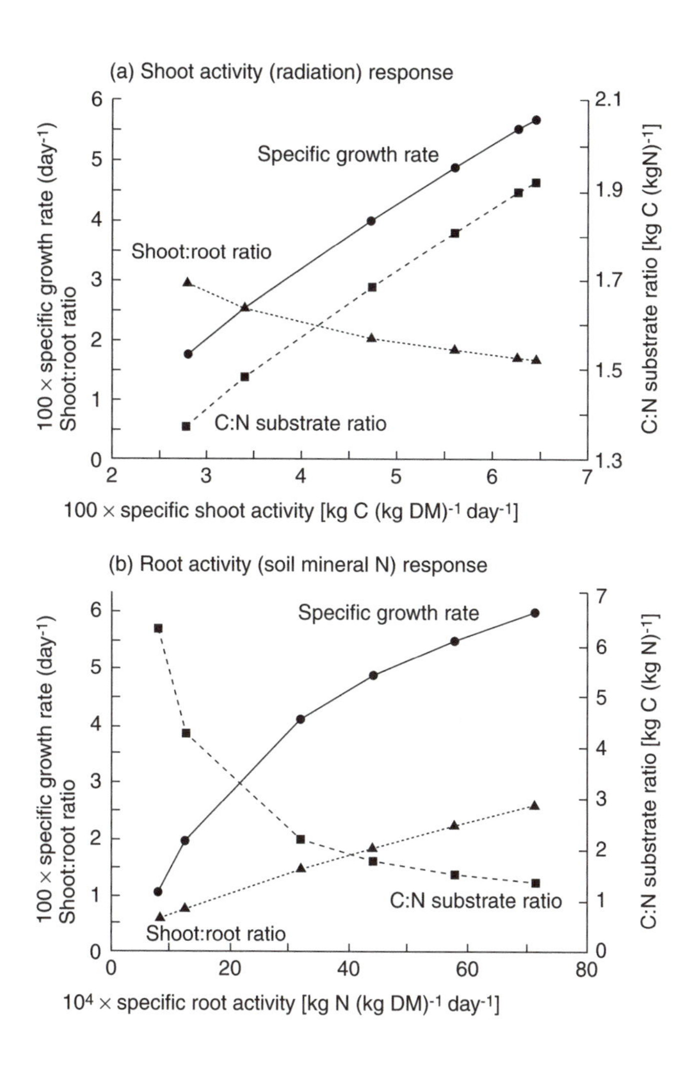

continued over

Fig. 3.10. Exponential growth responses. The simulations were begun from small initial values and halted when the leaf area index was 0.1 or less during the period of exponential growth. The variables shown are whole-plant specific (or relative) growth rate, the ratio of shoot:root total dry mass, and the ratio of the whole-plant C:N substrate masses (Fig. 3.1). (a) Specific shoot activity is canopy photosynthesis [Eqn (3.2n)] divided by total shoot dry mass [Eqn (3.1f)]; canopy photosynthetic rate was varied by assigning daily PAR radiation values of 1.92, 2.4, 3.6, 4.8, 7.2, 9.6 MJ PAR m^{-2} day^{-1} (1 MJ PAR m^{-2} day^{-1} = 106 μmol m^{-2} s^{-1} over a 12-hour day; Appendix). (b) Specific root activity is N uptake [Eqn (3.4h)] divided by total root dry mass [Eqn (3.1f)]; N uptake rate was varied using the soil mineral N concentration, N_{min} [Eqn (3.4f)], assigning N_{min} values of 10^{-3} × (0.05, 0.1, 0.5, 1, 2, 4) kg N m^{-2}. (c) Air and soil temperature are varied together as shown, with 4.8 MJ PAR day^{-1}, 12-hour day, and N_{min} = 0.001 kg N m^{-2}; the temperature function is as in Eqn (3.11b) (Fig. 3.6, cubic).

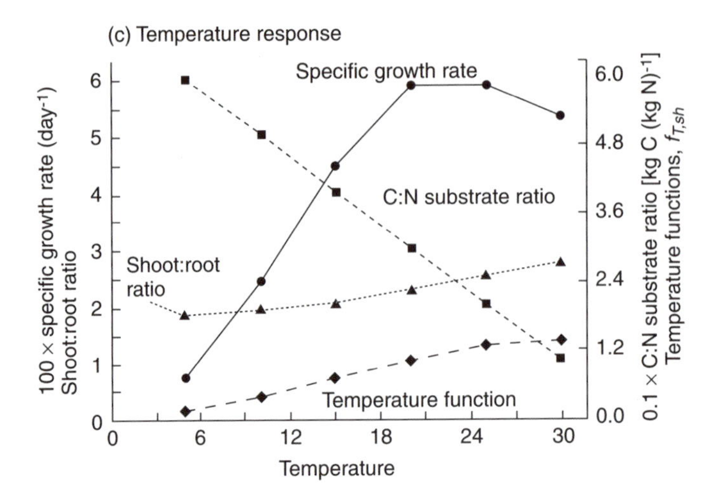

Fig. 3.10. *continued*

light response curve (Fig. 3.3) as these have different temperature dependencies. Note that the specific growth rate increases faster than the general temperature function [Eqn (3.11a)], and reaches its maximum at a lower temperature (20°C as opposed to 30°C for the temperature function; Fig. 3.6, cubic).

3.13.3 Steady-state responses

Steady-state responses are illustrated in Fig. 3.11. The simulations are continued until there is no appreciable change in any of the variables of the model. Canopy photosynthesis and leaf area index (Fig. 3.11a) increase in step with increasing radiation receipt. Shoot:root ratio decreases by 10% over the range examined and the C:N substrate ratio increases by 10%. It is interesting to note that if root uptake scales with root mass [achieved by setting the parameter $K_{MXrt,uN} = 10^{10}$ in Eqn (3.4h)], then the shoot:root ratio and all substrate concentrations remain constant in the steady state. This unexpected result is a consequence of the linear degradation/senescence processes assumed throughout [e.g. Eqns (3.5a)], and can be confirmed analytically using a simpler model (Thornley, 1997). Modellers frequently make linear assumptions for reasons of simplicity. Such assumptions are often unrealistic and can have major consequences on the responses given by a model. Fig. 3.11b shows that shoot:root ratio and the C:N substrate ratio are more

Fig. 3.11. (opposite) Steady-state responses. The simulations were continued until a constant state was achieved. The variables shown are leaf area index [Eqn (3.1b)], canopy photosynthetic rate [Eqn (3.2n)], shoot:root total dry matter ratio, whole-plant C:N substrate ratio [Eqn (3.1k)], root uptake rate [Eqn (3.4h)], temperature function [Eqn (3.11a), cubic]. (a) Response to irradiance (10 MJ PAR m^{-2} day^{-1} = 1065 μmol m^{-2} s^{-1} over a 12-hour day; Appendix). (b) Response to soil mineral N, N_{min} [Eqn (3.4f)]. (c) Air and soil temperature are varied together as shown.

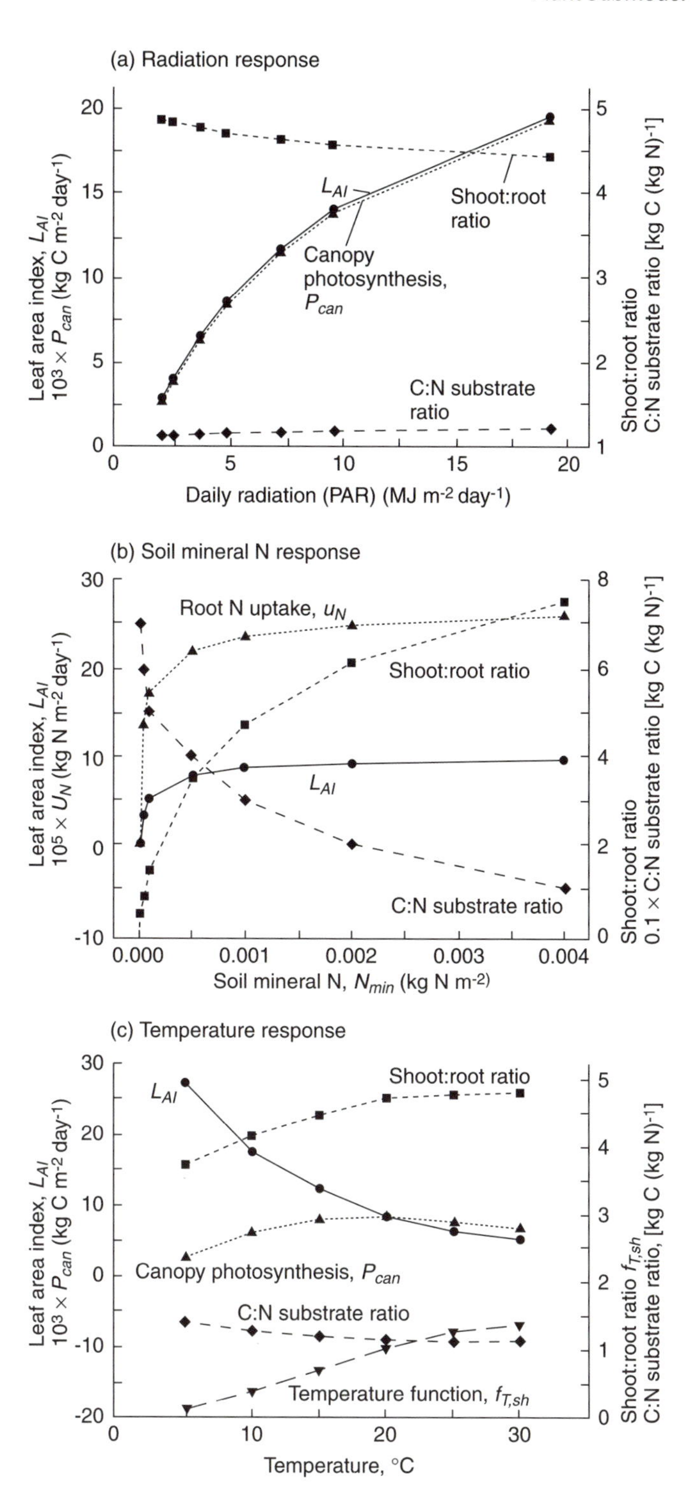
(a) Radiation response
Leaf area index, L_{AI}
$10^3 \times P_{can}$ (kg C m^{-2} day^{-1})
L_{AI}
Shoot:root ratio
Canopy photosynthesis, P_{can}
C:N substrate ratio
Daily radiation (PAR) (MJ m^{-2} day^{-1})
Shoot:root ratio
C:N substrate ratio [kg C (kg N)$^{-1}$]
(b) Soil mineral N response
Leaf area index, L_{AI}
$10^5 \times U_N$ (kg N m^{-2} day^{-1})
Root N uptake, u_N
Shoot:root ratio
L_{AI}
C:N substrate ratio
Soil mineral N, N_{min} (kg N m^{-2})
Shoot:root ratio
$0.1 \times$ C:N substrate ratio [kg C (kg N)$^{-1}$]
(c) Temperature response
Leaf area index, L_{AI}
$10^3 \times P_{can}$ (kg C m^{-2} day^{-1})
L_{AI}
Shoot:root ratio
Canopy photosynthesis, P_{can}
C:N substrate ratio
Temperature function, $f_{T,sh}$
Temperature, °C
Shoot:root ratio $f_{T,sh}$
C:N substrate ratio, [kg C (kg N)$^{-1}$]

strongly influenced by soil mineral N, N_{min}, than by radiation (Fig. 3.11a), although N uptake saturates with respect to N_{min} more readily than canopy photosynthesis saturates with respect to radiation. This occurs in spite of the N_{min} values examined lying below the Michaelis parameter $K_{N_{eff}}$ of Eqn (3.4h). Note however, that N uptake has negative feedback control [the $N_{rt}/J_{N,uN}$ term in Eqn (3.4h)] whereas canopy photosynthesis is not inhibited by high C substrate levels. The temperature responses of shoot:root ratio and whole-plant C:N substrate ratio are similar to those found in exponential growth (Fig. 3.10c). Leaf area index decreases continuously with increasing temperature, and this shifts the maximum of canopy photosynthesis to below 20°C (cf. specific growth rate in Fig. 3.10c), which is well below the maximum of 30°C of the temperature function (Fig. 3.6, cubic).

Animal Submodel 4

4.1 Introduction

The animal submodel (Fig. 4.1; see also Fig. 4.3) is designed to provide a simple method of calculating the rate of removal of plant tissues (C and N) during grazing, the consequent C and N fluxes to the soil as urine and faeces, and gaseous C fluxes (respiration and methane) to the atmosphere. Any N fluxes from animal to atmosphere are ignored. This enables us to consider the role of the animals in the cycling and fate of C and N from plants to the environment and soil, including the balance of C and N cycling via litter versus excreta.

As shown in Fig 4.1, the model has two state variables, $M_{C,an,ret}$ and $M_{N,an,ret}$ [kg C, N (m^2 ground)$^{-1}$]. It is stressed that the model is *not* intended to consider the growth or energy balance of animals, but merely how mature animals which are only growing very slowly [Eqn (4.1b)] with predetermined characteristics of intake in relation to pasture state will partition C and N fluxes between plant, soil and atmosphere. The state variables $M_{C,an,ret}$ and $M_{N,an,ret}$ are included so that the role of C and N retention in animals in the overall C and N balances can be examined. These are simple accumulators for the masses of C and N retained in the animal, and are not involved in any equations in the model.

Many detailed ruminant models have been reported, e.g. for the physical and behavioural constraints to intake (Parsons *et al.*, 1994; Newman *et al.*, 1995), and for mass, energy balance and metabolism (ARC, 1980; Gill *et al.*, 1989; Forbes and France, 1993).

The model is driven by the C and N inputs of grazed material from the plant submodel, and the outputs are C fluxes to the atmosphere, and C and N fluxes in excreta to the soil and litter submodel. There is no direct dependence of the animal submodel on the environment, the soil and litter submodel or the water submodel.

Three variables from the plant submodel are required for the operation of the animal submodel: leaf area index, L_{AI} [Eqn (3.1b)], and the C and N concentrations in shoot dry matter, $C_{tot,sh}$, $N_{tot,sh}$ [Eqns (3.1i)]. State variables, initial values, parameters and other variables are listed in Table 4.1.

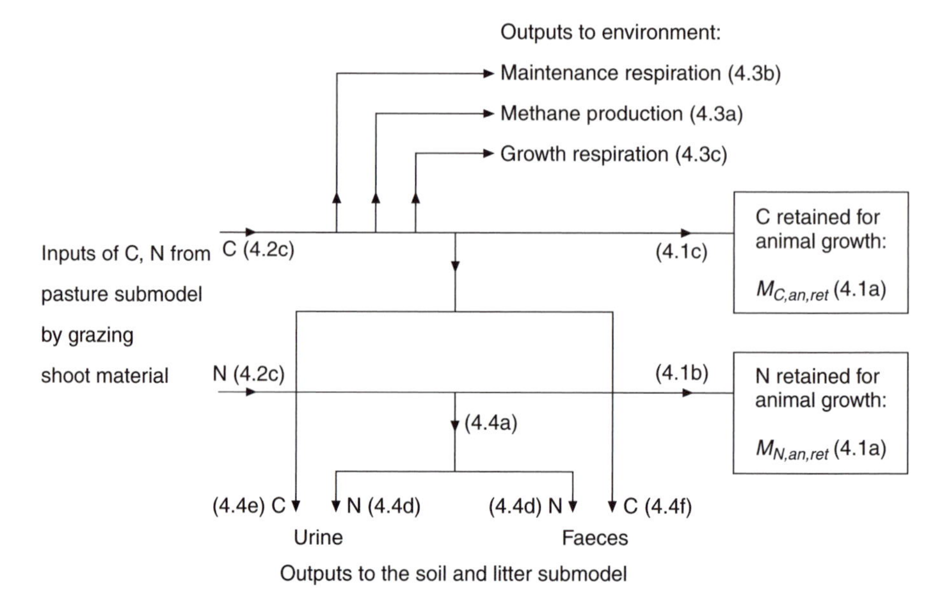

Fig. 4.1. Animal submodel. There are two state variables: the masses of C and N retained by the animals. Equation numbers refer to fluxes or to differential equations.

4.2 Retention of C and N in the Animal

The differential equations for the pools of C and N retained in the animals are

$$\frac{dM_{C,an,ret}}{dt} = I_{C,an,ret}, \qquad \frac{dM_{N,an,ret}}{dt} = I_{N,an,ret},$$
$$M_{C,an,ret}(t=0) = 0 \text{ kg C m}^{-2}, \qquad M_{N,an,ret}(t=0) = 0 \text{ kg N m}^{-2}. \tag{4.1a}$$

These pools are calculated on a ground area basis. The C and N inputs are given in Eqns (4.1c) and (4.1b) below.

4.2.1 Inputs of C and N

It is assumed that a fraction, $f_{N,an,ret}$, of the total N intake, $I_{N,pl\rightarrow an}$ [Eqn (4.2c)], is retained by the animals, so that the flux of N into the animal N retention pool, $I_{N,an,ret}$, is

$$I_{N,an,ret} = f_{N,an,ret} I_{N,pl\rightarrow an}, \qquad f_{N,an,ret} = 0.01. \tag{4.1b}$$

In adult non-lactating non-pregnant animals the retention of N is very small at 1% of total N intake, and this small amount of growth makes a correspondingly small contribution to the C and N fluxes and pools in the plant–animal–soil system. This highly simplified representation of C and N partition in the animal (Fig. 4.3) would not suffice for detailed considerations of animal digestion, metabolism and growth (e.g. Gill *et al.*, 1989), but does suffice for present purposes.

The input of C for retention by the animal, $I_{C,an,ret}$, is calculated from the rate of N retention by assuming a constant C:N ratio for retained material in the animal, $r_{C/N,an}$, with

Table 4.1. Symbols of animal submodel (m^2 refers to ground area unless otherwise specified).

State variables	Description	Initial value
$M_{C,an,ret}$, $M_{N,an,ret}$	Masses of C, N retained in animal (4.1a)	0, 0 kg C, N m^{-2}
Parameters	**Description**	**Value**
$f_{C,an,mai}$	Fraction of C input respired for maintenance (4.3b)	0.65
$f_{C,an,met}$	Fraction of C input converted to methane (4.3a)	0.05
$f_{N,an,ret}$	Fraction of N input retained in animal (4.1b)	0.01
$I_{DM,pl \to an,max}$	Maximum dry matter intake (4.2a)	1.5 kg DM $animal^{-1}$ day^{-1}
$K_{LAI,an}$	Intake response Michaelis–Menten parameter (4.2a)	1 m^2 leaf (m^2 ground)$^{-1}$
q_{LAI}	Intake response 'steepness' parameter (4.2a)	3
$r_{C:N,an}$	C:N ratio in animal (4.1c)	3.5 kg C (kg N)$^{-1}$
Y_{an}	C Growth efficiency (4.3c)	0.4
—	Urine/faeces N partition parameters (4.4c)	See Eqn (4.4c)
Other animal variables	**Description**	**Units**
$f_{N,urine}$	Fraction of excreted N in urine (4.4c)	
$I_{C,an,ret}$	Input to retained animal C pool (4.1c)	kg C m^{-2} day^{-1}
$I_{C,pl \to an}$	Gross C input to the animal (4.2c)	kg C m^{-2} day^{-1}
$I_{DM,pl \to an}$	Input of dry matter from plant to animal (4.2a)	kg DM $animal^{-1}$ day^{-1}
$I_{DM,pl \to an,gnd}$	Input of dry matter from plant to animal (4.2b)	kg DM m^{-2} day^{-1}
$I_{N,an,ret}$	Input to retained animal N pool (4.1b)	kg N m^{-2} day^{-1}
$I_{N,pl \to an}$	Gross N input to the animal (4.2c)	kg N m^{-2} day^{-1}
$O_{C,an \to env,G}$	Growth respiration (4.3c)	kg C m^{-2} day^{-1}
$O_{C,an \to env,mai}$	Maintenance respiration (4.3b)	kg C m^{-2} day^{-1}
$O_{C,an \to env,met}$	Methane production (4.3a)	kg C m^{-2} day^{-1}
$O_{C,an \to so,fa}$	C output to faeces (4.4f)	kg C m^{-2} day^{-1}
$O_{C,an \to so,ur}$	C output to urine (4.4e)	kg C m^{-2} day^{-1}
$O_{N,an \to exc}$	N output to excreta (4.4a)	kg N m^{-2} day^{-1}
$O_{N,an \to so,fa}$	N output to faeces (4.4d)	kg N m^{-2} day^{-1}
$O_{N,an \to so,ur}$	N output to urine (4.4d)	kg N m^{-2} day^{-1}
Management variables		
$n_{animals}$	Stocking density (7.6f)	Sheep m^{-2}
Plant variables		
$C_{tot,sh}$, $N_{tot,sh}$	Total C and N concentrations in shoot (3.1i)	kg C, N (kg dry matter)$^{-1}$
L_{AI}	Total leaf area index (3.1b)	m^2 leaf (m^2 ground)$^{-1}$

$$I_{C,an,ret} = r_{C:N,an} I_{N,an,ret}, \quad r_{C:N,an} = 3.5 \text{ kg C (kg N)}^{-1}. \tag{4.1c}$$

This C:N mass ratio is typical for animal protein (e.g. Fruton and Simmonds, 1958, pp. 27–28).

4.3 Grazing Intake

It is assumed that grazing intake is determined by the leaf area index of the sward, using a sigmoidal response as described by Johnson and Parsons (1985a) and as observed by Penning *et al.* (1991). The input flux of dry matter (DM) from the plants (*pl*) to the animals (*an*), $I_{DM,pl\rightarrow an}$ (kg DM animal^{-1} day^{-1}), is given by

$$\begin{aligned} I_{DM,pl\rightarrow an} &= I_{DM,pl\rightarrow an,max} \frac{\left(L_{AI} / K_{LAI,an}\right)^{q_{LAI,an}}}{1 + \left(L_{AI} / K_{LAI,an}\right)^{q_{LAI,an}}}, \\ I_{DM,pl\rightarrow an,max} &= 1.5 \text{ kg DM animal}^{-1} \text{ day}^{-1} \text{ (sheep)}, \\ K_{LAI,an} &= 1 \text{ m}^2 \text{ leaf (m}^2 \text{ ground)}^{-1}, \qquad q_{LAI,an} = 3. \end{aligned} \tag{4.2a}$$

This equation is drawn in Fig. 4.2, giving the response of animal intake to leaf area index (LAI), denoted by the variable L_{AI} which is obtained from the plant submodel [Eqn (3.1b)]. The maximum intake for large LAI is $I_{DM,pl\rightarrow an,max}$, typically 1.0–2.5 kg DM animal^{-1} day^{-1} (default 1.5) for sheep and 10–20 kg DM animal^{-1} day^{-1} for cattle, depending on the age and physiological state of the animals (e.g. see Leaver,

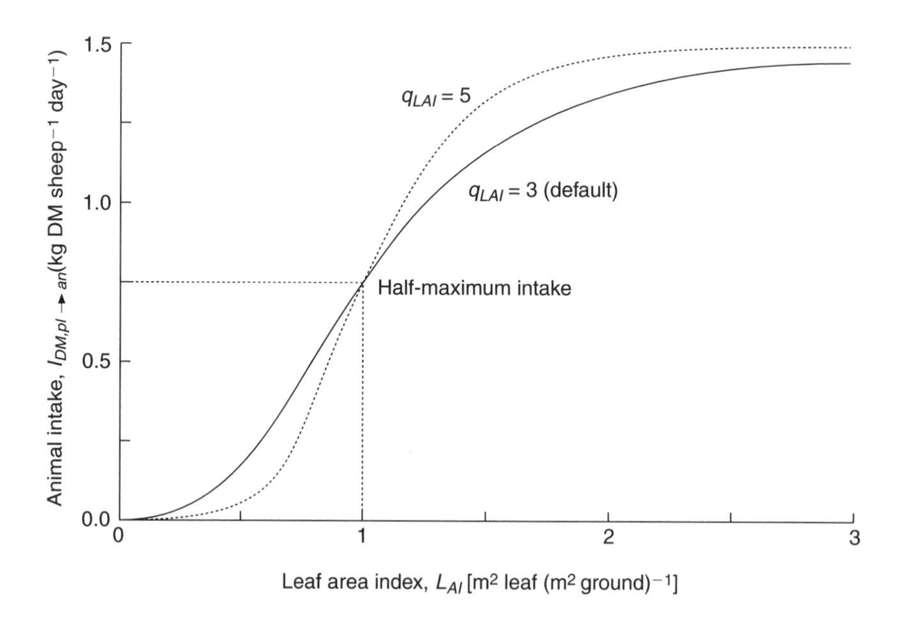

Fig. 4.2. Animal intake. Dry matter grazing intake [$I_{DM,pl\rightarrow an}$, Eqn (4.2a)] in response to leaf area index (L_{AI}). Parameters as given in Eqn (4.2a).

1985; Penning *et al.*, 1995). The LAI for half maximum intake is $K_{LAI,an}$. The response is sigmoidal with a steepness parameter (dimensionless) of $q_{LAI,an}$ (higher values of $q_{LAI,an}$ give a more steplike response – see Fig. 4.2).

The DM input to the animals per unit ground area is [kg DM (m^2 ground)$^{-1}$ day^{-1}]

$$I_{DM,pl\rightarrow an,gnd} = n_{animals} I_{DM,pl\rightarrow an}. \tag{4.2b}$$

The stocking density, $n_{animals}$ (sheep m^{-2}), is given in Eqn (7.6f).

The C and N inputs to the animal per unit ground area, $I_{C,pl\rightarrow an}$ and $I_{N,pl\rightarrow an}$ (kg C, N m^{-2} day^{-1}), and the C:N ratio in the ingested material, $r_{C:N,tot,sh}$, are obtained by

$$I_{C,pl\rightarrow an} = C_{tot,sh} I_{DM,pl\rightarrow an,gnd}, \qquad I_{N,pl\rightarrow an} = N_{tot,sh} I_{DM,pl\rightarrow an,gnd},$$
$$r_{C:N,tot,sh} = \frac{C_{tot,sh}}{N_{tot,sh}}. \tag{4.2c}$$

The C and N concentrations (total) of shoot dry matter, $C_{tot,sh}$ and $N_{tot,sh}$, are variables of the plant submodel and are calculated in Eqns (3.1i). The C:N ratio in the ingested material, $r_{C:N,tot,sh}$, is used below for determining the partition of excreted N between urine and faeces.

4.4 C Outputs to the Atmosphere

4.4.1 Methane production

It is assumed that a fraction, $f_{C,an,met}$, of the total C input is converted into methane, giving an output flux to the atmosphere of

$$O_{C,an\rightarrow env,met} = f_{C,an,met} I_{C,pl\rightarrow an}, \qquad f_{C,an,met} = 0.05. \tag{4.3a}$$

4.4.2 Respiration – maintenance and growth

It is assumed that a fraction, $f_{C,an,mai}$, of the total C input, $I_{C,pl\rightarrow an}$ [Eqn (4.2c)], is taken for maintenance and is respired as CO_2, giving an output flux of

$$O_{C,an\rightarrow env,mai} = f_{C,an,mai} I_{C,pl\rightarrow an}, \qquad f_{C,an,mai} = 0.65. \tag{4.3b}$$

For mature non-pregnant non-lactating animals maintenance takes a high proportion of the energy taken in.

Denoting the growth efficiency by Y_{an}, the growth respiration (the output of C from the animal as CO_2 arising from the growth process) is

$$O_{C,an\rightarrow env,G} = \frac{1-Y_{an}}{Y_{an}} I_{C,an,ret}, \qquad Y_{an} = 0.4. \tag{4.3c}$$

Here, $I_{C,an,ret}$ [Eqn (4.1c)] is the C retained in the animals due to growth. Per unit mass of C used for animal growth, a fraction Y_{an} is converted into product (i.e. is retained), and the remaining fraction, $1 - Y_{an}$, is respired. This value is consistent with the values reported by Gill *et al.* (1989, table 3.3, p. 109).

4.5 C and N Outputs to the Soil and Litter Submodel

The N flux to excreta, $O_{N,an\rightarrow exc}$, is obtained by difference, subtracting the retained N flux, $I_{N,an,ret}$, Eqn (4.1b), from the total N intake, $I_{N,pl\rightarrow an}$, Eqn (4.2c), giving

$$O_{N,an\rightarrow exc} = (1 - f_{N,an,ret}) I_{N,pl\rightarrow an}. \tag{4.4a}$$

It is assumed that the fractions of total excreted N found in urine and in faeces depend on the C:N ratio in the ingested material, $r_{C:N,tot,sh}$, derived in Eqn (4.2c). Experiments on ewes (Parsons *et al.*, 1991b; Orr *et al.*, 1995) suggest that the fraction of excreted N found in the urine, $f_{N,urine}$, has the values

$$\begin{aligned} f_{N,urine} &= 0.8 \qquad \text{when } r_{C:N,tot,sh} = 12, \\ f_{N,urine} &= 0.7 \qquad \text{when } r_{C:N,tot,sh} = 25. \end{aligned} \tag{4.4b}$$

Interpolating linearly between these values leads to

$$f_{N,urine} = 0.8 - (0.8 - 0.7)\left(\frac{r_{C:N,tot,sh} - 12}{25 - 12}\right). \tag{4.4c}$$

Using Eqn (4.4a) for the total excreted N, the outputs of N to urine and faeces are

$$\begin{aligned} O_{N,an\rightarrow so,ur} &= f_{N,urine} O_{N,an\rightarrow exc}, \\ O_{N,an\rightarrow so,fa} &= \left(1 - f_{N,urine}\right) O_{N,an\rightarrow exc}. \end{aligned} \tag{4.4d}$$

The urine N output carries with it a C flux, given by

$$O_{C,an\rightarrow so,ur} = \frac{12}{28} O_{N,an\rightarrow so,ur}. \tag{4.4e}$$

This is based on the chemical formula for urea, which is $H_2N.CO.NH_2$, with two atoms of N ($2 \times 14 = 28$) per atom of C (12). It is assumed that all urine N is in urea (but see Whitehead, 1995, table 4.6, p. 74).

The C output to the faeces is obtained by difference, with

$$\begin{aligned} O_{C,an\rightarrow so,fa} = I_{C,pl\rightarrow an} &- O_{C,an\rightarrow env,mai} - O_{C,an\rightarrow env,met} - I_{C,an,ret} \\ &- O_{C,an\rightarrow env,G} - O_{C,an\rightarrow so,ur}. \end{aligned} \tag{4.4f}$$

The terms subtracted from the gross C intake of $I_{C,pl\rightarrow an}$ [Eqn (4.2c)] are due to maintenance respiration [$O_{C,an\rightarrow env,mai}$, Eqn (4.3b)], methane production [$O_{C,an\rightarrow env,mai}$, Eqn (4.3a)], growth [retained C, $I_{C,an,ret}$, Eqn (4.1c)], growth respiration [$O_{C,an\rightarrow env,G}$, Eqn (4.3c)] and urine [$O_{C,an\rightarrow so,ur}$, Eqn (4.4e)].

Note that, with the default parameter settings, this implies an apparent digestibility for C in non-growing animals of about 0.7. This is similar to the organic matter digestibility of cool-temperate perennial grasses (e.g. Gill *et al.*, 1989, p. 96).

The C and N balances, relative to intakes of 100 units, as given by the above equations, are summarized in Fig. 4.3. The C:N ratio of the plant shoot is about 20 [Eqn (4.2c)], so these numbers are in agreement with the observation that the C:N ratio of cattle and sheep dung is about 23 (Kirchmann, 1991). C digestibility is an output of the submodel, depending to an extent on the C:N ratio in the intake material.

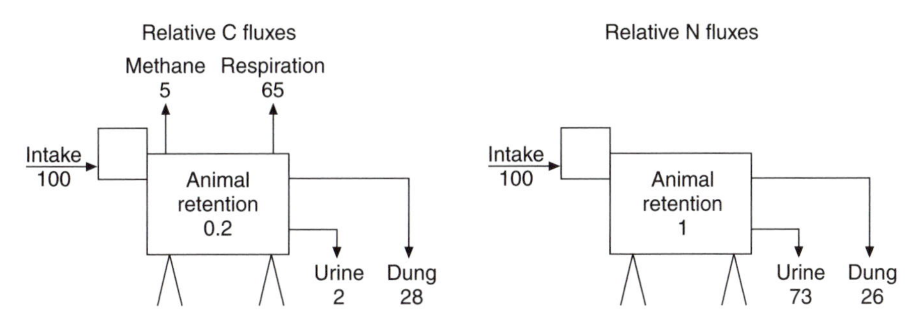

Fig. 4.3. Relative C and N fluxes for a mature non-lactating non-pregnant sheep, as given by the animal submodel.

Because most of the C intake is respired, note that net primary production, using the usual definition of gross photosynthetic rate less plant respiration [Section 8.7, Eqn (8.1a)], can be substantially different from the C input to the litter and soil submodel, the difference depending on the grazing intensity, which also affects plant litter production.

4.6 Impacts of Grazing on the Carbon Fluxes from the Plant Submodel

Grazing can have a profound influence on both the dynamic and stationary properties of the system. Here we consider some of the steady-state characteristics of a plant–animal system which has been decoupled from the water submodel (Chapter 6) by making all the water modifying functions [Eqns (6.7b)] equal to unity (program switch s_Water), and from the soil submodel (Chapter 5) by causing the soil ammonium and nitrate pools (Fig. 5.1) to remain constant (program switch soilvr). Also, the diurnal and seasonal components of the environment (Chapter 7) are removed using appropriate switches (pp. 140, 146), and photosynthesis is applied at a constant rate (p. 150).

4.6.1 The stability of grazed pastures

The stability of grazing systems was considered in a pioneering paper by Noy-Meir (1975), who used some simple relationships from prey–predator theory. More recently, Johnson and Parsons (1985a) explored the stability of a grazed pasture using a physiological model of the growth of the above-ground component of a grass crop together with the intake curve shown in Fig. 4.2. In this section, we demonstrate that present model gives rise to the same phenomena.

In Fig. 3.9c, it can be seen that net shoot growth rate (g_y) versus shoot dry mass (y, equivalent to leaf area index) is approximately parabolic, following the equation

$$g_y = by(y_m - y), \tag{4.5a}$$

where b and y_m are constants. The intake equation is [cf. Eqn (4.2a)]

$$i = i_m \frac{y^q}{y^q + K^q}, \tag{4.5b}$$

where i_m is the maximum intake rate, K and q are constants. It is the intersections where $g_y = i$ of these two equations that determine the grazing equilibria. There may be two intersections, in which case $y_1 = 0$ is unstable, and y_2 is a positive stable grazing equilibrium; or there may be four intersections, in increasing magnitude, y_1, … , y_4; in this case $y_1 = 0$ is unstable as before, y_2 is stable, y_3 is unstable, and y_4 is stable. The system settles at y_2 or y_4 depending upon where it starts from (the initial value of y), and there may be a region of y-space (shoot dry mass, or leaf area index), which may be unattainable as parameters such as stocking density are varied. An illustration of this simplified view of grazing stability is in Thornley and Johnson (1990, fig. 6.6, p. 151).

In Fig. 4.4, stocking density $n_{animals}$ has been varied with the plant submodel growing at two constant temperatures, 10 and 20°C. At 10°C with stocking densities between 3 and 5 sheep ha^{-1}, there are two steady states, with leaf area index (L_{AI}) of 1 or 5.5. If the simulation begins with a low L_{AI} starting value, then the lower branch solutions are obtained, and vice versa. Outside the stocking density range of 3–5 sheep ha^{-1}, there is only a single solution whatever the starting values of the simulation. As stocking density is varied continuously, there is a discontinuous jump in L_{AI}. The high L_{AI} branch is the more productive sward, both in terms of animal intake per hectare and in terms of net primary production. However, the system

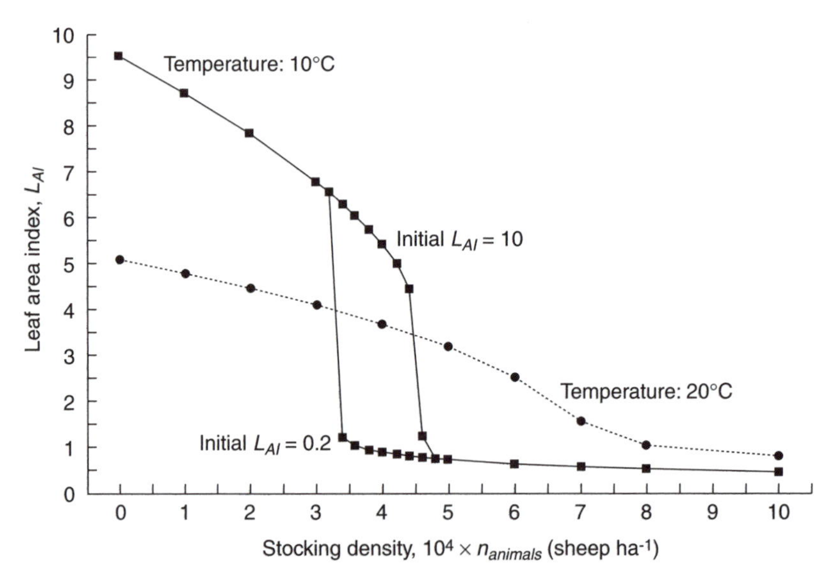

Fig. 4.4. Responses of leaf area index L_{AI} to stocking density $n_{animals}$. The water submodel is decoupled. The soil submodel is also decoupled so that the soil ammonium and nitrate pools are constant at 0.0007 and 0.0003 kg N m^{-2}. The environment is constant, with both seasonal and diurnal changes switched off (Chapter 7). Each point on the graph represents a steady state, reached by dynamic simulation from initial values of leaf area index as indicated for the 10°C response.

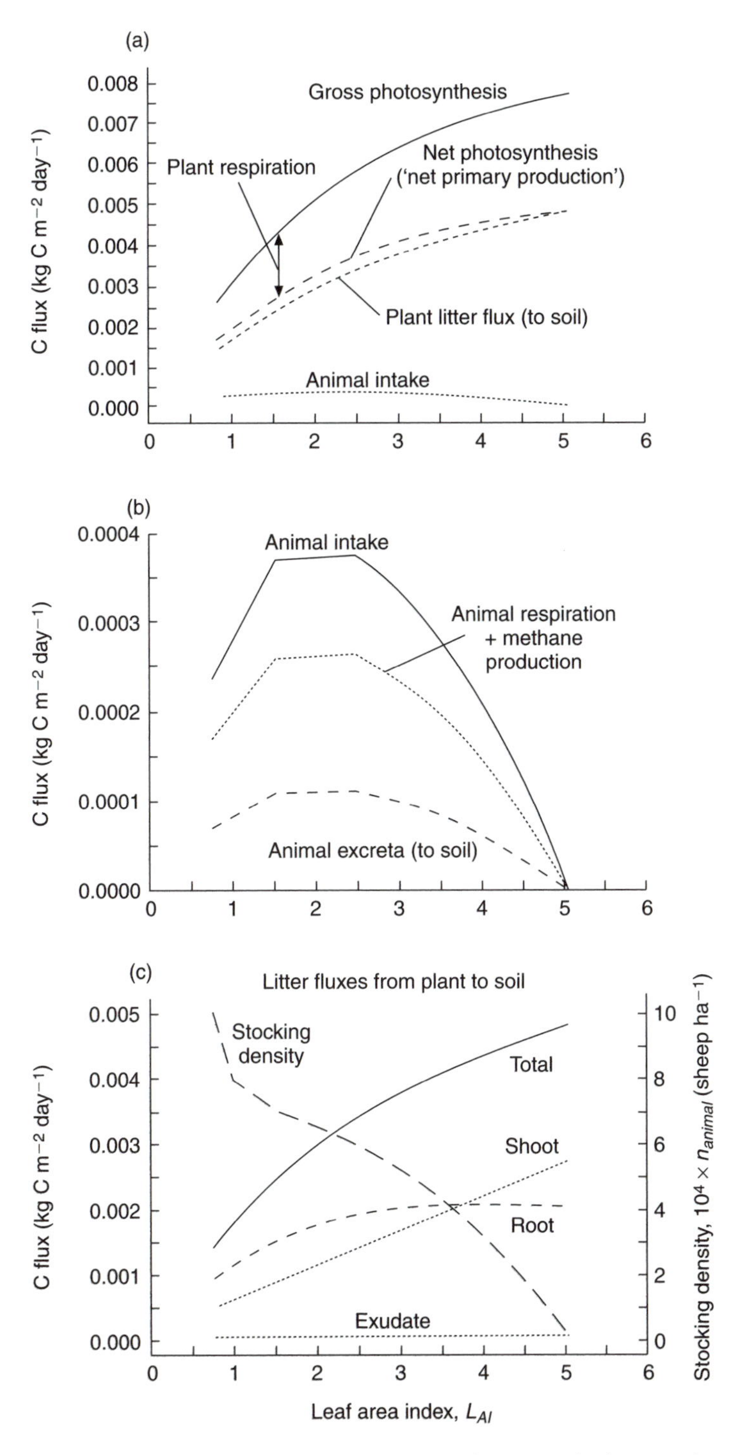

Fig. 4.5. Impact of grazing and consequent changes in leaf area index on the carbon fluxes from plant and animal to soil. These simulations were performed as in Fig. 4.4 in the 20°C response.

is quite easily flipped from one state to the other by changes in grazing management. At 20°C this bifurcation disappears, and varying the stocking density now gives continuous variation in L_{AI} over the whole range, and the steady state achieved is independent of the initial values. Both environment as well as plant and animal parameters can greatly influence grazing stability. In this instance, lower temperatures make the grassland ecosystem more prone to instability than higher temperatures. The mean annual temperature in central southern Britain is 10°C.

4.6.2 Carbon fluxes in the plant–animal system

The carbon fluxes corresponding to the 20°C steady states of Fig. 4.4 are shown in Fig. 4.5. The carbon fluxes to soil and litter decrease as a result of introducing grazing for two principal reasons: first, leaf area index (L_{AI}) is decreased and this tends to decrease gross photosynthesis (Fig. 4.5a), especially at lower leaf area indices where the canopy is not capturing all the incident radiation; and second, only a small fraction of animal intake is excreted to the soil the rest being respired (Fig. 4.5b). In a grazed system, the traditional definition of net primary production [gross photosynthesis less plant respiration: Section 8.7, Eqn (8.1a)] may be an inaccurate measure of the carbon input to the soil, especially for management regimes which maximize animal intake per unit ground area (Figs 4.5a,b). This has been emphasized by Parsons *et al.* (1983, particularly fig. 5). As grazing intensity increases with increasing stocking density (Fig. 4.5c), shoot litter is decreased far more than root litter, plant maintenance respiration decreases, and animal respiration increases.

Soil and Litter Submodel

5

5.1 Introduction

The soil and litter submodel is shown in Fig. 5.1. The inputs from plant litter and faeces are divided into three fractions:

1. metabolizable (*met*), a rapidly degraded fraction consisting mostly of carbohydrates and proteins;
2. cellulose (*cel*), a more slowly degraded fraction comprising mostly cellulose and hemicellulose;
3. lignin (*lig*), a resistant fraction of lignified material.

The *cel* and *lig* fractions are assumed to have constant C:N ratios. The animal and plant submodels generate fluxes of C and N in faeces [$O_{C,an \to so,fa}$, $O_{N,an \to so,fa}$, Eqns (4.4f), (4.4d)] and plant litter [$O_{C,sh \to so}$, $O_{N,sh \to so}$, $O_{C,rt \to so}$, $O_{N,rt \to so}$, Eqns (3.10b)]. Fractions of these C flows are assigned to the surface and soil *cel* and *lig* pools with the fixed C:N ratios shown in Fig. 5.1; the remainder of the C and N is assigned to the *met* C and N pools where the C and N contents can vary independently.

This approach has been widely followed (van Veen and Paul, 1981; Verberne *et al*., 1990; Parton *et al*., 1993), and, although others have used two categories for the inputs (e.g. Parton *et al*., 1987; Jenkinson, 1990), there is general agreement that this is a good starting point for litter characterization.

On the other hand, the characterization of soil organic matter is contentious. Chemical categories based on fulvic and humic acids have earlier been in vogue (e.g. Tate, 1987). However, more recently, a view which is increasingly supported is that the *physical* state of the material, in particular the degree to which it is in association with clay or other soil particles, may be more important than its chemical state (e.g. Christensen, 1996; see Hassink and Whitmore, 1997, for a model of the physical protection of organic matter in soils). We adopt here the latter position, although the criticism that schemes proposed for modelling purposes are often difficult to relate to data from fractionation and chemical analysis still has some validity.

Soil organic matter (SOM), apart from the soil microbe pool, is placed in three

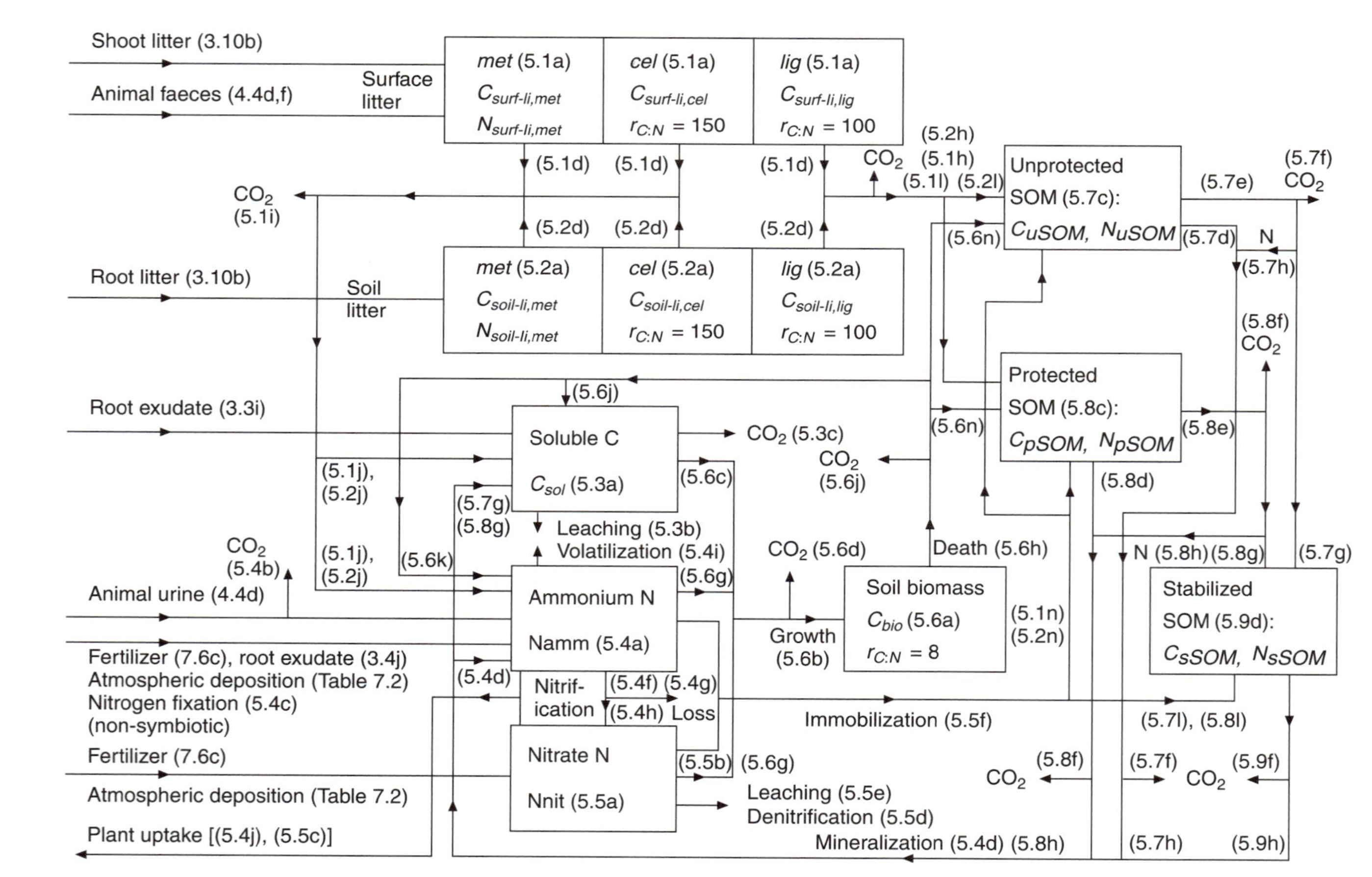

Fig. 5.1. Soil and litter submodel. The 18 state variables of the submodel (Table 5.1) are shown in the boxes. Equation numbers refer to the differential equations, or the fluxes, or expressions from which the flux can be calculated. *met, cel, lig* denote metabolic, cellulose and lignin, as the three categories into which the plant litter and animal faeces inputs are placed. SOM denotes soil organic matter.

categories: unprotected SOM (*uSOM*), protected SOM (*pSOM*) and stabilized SOM (*sSOM*). 'Protection' is against degradation (mineralization) and is provided by association with clay particles or other soil particles. Material from the two lignin (*lig*) pools suffers some respiratory loss to CO_2, and then enters the unprotected and protected SOM pools (*uSOM*, *pSOM*), where it is divided between the two according to clay content of the soil (f_{clay}). There are also inputs to the *uSOM* and *pSOM* pools from dying soil microbes, again with a division depending on soil clay content. A process of chemical stabilization can convert *uSOM* or *pSOM* to a stabilized form (*sSOM*). This process competes with possible mineralization of *uSOM* and *pSOM*. After Parton *et al.* (1993), it is assumed that *uSOM*, *pSOM* and *sSOM* have variable C:N ratios depending on soil mineral N, N_{min}. Stabilized SOM (*sSOM*) has a C:N ratio which also depends on clay content.

The definitions of 'mineralization' and 'immobilization' are partly arbitrary. Here N mineralization is the gross flux of N from the three SOM pools to the ammonium pool. The fluxes of N from decomposing litter are not included in our definition of mineralization. The C fluxes out of unprotected SOM (Fig. 5.1) are to stabilized SOM, to the environment (CO_2), and to the soluble C pool. It is assumed that the N associated with the C flux into the soluble C pool, and the N associated with the respired fraction of the C flux to stabilized SOM, are returned to the ammonium N pool, N_{amm}. To satisfy the lower C:N ratio in stabilized SOM then requires an immobilizing flux taken from both mineral N pools. Alternatively it could have been assumed that the mineralizing fluxes, which arguably comprise N returned to the bulk soil ammonium pool, were smaller, and then the corresponding immobilizing N flux would be less.

Jenkinson (1990), Parton *et al.* (1993) and Verberne *et al.* (1990) all represent microbes, although none of these authors involve the size or activity of the microbial pool in any of the transformations occurring, and assume linear kinetics throughout (Section 1.6). The microbial pool is assumed to be involved in some of the processes shown in Fig. 5.1, namely nitrification, denitrification and microbial growth. However, the three mineralization reactions, of *uSOM*, *pSOM* and *sSOM*, are linear. Non-linear processes may give a more accurate representation of the actual mechanisms, but difficulties were encountered in mixing linear and non-linear processes. Non-linear processes can lead to more complex responses (p. 62), which are generally more realistic (e.g. Murray, 1977), but could be misleading if the non-linear representation is not well-chosen. However, excessive reliance on linearity can give models with very unrealistic behaviour. These difficulties point to a need for more mechanistic information at the process level.

A different class of decay models has been developed, e.g. by Carpenter (1981), who considers the decay of heterogenous detritus. In this approach there is a continuous scale of decomposability, and fractions of detritus can move up or down the scale. The method is more mathematical and perhaps less accessible than traditional SOM models of the type used here. However, more general techniques such as this may be needed in order to make long-term progress.

We depart from tradition in using 'substrate' pools for C and N substrates in the soil, labelled C_{sol}, N_{amm} and N_{nit} in Fig. 5.1. Modellers often try to simplify their models by ignoring small and relatively labile pools. However, experience in modelling plants suggests that substrate pools can have quite a decisive influence in determining activity (e.g. of shoot and root), and allocating growth between shoot and root (Thornley, 1997). It seems more realistic to represent these small pools rather than using awkward empirical devices which often are too rigid, and may block off development routes for the model. The inclusion of soil substrate pools means that processes such as root exudation, microbial growth and death, and a role of microbial biomass in determining rates of some processes, can be explicitly represented; these possibilities seem to be worth having.

Non-symbiotic N fixation is included at a low but variable level as it seems likely to be a significant process for an important class of plant ecosystems: N-poor grassland and forest under increasing ambient CO_2 concentrations. The plant submodel (Chapter 3) is for a monoculture, and does not take explicit account of grass–legume competition and symbiotic N fixation by a legume sward component. However, symbiotic N fixation may be crudely introduced by adding N inputs to the system, possibly dependent on the C:N status of the plants, e.g. extra N inputs could be applied to the substrate N pool in the plant root (Fig. 3.1), or to the soil ammonium N pool (Fig. 5.1). Grass–legume competition poses a difficult problem which is best first addressed with simpler models rather than within the framework of a plant ecosystem simulator (e.g. Thornley *et al.*, 1995; Schwinning and Parsons, 1996a, b).

The soil and litter submodel presented here has many simplifications: the inclusion of only C and N (omitting phosphorus), the use of a single microbial pool, and an analysis based on a single soil horizon, are among the more important. The model assumes a spatially uniform distribution of all processes, including dung and urine deposition: in a temperate environment with adequate moisture this assumption could be reasonable, but see Schwinning and Parsons (1996b). Note also that the model does not include any effect of soil organic matter on the soil water characteristic, which could be important in long-term responses. Recent results reported by Fisher *et al.* on C sequestration in grassland soils (1994, 1995; see also Davidson *et al.*, 1995) indicate how much remains to be done, and how far we are from being able to make credible long-term predictions of ecosystem responses.

It is worth remarking that the scheme of Fig. 5.1 is considerably more complex than a scheme Verberne and I proposed earlier (Thornley and Verberne, 1989). However, in many respects the increase in complexity made little difference, other than changing the speed of some of the dynamic responses and the positions of the long-term asymptotes. The more detailed soil model can of course be used to address more problems, such as exudation and C leaching, but in many ways, the soil appears to be relatively passive in its responses compared with the very plastic responses of the plant to the environment.

The different pools with differential equations for the state variables, and expressions for the input/output terms, are presented in separate sections. State variables, parameters and other variables used in this Chapter are listed in Table 5.1. Many parameter values are close to those used by Verberne *et al.* (1990) or Parton *et al.* (1993), but as is usual (Section 1.3), some parameters have been adjusted to obtain reasonable behaviour.

Table 5.1. Symbols of soil and litter submodel. SOM denotes soil organic matter. Equation references are: (i) for state variables, to their differential equation; (ii) for parameters, to the equation where the parameter is introduced and explained; (iii) for other soil and litter variables, to the equation where the variable is defined; (iv) for variables from other submodels, to the equation in this chapter where the variable is used, and to the equation where the variable is defined. Rate constants [e.g. k_{bioG} of Eqn (5.6b)] are not listed under other variables; the standard value (e.g. of k_{bioG20}) and definition is given under parameters.

State variables	Description	Initial value
C_{bio}	C content of soil microbial biomass pool (5.6a)	0.2 kg C m^{-2}
$C_{soil\text{-}li,i}$, i = *met, cel, lig*	C content of soil litter pools, *i* = *met*abolic, *cel*lulose, *lig*nin (5.2a)	0.0025, 0.1, 0.15 kg C m^{-2}
C_{sol}	Soil soluble C (C substrate) pool (5.3a)	0.0005 kg C m^{-2}
$C_{surf\text{-}li,i}$, i = *met, cel, lig*	C content of surface litter pools, *i* = *met*abolic, *cel*lulose, *lig*nin (5.1a)	0.0025, 0.1, 0.15 kg C m^{-2}
C_{pSOM}, C_{sSOM}, C_{uSOM}	C content of protected, stabilized and unprotected SOM pools [(5.8c), (5.9d), (5.7c)]	4, 6, 0.4 kg C m^{-2}
N_{amm}	Ammonium N pool (5.4a)	0.0007 kg N m^{-2}
N_{nit}	Nitrate N pool (5.5a)	0.0003 kg N m^{-2}
N_{pSOM}, N_{sSOM}, N_{uSOM}	N content of protected, stabilized and unprotected SOM pools [(5.8c), (5.9d), (5.7c)]	0.4, 0.5, 0.04 kg N m^{-2}
$N_{soil\text{-}li,met}$, $N_{surf\text{-}li,met}$	Soil and surface litter metabolic N pools [(5.2a), (5.1a)]	0.0005, 0.0005 kg N m^{-2}

Parameters	Description	Value
$C_{bio,min}$	Minimum value applied to the microbial biomass population (5.6h)	0.001 kg C in biomass m^{-2}
$c_{bioD,clay}$	Parameter determining fate of dying microbial biomass as affected by clay content (5.6n)	3
$c_{lit\rightarrow prot}$	Constant affecting dependence on clay content of fraction of decaying litter assigned to protected SOM (5.1l)	3
$c_{N,bio}$	Constant in microbial biomass growth N use (5.6b)	1 kg ammonium N (kg nitrate N)$^{-1}$
$c_{rC:N,li}$	Constant affecting dependence of litter decay rates on C:N ratio (5.1f)	25 kg C (kg N)$^{-1}$
$f_{CbioD,R}$	Fraction of microbial biomass death C flux respired (5.6j)	0.5
$f_{CbioD\rightarrow Csol}$	Fraction of microbial biomass death C flux returned to C_{sol} pool (5.6j)	0.1
$f_{CbioD\rightarrow uSOM,clay}$, $f_{CbioD\rightarrow uSOM,clay}$	Fraction of dying microbial biomass C flux which is unprotected in clay, sandy soil (5.6n)	0.3, 0.7
$f_{C,fa,i}$, i = *met, cel, lig*	Fractions of faeces C flux which is *met*abolic, *cel*lulose, *lig*nin (5.1b)	0.2, 0.65, 0.15

Continued over

Table 5.1. Continued

Parameters	Description	Value
$f_{C,pSOM\to minR}$, $f_{C,pSOM\to sSOM,R}$	Fractions of C fluxes from protected SOM to mineralization and stabilization which are respired (5.8f)	0.4, 0.2
$f_{C,rt\text{-}li,i}$, $i = met, cel, lig$	Fractions of root litter C flux which is *met*abolic, *cel*luose, *lig*nin (5.2b)	0.2, 0.65, 0.15
$f_{C,sh\text{-}li,i}$, $i = met, cel, lig$	Fractions of shoot litter C flux which is *met*abolic, *cel*luose, *lig*nin (5.1b)	0.2, 0.65, 0.15
$f_{C,soil\text{-}li,iR}$, $i = met, cel, lig$	Fractions of soil litter C fluxes from *met*abolic, *cel*luose, *lig*nin pools which are respired (5.2h)	0.6, 0.4, 0.2
$f_{C,sSOM\to minR}$	Fraction of C flux from stabilized SOM to mineralization which is respired (5.9f)	0.4
$f_{C,surf\text{-}li,iR}$, $i = met, cel, lig$	Fractions of surface litter C fluxes from *met*abolic, *cel*luose, *lig*nin pools which are respired (5.1h)	0.6, 0.4, 0.2
$f_{C,uSOM\to minR}$, $f_{C,uSOM\to sSOM,R}$	Fractions of C fluxes from unprotected SOM to mineralization and stabilization which are respired (5.7f)	0.4, 0.2
f_{clay}	Clay content (5.1l)	0.1
f_{nitrif}	Fraction of nitrification flux lost to environment (5.4g)	0.1
$f_{SOM,bio}$	Constant determining maximum in soil microbial biomass (5.6b)	0.05 kg C in biomass (kg C in SOM)$^{-1}$
$J_{Nfix,bio}$	Inhibition constant for non-symbiotic N fixation (5.4c)	0.001 kg N m^{-2}
$K_{C,bioG}$, $K_{N,bioG}$	Michaelis–Menten parameters in microbial biomass growth (5.6b)	0.001, 0.0002 kgC, N m^{-2}
$K_{Csol,Nfix}$	Michaelis–Menten constant for non-symbiotic N fixation (5.4c)	0.0005 kg C m^{-2}
$K_{rC:N,pSOM}$, $K_{rC:N,sSOM}$, $K_{rC:N,uSOM}$	Michaelis–Menten parameters for dependence of new protected, stabilized, unprotected SOM on mineral N [5.8b), (5.9b), (5.7b)]	0.002 kg N m^{-2}
k_{bioD20}	Microbial biomass death rate parameter (5.6h)	0.01 day^{-1}
k_{bioG20}	Microbial biomass growth rate parameter (5.6b)	0.1 (kg C)$^{1-q_{bio}}$ day^{-1}
$k_{CsolR20}$	Respiration constant (5.3c)	0.1 day^{-1}
$k_{denit20}$	Denitrification rate constant (5.5d)	0.5 (kg biomass C m^{-2})$^{-1}$ day^{-1}
$k_{Nfix,bio}$	Rate constant for non-symbiotic N fixation (5.4c)	50×10^{-6} kg N (kg biomass C)$^{-1}$ day^{-1}
k_{nit20}	Nitrification rate constant (5.4f)	0.4 (kg biomass C m^{-2})$^{-1}$ day^{-1}
$k_{soil\text{-}li,i20}$, $i = met, cel, lig$	Rate constants for soil litter degradation (5.2e)	0.2, 0.1, 0.02 day^{-1}
$k_{surf\text{-}li,i20}$, $i = met, cel, lig$	Rate constants for surface litter degradation (5.1e)	0.2, 0.1, 0.02 day^{-1}
$k_{iSOM\to min20}$, $i = p, s, u$	Mineralization rate constants for protected (*p*), stabilized (*s*), unprotected (*u*) SOM [(5.8d), (5.9e), (5.7d)]	0.00015, 0.00003, 0.003 (kg C in biomass m^{-2})$^{-1}$ day^{-1}

Table 5.1. Continued

Parameters	Description	Value
$k_{pSOM \to sSOM20}$, $k_{uSOM \to sSOM20}$	Stabilization rate constants for protected and unprotected SOM [(5.8e), (5.7e)]	0.00002, 0.0002 day^{-1}
k_{vol20}	Volatilization rate constant (5.4i)	0.02 day^{-1}
q_{bio}	Allometric constant for growth of microbial biomass (5.6b)	2/3
q_{denit}	Denitrification parameter (5.5d)	10
$q_{lit,decay,lig}$	Constant affecting dependence of litter decay on lignin (5.1f)	3
$q_{W,soil}$, $q_{W,surf}$	Exponents governing effects of water stress on soil and surface processes (5.11b)	20, 30
$r_{C:N,bio}$	C:N ratio is soil microbial biomass (5.6e)	8 kg C (kg N)$^{-1}$
$r_{C:N,cel}$, $r_{C:N,lig}$	C:N ratios in *cel*lulose and *lig*nin pools (5.1c)	150, 100 kg C (kg N)$^{-1}$
$r_{C:N,pSOM,max}$, $r_{C:N,pSOM,min}$, $r_{C:N,uSOM,max}$, $r_{C:N,uSOM,min}$	Maximum, minimum C:N ratios of protected, unprotected SOM (5.8b) (5.7b)	12, 6, 12, 6 kg C (kg N)$^{-1}$
Y_{bio}	Yield constant for microbial biomass growth process (5.6c)	0.5

Other soil and litter variables	Description	Units
$C_{soil\text{-}li}$, $C_{surf\text{-}li}$	Total C in soil, surface litter [(5.2g), (5.1g)]	kg C m^{-2}
C_{SOM}	Total C content of the three SOM pools (5.10a)	kg C m^{-2}
D_{bio}	Death rate of soil microbial biomass (5.6h)	kg C m^{-2} day^{-1}
$f_{bio,max}$	Maximum microbial biomass density modifier of growth rate of biomass, set by SOM (5.6b)	
$f_{bio,min}$	Minimum microbial biomass density, modifier of death rate of biomass (5.6h)	
$f_{CbioD \to pSOM}$, $f_{CbioD \to uSOM}$	Fractions of C from microbial biomass death assigned to protected and unprotected SOM pools (5.6n)	
$f_{CN,bio}$	Dependence of microbial biomass growth rate on supply of C and N (5.6b)	
$f_{C,lit\text{-}prot}$	Fraction of lignin C entering protected SOM pool as affected by clay content (5.1l)	
$f_{decay,lig,soil}$, $f_{decay,lig,surf}$	Modifier of soil, surface litter degradation according to lignin content [(5.2f), (5.1f)]	
$f_{denit,W}$	Modifier of denitrification as affected by soil aerobicity (5.5d)	
$f_{lig,soil}$, $f_{lig,surf}$	Fractions of lignin in soil, surface litter [(5.2f), (5.1f)]	
$f_{Namm \to bio}$, $f_{Nnit \to bio}$	Fractions of microbial biomass growth N reqirements taken from ammonium and nitrate pools (5.6f)	
$f_{rC:N,soil\text{-}li}$, $f_{rC:N,surf\text{-}li}$	Modifiers of soil, surface litter degradation according to C:N ratio [(5.2f), (5.1f)]	
$f_{T,soil}$, $f_{T,surf}$	Soil and surface temperature functions (5.11a)	
$f_{W,soil}$, $f_{W,surf}$	Soil and surface water functions (5.11b)	
G_{bio}	Growth rate of soil microbial biomass (5.6b)	kg C m^{-2} day^{-1}
$I_{C,bioD \to Csol}$	Input of C from microbial biomass death to soluble C pool (5.6j)	kg C m^{-2} day^{-1}

Continued over

Table 5.1. Continued

Other soil and litter variables	Description	Units
$I_{Cbio \to SOM}$	Flux of C from microbial biomass death into unprotected and protected SOM pools (5.6m)	kg C m^{-2} day^{-1}
$I_{Cbio \to pSOM}$, $I_{Cbio \to uSOM}$	Fluxes of C from microbial biomass death into protected, unprotected SOM pools (5.6n)	kg C m^{-2} day^{-1}
$I_{C,pSOM \to Csol}$, $I_{C,pSOM \to sSOM}$, $I_{C,uSOM \to Csol}$, $I_{C,uSOM \to sSOM}$	Fluxes of C from protected, unprotected SOM pools to soluble C (C_{sol}) and stabilized SOM [(5.8g), (5.7g)]	kg C m^{-2} day^{-1}
$I_{C,soil\text{-}li \to Csol}$, $I_{C,surf\text{-}li \to Csol}$	Inputs of C to soluble C (C_{sol}) substrate pool from soil, surface litter [5.2j), (5.1j)]	kg C m^{-2} day^{-1}
$I_{C,soil\text{-}li \to SOM}$, $I_{C,surf\text{-}li \to SOM}$	C inputs to SOM pools from soil, surface litter [(5.2k), (5.1k)]	kg C m^{-2} day^{-1}
$I_{C,soil\text{-}li \to pSOM}$, $I_{C,soil\text{-}li \to uSOM}$, $I_{C,surf\text{-}li \to pSOM}$, $I_{C,surf\text{-}li \to uSOM}$	C inputs to protected and unprotected SOM pools from soil, surface litter [(5.2l), (5.1l)]	kg C m^{-2} day^{-1}
$I_{C,soil\text{-}li,i}$, $I_{C,surf\text{-}li,i}$, $i = met, cel, lig$	Inputs of C into soil, surface litter *met*abolic, *cel*lulose, *lig*nin pools [(5.2b), (5.1b)]	kg C m^{-2} day^{-1}
$I_{C,sSOM \to Csol}$	C input to soluble C pool (C_{sol}) from stabilized SOM (5.9g)	kg C m^{-2} day^{-1}
$I_{Namm \to Nnit}$	N input to nitrate N pool from ammonium N pool (5.4h)	kg N m^{-2} day^{-1}
$I_{Nbio \to Namm}$	Total N input to ammonium N pool from microbial biomass death (5.6q)	kg N m^{-2} day^{-1}
$I_{Nbio \to SOM}$	N input to SOM reqired by flow of C from microbial biomass death to SOM (5.6o)	kg N m^{-2} day^{-1}
$I_{N,bioD \to Namm}$	Input of N from microbial biomass death to ammonium pool (5.6k)	kg N m^{-2} day^{-1}
$I_{N,bioD\text{-}SOM \to Namm}$	Input of N to ammonium N from fraction of dying microbial biomass transformed into SOM (5.6p)	kg N m^{-2} day^{-1}
$I_{Nfix \to Namm}$	Non-symbiotic N fixation by soil microorganisms (5.4c)	kg N m^{-2} day^{-1}
$I_{N,nitrif \to env}$	Nitrification N input to environment (5.4g)	kg N m^{-2} day^{-1}
$I_{N,pSOM \to Namm}$, $I_{N,uSOM \to Namm}$	Mineralizing N inputs from protected, unprotected SOM to ammonium N [(5.8h), (5.7h)]	kg N m^{-2} day^{-1}
$I_{N,soil\text{-}li \to Namm}$, $I_{N,surf\text{-}li \to Namm}$	Inputs of N to ammonium N (N_{amm}) pool from soil, surface litter [(5.2j), (5.1j)]	kg N m^{-2} day^{-1}
$I_{N,soil\text{-}li,met}$, $I_{N,surf\text{-}li,met}$	Inputs of N into soil, surface litter *met*abolic N pools [(5.2c), (5.1c)]	kg N m^{-2} day^{-1}
$I_{N,SOM \to Namm}$	N input to ammonium N from mineralization of SOM (5.4d)	kg N m^{-2} day^{-1}
$I_{N,sSOM \to Namm}$	Mineralizing N flux from stabilized SOM into ammonium N (5.9h)	kg N m^{-2} day^{-1}
$N_{eff,bio}$	Effective mineral N concentration seen by microbial biomass (5.6b)	kg N m^{-2}
N_{min}	Total mineral N pool in soil (5.5g)	kg N m^{-2}
$N_{soil\text{-}li}$, $N_{surf\text{-}li}$	Total N in soil, surface litter [(5.2g), (5.1g)]	kg N m^{-2}
N_{SOM}	Total N in three SOM pools (5.10a)	kg N m^{-2}
$O_{C,pSOM \to min}$, $O_{C,uSOM \to min}$	C outputs from protected, unprotected SOM pools to mineralization [(5.8d), (5.7d)]	kg C m^{-2} day^{-1}

Table 5.1. Continued

Other soil and litter variables	Description	Units
$O_{C,pSOM \to sSOM}$, $O_{C,uSOM \to sSOM}$	C outputs from protected, unprotected SOM pools to stabilized SOM [(5.8e), (5.7e)]	kg C m^{-2} day^{-1}
O_{Cbio}	Total C flux associated with microbial biomass death (5.6i)	kg C m^{-2} day^{-1}
$O_{Csol \to bioG}$, $O_{Csol \to lch}$, $O_{Csol \to R}$	C output from C_{sol} pool to microbial biomass growth (5.6c), to leaching (5.3b), and to direct respiration (5.3c)	kg C m^{-2} day^{-1}
$O_{C,soil\text{-}li,i}$, $O_{C,surf\text{-}li,i}$, i = *met*, *cel*, *lig*	Outputs of C from soil, surface litter *met*abolic, *cel*lulose, *lig*nin pools [(5.2d), (5.1d)]	kg C m^{-2} day^{-1}
$O_{C,sSOM \to min}$	C output from stabilized SOM to mineralization (5.9e)	kg C m^{-2} day^{-1}
$O_{Namm \to bioG}$, $O_{Nnit \to bioG}$	Outputs of N from ammonium and nitrate pools for microbial biomass growth (5.6g)	kg N m^{-2} day^{-1}
$O_{Namm \to imm}$, $O_{Nnit \to imm}$, $O_{Nmin \to imm}$	Total immobilizing fluxes from ammonium N, nitrate N, and combined pools [(5.4e), (5.5b), (5.5f)]	kg N m^{-2} day^{-1}
$O_{Namm \to nitrif}$, $O_{Namm \to rt}$, $O_{Namm \to vol}$	Output of N from ammonium N pool to nitrification (5.4f), root uptake (5.4j), volatilization (5.4i)	kg N m^{-2} day^{-1}
$O_{Namm,pSOM \to sSOM}$, $O_{Nnit,pSOM \to sSOM}$, $O_{Namm,uSOM \to sSOM}$, $O_{Nnit,uSOM \to sSOM}$	Mineralizing N fluxes from ammonium and nitrate pools associated with transformations of protected, unprotected SOM to stabilized SOM [(5.8l), (5.7l)]	kg N m^{-2} day^{-1}
O_{Nbio}	Total N output associated with microbial biomass death (5.6i)	kg N m^{-2} day^{-1}
$O_{Nbio \to SOM}$	N output from microbial biomass death towards unprotected and protected SOM pools (5.6l)	kg N m^{-2} day^{-1}
$O_{Namm,soil\text{-}li \to SOM}$, $O_{Nnit,soil\text{-}li \to SOM}$, $O_{Namm,surf\text{-}li \to SOM}$, $O_{Nnit,surf\text{-}li \to SOM}$	N outputs required from ammonium and nitrate N pools as soil, surface litter lignin is transformed to SOM [(5.2o), (5.1o)]	kg N m^{-2} day^{-1}
$O_{Nmin \to bioG}$	Output from mineral N pools (N_{amm} and N_{nit}) to microbial biomass growth (5.6e)	kg N m^{-2} day^{-1}
$O_{Nmin,pSOM \to sSOM}$, $O_{Nmin,uSOM \to sSOM}$	Immobilizing N fluxes taken from mineral N pools associated with tranformation of protected, unprotected SOM to stabilized SOM [(5.8k), (5.7k)]	kg N m^{-2} day^{-1}
$O_{Nmin,soil\text{-}li \to SOM}$, $O_{Nmin,surf\text{-}li \to SOM}$	N outputs required from mineral N pools as soil, surface litter lignin is transformed to SOM [(5.2n), (5.1n)]	kg N m^{-2} day^{-1}
$O_{Nnit \to denit}$, $O_{Nnit \to lch}$, $O_{Nnit \to rt}$	Output of N from nitrate N pool to denitrification (5.5d), leaching (5.5e), root uptake (5.5c)	kg N m^{-2} day^{-1}
$O_{N,soil\text{-}li,lig}$, $O_{N,soil\text{-}li,met}$, $O_{N,surf\text{-}li,lig}$, $O_{N,surf\text{-}li,met}$	Outputs of N from soil, surface litter *lig*nin and *met*abolic pools [(5.2m), (5.2d),(5.1m), (5.1d)]	kg N m^{-2} day^{-1}
R_{bioD}	Respiration associated with microbial biomass death (5.6j)	kg C m^{-2} day^{-1}

Continued over

Table 5.1. Continued

Other soil and litter variables	Description	Units
R_{bioG}	Microbial biomass growth respiration (5.6d)	kg C m^{-2} day^{-1}
$R_{pSOM \to min}$, $R_{uSOM \to min}$, $R_{pSOM \to sSOM}$, $R_{uSOM \to sSOM}$	Respiration associated with mineralization and stabilization of protected and unprotected SOM [(5.8f), (5.7f)]	kg C m^{-2} day^{-1}
$R_{sSOM \to min}$	Respiration associated with mineralization of stabilized SOM (5.9f)	kg C m^{-2} day^{-1}
$R_{soil\text{-}li}$, $R_{surf\text{-}li}$	Total soil, surface litter respiration [(5.2i), (5.1i)]	kg C m^{-2} day^{-1}
$R_{soil\text{-}li,i}$, $R_{surf\text{-}li,i}$, $i = met, cel, lig$	Respiration from soil, surface litter *met*abolic, *cel*lulose, *lig*nin pools [(5.2h), (5.1h)]	kg C m^{-2} day^{-1}
R_{urine}	Respiratory C flux associated with urine hydrolysis (5.4b)	kg C m^{-2} day^{-1}
$r_{C:N,pSOM,new}$, $r_{C:N,sSOM,new}$, $r_{C:N,uSOM,new}$	C:N ratio in new protected, stabilized, unprotected SOM [(5.8b), (5.9b), (5.7b)]	kg C (kg N)$^{-1}$
$r_{C:N,soil\text{-}li}$, $r_{C:N,surf\text{-}li}$	Soil, surface litter C:N ratios [(5.2f), (5.1f)]	kg C (kg N)$^{-1}$
$r_{C:N,SOM}$	Ratio of C:N in total SOM (5.10b)	kg C (kg N)$^{-1}$
$r_{C:N,iSOM}$, $i = p,s,u$	C:N ratio is in SOM pools [(5.8a), (5.9a), (5.7a)]	kg C (kg N)$^{-1}$
Animal submodel variables		
$O_{C,an \to so,fa}$, $O_{C,an \to so,ur}$	Outputs of C from animal in faeces [(5.1b), (4.4f)], urine [(5.4b), (4.4e)]	kg C m^{-2} day^{-1}
$O_{N,an \to so,fa}$, $O_{N,an \to so,ur}$	Outputs of N from animal in faeces [(5.1c), (4.4d)], urine [(5.4a), (4.4d)]	kg N m^{-2} day^{-1}
Environment and management		
$I_{N,env \to Namm}$, $I_{N,env \to Nnit}$	Inputs of atmospheric ammonium N, nitrate N [(5.4a), (5.5a), Table 7.2]	kg N m^{-2} day^{-1}
$I_{N,fert \to Namm}$, $I_{N,fert \to Nnit}$	Fertilizer application of ammonium N, nitrate N [(5.4a), (5.5a), (7.6c)]	kg N m^{-2} day^{-1}
T_{air}, T_{soil}	Air and soil temperatures [(5.11a), (7.5e); (7.4a), Table 7.3]	°C
Plant submodel variables		
$O_{C,rt \to so}$, $O_{C,sh \to so}$	Output of C from root, shoot litter [(5.2b), (5.1b), (3.10b)]	kg C m^{-2} day^{-1}
$O_{CS,rt \to so,ex}$, $O_{NS,rt \to so,ex}$	Outputs of C, N substrates from root by exudation [(5.3a), (3.3i); (5.4a), (3.4j)]	kg C, N m^{-2} day^{-1}
$O_{N,rt \to so}$, $O_{N,sh \to so}$	Outputs of N from root, shoot litter [(5.2c), (5.1c), (3.10b)]	kg N m^{-2} day^{-1}
u_{Namm}, u_{Nnit}	Plant uptake of N from ammonium, nitrate pools [(5.4j), (5.5c); (3.4i)]	kg N m^{-2} day^{-1}
Water submodel variables		
$a_{W,sh}$, $a_{W,so}$	Shoot, soil water activity [(5.11b), (6.7a)]	
k_{leach}	Leaching rate constant [(5.3b), (5.5e); (6.2e)]	day^{-1}

5.2 Surface Litter Pools, $C_{surf\text{-}li,met}$, $N_{surf\text{-}li,met}$, $C_{surf\text{-}li,cel}$, $C_{surf\text{-}li,lig}$

5.2.1 Differential equations

The surface and soil litter pools are treated similarly, with each having the three categories of metabolizable (*met*), cellulose (*cel*) and lignin (*lig*). For the surface litter pool state variables, the differential equations and initial values of the state variables are

$$\begin{aligned}
&\frac{dC_{surf\text{-}li,met}}{dt} = I_{C,surf\text{-}li,met} - O_{C,surf\text{-}li,met},\\
&\frac{dN_{surf\text{-}li,met}}{dt} = I_{N,surf\text{-}li,met} - O_{N,surf\text{-}li,met},\\
&\frac{dC_{surf\text{-}li,cel}}{dt} = I_{C,surf\text{-}li,cel} - O_{C,surf\text{-}li,cel},\\
&\frac{dC_{surf\text{-}li,lig}}{dt} = I_{C,surf\text{-}li,lig} - O_{C,surf\text{-}li,lig},\\
&C_{surf\text{-}li,met}(t=0) = 0.0025 \text{ kg C m}^{-2}, \quad N_{surf\text{-}li,met}(t=0) = 0.0005 \text{ kg N m}^{-2}\\
&C_{surf\text{-}li,cel}(t=0) = 0.1, \quad C_{surf\text{-}li,lig}(t=0) = 0.15 \text{ kg C m}^{-2}.
\end{aligned} \tag{5.1a}$$

The inputs and outputs are described immediately below.

5.2.2 Inputs

There are two inputs of C to surface litter: $O_{C,sh\to so}$ of Eqn (3.10b) from shoot litter, and $O_{C,an\to so,fa}$ of Eqn (4.4f) from faeces. The substrate C component ($O_{CS,sh\to so}$) of the shoot litter flux is placed directly in the *met* C pool (Fig. 5.1). The structural C flux from the shoot ($O_{CX,sh\to so}$) and the C flux from faeces ($O_{an\to so,fa}$) are divided in constant fractions $f_{C,sh\text{-}li,i}$ and $f_{C,fa,i}$ between the (i =) *met*, *cel* and *lig* pools. This results in C inputs to *met*, *cel* and *lig* of

$$\begin{aligned}
&I_{C,surf\text{-}li,met} = O_{CS,sh\to so} + f_{C,sh\text{-}li,met} O_{CX,sh\to so} + f_{C,fa,met} O_{C,an\to so,fa},\\
&I_{C,surf\text{-}li,cel} = f_{C,sh\text{-}li,cel} O_{CX,sh\to so} + f_{C,fa,cel} O_{C,an\to so,fa},\\
&I_{C,surf\text{-}li,lig} = f_{C,sh\text{-}li,lig} O_{CX,sh\to so} + f_{C,fa,lig} O_{C,an\to so,fa},\\
&f_{C,sh\text{-}li,met} = 0.2, \quad f_{C,sh\text{-}li,cel} = 0.65, \quad f_{C,sh\text{-}li,lig} = 0.15,\\
&f_{C,fa,met} = 0.2, \quad f_{C,fa,cel} = 0.65, \quad f_{C,fa,lig} = 0.15.
\end{aligned} \tag{5.1b}$$

As the *cel* and *lig* pools have fixed C:N ratios, the N input to *met* is

$$\begin{aligned}
&I_{N,surf\text{-}li,met} = O_{N,sh\to so} + O_{N,an\to so,fa} - \frac{I_{C,surf\text{-}li,cel}}{r_{C:N,cel}} - \frac{I_{C,surf\text{-}li,lig}}{r_{C:N,lig}},\\
&r_{C:N,cel} = 150 \text{ kg C (kg N)}^{-1}, \quad r_{C:N,lig} = 100 \text{ kg C (kg N)}^{-1}.
\end{aligned} \tag{5.1c}$$

The shoot litter N flux $O_{N,sh\to so}$ and the faeces N flux $O_{N,an\to so,fa}$ are given in Eqns (3.10b) and (4.4d). $O_{N,sh\to so}$ includes both substrate N ($O_{NS,sh\to so}$) and the structural N ($O_{NX,sh\to so}$) components. Note that the ratio of C:N in the inputs in the *met* pool depends on the C and N contents of shoot litter and faeces, and this can cause the

C:N ratio in the *met* pool to vary. The faeces C:N ratio is about 23 (see Fig. 4.3 and Section 4.5); the shoot litter C:N ratio is *c*. 25 but can vary considerably with growing conditions [Eqns (3.10b) and Section 4.5]. Verberne *et al.* (1990, p.227) assume that the C:N ratio in their decomposable fraction is six. For comparison, proteins typically contain 50–55% C and 12–19% N (Fruton and Simmonds, 1958, p. 27), giving a C:N ratio between 2.6 and 4.6. With default values and no grazing, the model gives a C:N ratio in the *met* pool which varies between 6.9 and 8.4 over the second year of simulation.

5.2.3 Outputs

The output terms are calculated assuming linear kinetics (Section 1.6), and are

$$\begin{aligned} O_{C,surf\text{-}li,met} &= k_{surf\text{-}li,met} C_{surf\text{-}li,met}, \quad & O_{N,surf\text{-}li,met} &= k_{surf\text{-}li,met} N_{surf\text{-}li,met} \\ O_{C,surf\text{-}li,cel} &= k_{surf\text{-}li,cel} C_{surf\text{-}li,cel}, \quad & O_{C,surf\text{-}li,lig} &= k_{surf\text{-}li,lig} C_{surf\text{-}li,lig}. \end{aligned} \tag{5.1d}$$

The rate constants (day^{-1}) are given by

$$\begin{aligned} &k_{surf\text{-}li,met} = k_{surf\text{-}li,met20} f_{T,surf} f_{W,surf} f_{rC:N,surf\text{-}li}, \\ &k_{surf\text{-}li,cel} = k_{surf\text{-}li,cel20} f_{T,surf} f_{W,surf} f_{rC:N,surf\text{-}li} f_{decay,lig,surf}, \\ &k_{surf\text{-}li,lig} = k_{surf\text{-}li,lig20} f_{T,surf} f_{W,surf} f_{rC:N,surf\text{-}li} f_{decay,lig,surf}, \\ &k_{surf\text{-}li,met20} = 0.2, \qquad k_{surf\text{-}li,cel20} = 0.1, \qquad k_{surf\text{-}li,lig20} = 0.02 \text{ day}^{-1}. \end{aligned} \tag{5.1e}$$

The three rate constants $k_{..20}$ at 20°C are modified according to a surface temperature function ($f_{T,surf}$) and a surface water function ($f_{W,surf}$) [Eqns (5.11a), (5.11b)]. $f_{rC:N,surf\text{-}i}$ gives the dependence of all three decay rates on the total surface litter C:N ratio [Eqns (5.1f), (5.1g)], decreasing with increasing C:N; $f_{decay,lig,surf}$ describes the dependence of the *cel* and *lig* rates on lignin fraction, $f_{lig,surf}$ (Parton *et al.*, 1987; Verberne *et al.*, 1990). If all the modifiers are unity, the half-lives of the *met*, *cel* and *lig* pools are 3, 7 and 35 days. The modifiers are given by

$$\begin{aligned} &f_{rC:N,surf\text{-}li} = \frac{c_{rC:N,li}}{r_{C:N,surf-li}}, \qquad r_{C:N,surf\text{-}li} = \frac{C_{surf\text{-}li}}{N_{surf\text{-}li}}, \\ &f_{decay,lig,surf} = e^{-q_{\text{lit,decay,lig}} f_{lig,surf}}, \qquad f_{lig,surf} = \frac{C_{surf\text{-}li,lig}}{C_{surf\text{-}li,lig} + C_{surf\text{-}li,cel}}, \\ &c_{rC:N,li} = 25 \text{ kg C (kg N)}^{-1}, \qquad q_{lit,decay,lig} = 3. \end{aligned} \tag{5.1f}$$

A total litter C:N ratio ($r_{C:N,surf\text{-}li}$) of 25 (specified by $c_{rC:N,li}$) gives an $f_{rC:N,surf\text{-}li}$ of unity. A lower C:N ratio causes the rate constants to increase. $f_{decay,lig,surf}$ describes the dependence of the decay rates of *cel* and *lig* on the lignin fraction $f_{lig,surf}$. As the lignin fraction increases, decay rates decrease. $f_{lig,surf}$ has a maximum value of unity for no lignin giving $f_{decay,lig,surf} = 1$; $f_{decay,lig,surf}$ has a notional minimum value of e^{-3} or 0.05, reducing the decay rates to 5% in a highly lignified substrate where $f_{lig,surf} = 1$.

The total quantities of C and N in the surface litter pools are

$$C_{surf\text{-}li} = C_{surf\text{-}li,met} + C_{surf\text{-}li,cel} + C_{surf\text{-}li,lig},$$
$$N_{surf\text{-}li} = N_{surf\text{-}li,met} + \frac{C_{surf\text{-}li,cel}}{r_{C:N,cel}} + \frac{C_{surf\text{-}li,lig}}{r_{C:N,lig}}. \tag{5.1g}$$

The C:N ratio parameters are given in Eqns (5.1c).

5.2.4 Outputs into other pools

Respiratory losses to the environment (kg C m^{-2} day^{-1}) from the outputs are

$$R_{surf\text{-}li,met} = f_{C,surf\text{-}li,metR} O_{C,surf\text{-}li,met}, \quad R_{surf\text{-}li,cel} = f_{C,surf\text{-}li,celR} O_{C,surf\text{-}li,cel},$$
$$R_{surf\text{-}li,lig} = f_{C,surf\text{-}li,ligR} O_{C,surf\text{-}li,lig}, \tag{5.1h}$$
$$f_{C,surf\text{-}li,metR} = 0.6, \quad f_{C,surf\text{-}li,celR} = 0.4, \quad f_{C,surf\text{-}li,ligR} = 0.2.$$

Constant respiratory loss fractions, $f_{C,surf\text{-}li,..R}$, are assumed, with different values for the three surface litter categories (Verberne *et al.*, 1990).

Total surface litter respiration is

$$R_{surf\text{-}li} = R_{surf\text{-}li,met} + R_{surf\text{-}li,cel} + R_{surf\text{-}li,lig}. \tag{5.1i}$$

C and N inputs to the soluble C (C_{sol}) and ammonium N (N_{amm}) pools are

$$I_{C,surf\text{-}li \to Csol} = O_{C,surf\text{-}li,met} - R_{surf\text{-}li,met} + O_{C,surf\text{-}li,cel} - R_{surf\text{-}li,cel},$$
$$I_{N,surf\text{-}li \to Namm} = O_{N,surf\text{-}li,met} + \frac{O_{C,surf\text{-}li,cel}}{r_{C:N,cel}}. \tag{5.1j}$$

The C:N ratio parameter is given in Eqn (5.1c).

There is a C input to the SOM pools from the surface litter lignin pool:

$$I_{C,surf\text{-}li \to SOM} = O_{C,surf\text{-}li,lig} - R_{surf\text{-}li,lig}. \tag{5.1k}$$

This is divided between the protected and unprotected SOM pools according to clay content, with

$$I_{C,surf\text{-}li \to pSOM} = f_{C,lit \to prot} I_{C,surf\text{-}li \to SOM},$$
$$I_{C,surf\text{-}li \to uSOM} = (1 - f_{C,lit \to prot}) I_{C,surf\text{-}li \to SOM}, \tag{5.1l}$$
$$f_{C,lit \to prot} = c_{\text{lit} \to \text{prot}} f_{clay}.$$
$$c_{lit \to prot} = 3, \qquad f_{clay} = 0.1.$$

A high clay content, f_{clay}, directs more of the output from the surface litter lignin pool into the protected SOM pool. With the constant $c_{lit \to prot} = 3$, then f_{clay} may not exceed $\frac{1}{3}$ otherwise the fraction $f_{C,lit \to prot}$ exceeds unity.

It is assumed that new material entering the unprotected and protected SOM pools has variable C:N ratios of $r_{C:N,uSOM,new}$ and $r_{C:N,pSOM,new}$ [Eqns (5.7b), (5.8b)] which depend on the concentration of the soil mineral N pools N_{min} [Eqn (5.5g)]. Additional N is needed to sustain the flux of C from lignin with a C:N ratio of 100 [Eqn (5.1c)] into pools of SOM with C:N ratios which are much lower. This additional N flux is taken from the N_{amm} and N_{nit} pools (labelled immobilization in Fig. 5.1).

Accompanying the C flux from lignin of $O_{C,surf\text{-}li,lig}$ [Eqn (5.1d)] is an N flux of [see Eqn (5.1c) for the C:N ratio]

$$O_{N,surf\text{-}li,lig} = \frac{O_{C,surf\text{-}li,lig}}{r_{C:N,lig}}. \tag{5.1m}$$

The N flux required from the two mineral pools (denoted N_{min}) is therefore

$$O_{N\min,surf\text{-}li\to SOM} = \frac{I_{C,surf\text{-}li\to uSOM}}{r_{C:N,uSOM,new}} + \frac{I_{C,surf\text{-}li\to pSOM}}{r_{C:N,pSOM,new}} - O_{N,surf\text{-}li,lig}. \tag{5.1n}$$

The first two terms on the right-hand side of this equation are the C inputs to unprotected and protected SOM given by Eqns (5.1l), divided by the C:N ratios to turn this into an N requirement. Then the (usually small) flux of N from lignin [Eqn (5.1m)] is subtracted. The resulting N requirement is taken from the N_{amm} and N_{nit} pools in the same ratio in which microbial biomass takes ammonium N and nitrate N, to give outputs from the N_{amm} and N_{nit} pools of

$$\begin{aligned} O_{Namm,surf\text{-}li\to SOM} &= f_{Namm\to bio} O_{Nmin,surf\text{-}li\to SOM}, \\ O_{Nnit,surf\text{-}li\to SOM} &= f_{Nnit\to bio} O_{Nmin,surf\text{-}li\to SOM}. \end{aligned} \tag{5.1o}$$

The f fractions are given in Eqns (5.6f).

5.3 Soil Litter Pools, $C_{soil\text{-}li,met}$, $N_{soil\text{-}li,met}$, $C_{soil\text{-}li,cel}$, $C_{soil\text{-}li,lig}$

5.3.1 Differential equations

The treatment of the soil litter pools is similar to that of the surface litter pools in Section 5.2 above, where the explanation is more complete. Eqns (2.2a) to (2.2l) correspond to (2.1a) to (2.1l). For the soil litter pool state variables, the differential equations and initial values of the state variables are

$$\begin{aligned} &\frac{dC_{soil\text{-}li,met}}{dt} = I_{C,soil\text{-}li,met} - O_{C,soil\text{-}li,met}, \\ &\frac{dN_{soil\text{-}li,met}}{dt} = I_{N,soil\text{-}li,met} - O_{N,soil\text{-}li,met}, \\ &\frac{dC_{soil\text{-}li,cel}}{dt} = I_{C,soil\text{-}li,cel} - O_{C,soil\text{-}li,cel}, \quad \frac{dC_{soil\text{-}li,lig}}{dt} = I_{C,soil\text{-}li,lig} - O_{N,soil\text{-}li,lig}, \\ &C_{soil\text{-}li,met}(t=0) = 0.0025 \text{ kg C m}^{-2}, \quad N_{soil\text{-}li,met}(t=0) = 0.0005 \text{ kg N m}^{-2}, \\ &C_{soil\text{-}li,cel}(t=0) = 0.1, \quad C_{soil\text{-}li,lig}(t=0) = 0.15 \text{ kg C m}^{-2}. \end{aligned} \tag{5.2a}$$

5.3.2 Inputs

The C input to soil litter, $O_{C,rt\to so}$ of Eqn (3.10b) comes from root litter. It has two components: a substrate C flux ($O_{CS,rt\to so}$) and a structural C flux ($O_{CX,rt\to so}$). The substrate C flux is placed directly in the *met* C pool (Fig. 5.1). The structural C flux is divided in constant fractions $f_{C,rt\text{-}li,i}$ and $f_{C,lig,i}$ between the (i =) *met*, *cel* and *lig* pools, to give inputs to *met*, *cel* and *lig* of

$$I_{C,soil\text{-}li,met} = O_{CS,rt\to so} + f_{C,rt\text{-}li,met}O_{CX,rt\to so},$$
$$I_{C,soil\text{-}li,cel} = f_{C,rt-li,cel}O_{CX,rt\to so}, \qquad I_{C,soil\text{-}li,lig} = f_{CX,rt\text{-}li,lig}O_{CX,rt\to so}, \tag{5.2b}$$
$$f_{C,rt\text{-}li,met} = 0.2, \qquad f_{C,rt\text{-}li,cel} = 0.65, \qquad f_{C,rt\text{-}li,lig} = 0.15.$$

The *cel* and *lig* pools have fixed C:N ratios. Therefore the N input to *met* is

$$I_{N,soil\text{-}li,met} = O_{N,rt\to so} - \frac{I_{C,soil\text{-}li,cel}}{r_{C:N,cel}} - \frac{I_{C,soil\text{-}li,lig}}{r_{C:N,lig}}. \tag{5.2c}$$

The C:N ratios are given in Eqns (5.1c). The root litter N flux $O_{N,rt\to so}$ is given in Eqn (3.10b).

5.3.3 Outputs

The output terms are

$$O_{C,soil\text{-}li,met} = k_{soil\text{-}li,met}C_{soil\text{-}li,met}, \qquad O_{N,soil\text{-}li,met} = k_{soil\text{-}li,met}N_{soil\text{-}li,met},$$
$$O_{C,soil\text{-}li,cel} = k_{soil\text{-}li,cel}C_{soil\text{-}li,cel}, \qquad O_{C,soil\text{-}li,lig} = k_{soil\text{-}li,lig}C_{soil\text{-}li,lig}, \tag{5.2d}$$

with the rate constants (day^{-1}) given by

$$k_{soil\text{-}li,met} = k_{soil\text{-}li,met20}f_{T,soil}f_{W,soil}f_{rC:N,soil\text{-}li},$$
$$k_{soil\text{-}li,cel} = k_{soil\text{-}li,cel20}f_{T,soil}f_{W,soil}f_{rC:N,soil\text{-}li}f_{decay,lig,soil},$$
$$k_{soil\text{-}li,lig} = k_{soil\text{-}li,lig20}f_{T,soil}f_{W,soil}f_{rC:N,soil\text{-}li}f_{decay,lig,soil}, \tag{5.2e}$$
$$k_{soil\text{-}li,met20} = 0.2, \qquad k_{soil\text{-}li,cel20} = 0.1, \qquad k_{soil\text{-}li,lig20} = 0.02\ \text{day}^{-1}.$$

The three rate constants $k_{..20}$ at 20°C are modified according to a soil temperature function ($f_{T,soil}$) and a soil water function ($f_{W,soil}$) [Eqns (5.11a), (5.11b)]. $f_{rC:N,soil\text{-}li}$ gives the dependence of all three decay rates on the total soil litter C:N ratio, decreasing with increasing C:N; $f_{decay,lig,soil}$ describes the dependence of the *cel* and *lig* rates on lignin fraction, $f_{lig,soil}$. These are

$$f_{rC:N,soil\text{-}li} = \frac{c_{rC:N,li}}{r_{C:N,soil\text{-}li}}, \qquad r_{C:N,soil\text{-}li} = \frac{C_{soil\text{-}li}}{N_{soil\text{-}li}},$$
$$f_{decay,lig,soil} = \mathrm{e}^{-q_{\mathrm{lit,decay,lig}}f_{lig,soil}}, \qquad f_{lig,soil} = \frac{C_{soil\text{-}li,lig}}{C_{soil\text{-}li,lig} + C_{soil\text{-}li,cel}}. \tag{5.2f}$$

See Eqn (5.1f) and after for parameter values and further explanation.

The total quantities of C and N in the soil litter pools are

$$C_{soil\text{-}li} = C_{soil\text{-}li,met} + C_{soil\text{-}li,cel} + C_{soil\text{-}li,lig},$$
$$N_{soil\text{-}li} = N_{soil\text{-}li,met} + \frac{C_{soil\text{-}li,cel}}{r_{C:N,cel}} + \frac{C_{soil\text{-}li,lig}}{r_{C:N,lig}}. \tag{5.2g}$$

The C:N ratio parameters are given in Eqns (5.1c).

5.3.4 Outputs into other pools

Respiratory losses to the environment (kg C m^{-2} day^{-1}) from the outputs are:

$$R_{soil\text{-}li,met} = f_{C,soil\text{-}li,metR}O_{C,soil\text{-}li,met}, \quad R_{soil\text{-}li,cel} = f_{C,soil,li,celR}O_{C,soil\text{-}li,cel},$$
$$R_{soil\text{-}li,lig} = f_{C,soil\text{-}li,ligR}O_{C,soil\text{-}li,lig}, \tag{5.2h}$$
$$f_{C,soil\text{-}li,metR} = 0.6, \quad f_{C,soil,li,celR} = 0.4, \quad f_{C,soil\text{-}li,ligR} = 0.2.$$

Again, constant respiratory loss fractions, $f_{C,soil\text{-}li,..R}$, are assumed, with different values for the three soil litter categories.

Total soil litter respiration is

$$R_{soil\text{-}li} = R_{soil\text{-}li,met} + R_{soil\text{-}li,cel} + R_{soil\text{-}li,lig}. \tag{5.2i}$$

C and N inputs to the soluble C (C_{sol}) and ammonium N (N_{amm}) pools are

$$I_{C,soil\text{-}li\rightarrow Csol} = O_{C,soil\text{-}li,met} - R_{soil,li,met} + O_{C,soil\text{-}li,cel} - R_{soil\text{-}li,cel},$$
$$I_{N,soil\text{-}li\rightarrow Namm} = O_{N,soil\text{-}li,met} + \frac{O_{C,soil\text{-}li,cel}}{r_{C:N,cel}} \tag{5.2j}$$

The C:N ratio $r_{C:N,cel}$ is given in Eqn (5.1c).

There is a C input to the SOM pools from the soil litter lignin pool:

$$I_{C,soil\text{-}li\rightarrow SOM} = O_{C,soil\text{-}li,lig} - R_{soil\text{-}li,lig}. \tag{5.2k}$$

This is divided between the protected and unprotected SOM pools according to clay content, with

$$I_{C,soil\text{-}li\rightarrow pSOM} = f_{C,lit\rightarrow prot}I_{C,soil\text{-}li\rightarrow SOM},$$
$$I_{C,soil\text{-}li\rightarrow uSOM} = \left(1 - f_{C,lit\rightarrow prot}\right)I_{C,soil\text{-}li\rightarrow SOM}. \tag{5.2l}$$

See Eqn (5.1l) for parameters and explanation.

Additional N is needed to sustain the flux of C from lignin with a C:N ratio of 100 to the unprotected and protected SOM pools with much lower C:N ratios [Eqns (5.7b), (5.8b)]. This additional N flux comes from the N_{amm} and N_{nit} pools, and is labelled immobilization in Fig. 5.1. Accompanying the C flux from lignin of $O_{C,soil\text{-}li\rightarrow SOM}$ is an N flux of

$$O_{N,soil\text{-}li,lig} = \frac{O_{C,soil\text{-}li\rightarrow lig}}{r_{C:N,lig}}. \tag{5.2m}$$

The N flux required from the two mineral pools (N_{min}) is therefore

$$O_{Nmin,soil\text{-}li\rightarrow SOM} = \frac{I_{C,soil\text{-}li\rightarrow uSOM}}{r_{C:N,uSOM,new}} + \frac{I_{C,soil\text{-}li\rightarrow pSOM}}{r_{C:N,pSOM,new}} - O_{N,soil\text{-}li,lig}. \tag{5.2n}$$

The variable C:N ratios in the unprotected and protected SOM pools are defined in Eqns (5.7b) and (5.8b). The flux in Eqn (5.2n) is taken from the N_{amm} and N_{nit} pools in the same ratio in which microbial biomass take ammonium N and nitrate N [Eqns (5.6f)], to give outputs from the N_{amm} and N_{nit} pools of

$$O_{Namm,soil\text{-}li\rightarrow SOM} = f_{Namm\rightarrow bio}O_{Nmin,soil\text{-}li\rightarrow SOM},$$
$$O_{Nnit,soil\text{-}li\rightarrow SOM} = f_{Nnit\rightarrow bio}O_{Nmin,soil\text{-}li\rightarrow SOM}. \tag{5.2o}$$

The f fractions are given in Eqns (5.6f).

5.4 Carbon Substrate in Soil, C_{sol}

It is envisaged that this pool consists of labile C metabolites, such as glucose, other sugars and organic acids. Its inclusion enables processes such as C leaching, microbial biomass growth and activity to be represented.

5.4.1 Differential equation (kg C m^{-2} day^{-1})

This is

$$\frac{dC_{sol}}{dt} = I_{C,surf\text{-}li \to Csol} + I_{C,soil\text{-}li \to Csol} + O_{CS,rt \to so,ex} + I_{C,bioD \to Csol} + I_{C,uSOM \to Csol} + I_{C,pSOM \to Csol} + I_{C,sSOM \to Csol} - \left(O_{Csol \to bioG} + O_{Csol \to lch} + O_{Csol \to R}\right), \quad (5.3a)$$

$$C_{sol}(t = 0) = 0.0005 \text{ kg C m}^{-2}.$$

5.4.2 Inputs

There are inputs of $I_{C,surf\text{-}li \to Csol}$ and $I_{C,soil\text{-}li \to Csol}$, from the surface and soil litter metabolic and cellulose pools [Eqns (5.1j), (5.2j)]. There is root exudation of carbohydrate from the root C substrate pool [$O_{CS,rt \to so,ex}$ of Eqn (3.3i)]. The dying microbes give a flux of $I_{C,bioD \to Csol}$ into the C_{sol} pool [Eqn (5.6j)]. Finally, there are three mineralization fluxes from the three (*u*, *p*, *s*) SOM pools, $I_{C,uSOM \to Csol}$, $I_{C,pSOM \to Csol}$, $I_{C,sSOM \to Csol}$, given by Eqns (5.7g), (5.8g) and (5.9g).

5.4.3 Outputs

The C flux required for microbial growth, $O_{Csol \to bioG}$ [Eqn (5.6c)], is calculated in the microbial biomass section. This is the dominant flux out of the C_{sol} pool.

There is leaching of C at a rate of

$$O_{Csol \to lch} = k_{leach} C_{sol}. \quad (5.3b)$$

The leaching rate constant, k_{leach} (day^{-1}), is calculated in the water submodel [Eqn (6.2e)].

The C concentration of the leachate is $C_{sol}/(d_{rt}\theta_{so})$ where d_{rt} is root depth [Eqn (3.9c)] and θ_{so} is soil relative water content [Eqn (6.2f)]. With d_{rt} = 0.2 m, and θ_{so} = 0.4 m^3 water (m^3 soil)$^{-1}$, then C_{sol} = 0.0005 kg C m^{-2} is equivalent to a solution concentration of 0.0005/(0.2 × 0.4) = 0.006 kg C m^{-3} = 6 mg C litre^{-1}. In a mature spruce forest in North Wales, Stevens *et al.* (1989) measured values of between 0 and 88 mg litre^{-1}.

A further direct loss of C_{sol} to respiration is assumed, given by

$$O_{Csol \to R} = k_{CsolR} C_{sol}, \qquad k_{CsolR} = f_{T,soil} f_{W,soil} k_{CsolR20}, \qquad k_{CsolR20} = 0.1 \text{ day}^{-1}. \quad (5.3c)$$

The temperature and water functions $f_{T,soil}$ and $f_{W,soil}$ are defined in Eqns (5.11a) and (5.11b). The value of the rate constant k_{CsolR} corresponds to a half-life of 7 day at 20°C. This flux is additional to the respired component of the C flux required for microbial biomass growth [R_{bioG}, Eqn (5.6d)].

5.5 Ammonium N Pool, N_{amm}

5.5.1 Differential equation (kg N m^{-2} day^{-1})

This is

$$\begin{aligned}\frac{dN_{amm}}{dt} &= O_{NS,rt\to so,ex} + O_{N,an\to so,ur} + I_{N,fix\to Namm} \\ &+ I_{N,fert\to Namm} + I_{N,env\to Namm} + I_{N,surf\text{-}li\to Namm} \\ &+ I_{N,soil\text{-}li\to Namm} + I_{N,bioD\to Namm} + I_{N,bioD\text{-}SOM\to Namm} + I_{N,SOM\to Namm} \\ &- \left(O_{Namm\to imm} + O_{Namm\to bioG} + O_{Namm\to nitrif} + O_{Namm\to vol} + O_{Namm\to rt}\right), \\ N_{amm}(t=0) &= 0.0007 \text{ kg N m}^{-2}.\end{aligned} \tag{5.4a}$$

5.5.2 Inputs

Exudation of N by the plant roots gives a flux of $O_{NS,rt\to so,ex}$ [Eqn (3.4j)]. Nitrogen excreted by grazing animals in urine is assumed to be immediately hydrolysed to give ammonium N and CO_2 (Whitehead, 1995, p. 153). This generates an N flux of $O_{N,an\to so,ur}$ given by Eqn (4.4d). The urine-associated respiration is

$$R_{urine} = O_{C,an\to so,ur}. \tag{5.4b}$$

The C content of the urine flux is given by Eqn (4.4e).

It is assumed that non-symbiotic N fixation by soil microorganisms can occur, producing an N flux into the soil N ammonium pool alone. The flux is given by

$$\begin{aligned}I_{N,fix\to Namm} &= \frac{k_{Nfix,bio}C_{bio}}{\left(1+N_{eff,bio}/J_{Nfix,bio}\right)\left(1+K_{Csol,Nfix}/C_{Csol}\right)}, \\ k_{Nfix,bio} &= k_{Nfix,bio20}f_{T,soil}f_{W,soil}, \\ k_{Nfix,bio20} &= 50\times10^{-6} \text{ kg N (kg biomass C)}^{-1} \text{ day}^{-1}, \\ K_{Csol,Nfix} &= 0.0005 \text{ kg C m}^{-2}, \qquad J_{Nfix,bio} = 0.001 \text{ kg N m}^{-2}.\end{aligned} \tag{5.4c}$$

In this equation, the rate constant $k_{Nfix,bio}$ depends on soil temperature ($f_{T,soil}$) and soil water ($f_{W,soil}$) [Eqns (5.11a), (5.11b)]. The fixation rate is proportional to the soil microbial biomass pool. There is a Michaelis–Menten dependence on the soluble C pool C_{sol} with Michaelis–Menten constant $K_{Csol,Nfix}$. $N_{eff,bio}$ is the effective mineral N level seen by the soil microbial biomass [Eqn (5.6b)]; a high level of $N_{eff,bio}$ relative to the parameter $J_{Nfix,bio}$ inhibits N fixation (Svenning *et al.*, 1996). If C_{bio} = 0.2 kg dry mass C m^{-2}, C_{sol} is large, $N_{eff,bio}$ = 0, then at 20°C the rate of N fixation is 10×10^{-6} kg N m^{-2} day^{-1}, which is equivalent to 37 kg N ha^{-1} year^{-1} (Sheehy, 1989). Note that symbiotic N fixation, given by legumes, can give rise to substantially larger N fluxes into the system (Whitehead, 1995, p. 43), and it is not represented explicitly in the plant submodel (Chapter 3).

The ammonium N fertilizer application rate, $I_{N,fert\to Namm}$, is defined in Eqn (7.6c) . The atmospheric ammonium N input, $I_{N,env\to Namm}$, is defined in Table 7.2 and is equivalent to 50 kg N ha^{-1} year^{-1}; this value may be considered as including a contribution from symbiotic N fixation as well as the usual N deposition.

The inputs from the surface and soil litter pools, $I_{N,surf\text{-}li \to Namm}$ and $I_{N,soil\text{-}li \to Namm}$ are calculated in Eqns (5.1j) and (5.2j).

There is an input to N_{amm} from a fraction of the dying microbes, $I_{N,bioD \to Namm}$. This is calculated in Eqn (5.6k).

After respiratory loss of CO_2 the remainder of the dying microbes are assigned to the unprotected and protected SOM pools (*uSOM*, *pSOM*), in a ratio depending on clay content [Eqn (5.6n)]. Because these two pools usually have higher C:N ratios [Eqns (5.7b), (5.8b)] than the microbial biomass (Fig. 5.1), there is a release of N into N_{amm} of $I_{N,bioD\text{-}SOM \to Namm}$ given in Eqn (5.6p).

The last input, $I_{N,SOM \to Namm}$, is the total mineralizing N flux from the three SOM pools. This is given by

$$I_{N,SOM \to Namm} = I_{N,uSOM \to Namm} + I_{N,pSOM \to Namm} + I_{N,sSOM \to Namm}. \tag{5.4d}$$

The individual mineralizing N fluxes are given in Eqns (5.7h), (5.8h) and (5.9h).

5.5.3 Outputs

$O_{Namm \to imm}$ is the total immobilizing flux, with

$$\begin{aligned} O_{Namm \to imm} = {} & O_{Namm,surf-li \to SOM} + O_{Namm,soil-li \to SOM} \\ & + O_{Namm,uSOM \to sSOM} + O_{Namm,pSOM \to sSOM}. \end{aligned} \tag{5.4e}$$

$O_{Namm,surf\text{-}li \to SOM}$ and $O_{Namm,soil\text{-}li \to SOM}$ [Eqns (5.1o), (5.2o)] are immobilizing fluxes of N required as surface and soil litter lignin with a high C:N ratio is converted into SOM with a low C:N ratio in the unprotected and protected SOM pools [Eqns (5.7b), (5.8b)]. $O_{Namm,uSOM \to sSOM}$ and $O_{Namm,pSOM \to sSOM}$ are calculated as immobilizing fluxes of N (i.e. the fluxes are positive if immobilizing) in Eqns (5.7l) and (5.8l), to maintain the N balance sheet when unprotected and protected SOM are converted to stabilized SOM. Depending on the C:N ratios in the SOM pools, and the fractions of C lost to respiration, the fluxes could be negative, giving an apparent mineralizing N flux to both the N_{amm} and N_{nit} pools (normally, mineralizing N fluxes are input to the N_{amm} pool alone).

The $O_{Namm \to bioG}$ flux is the N required for soil microbial biomass growth [Eqn (5.6g)].

Nitrification is the process converting ammonium N into nitrate N (Parton *et al.*, 1996), with flux given by

$$\begin{aligned} & O_{Namm \to nitrif} = k_{nit} C_{bio} N_{amm}, \qquad k_{nit} = k_{nit20} f_{T,soil} f_{W,soil}, \\ & k_{nit20} = 0.4 \left(\text{kg bio C m}^{-2}\right)^{-1} \text{day}^{-1}. \end{aligned} \tag{5.4f}$$

The specific rate is assumed to be proportional to the microbial population (C_{bio}). The soil temperature ($f_{T,soil}$) and soil water ($f_{W,soil}$) modifying functions are as usual [Eqns (5.11a), (5.11b)]. The rate constant is such that at 20°C, with C_{bio} = 0.2 kg C m^{-2} and N_{amm} = 0.001 kg N m^{-2}, the nitrification flux is $0.4 \times 0.2 \times 0.001 = 80 \times 10^{-6}$ kg N m^{-2} day^{-1}, equivalent to *c.* 300 kg N ha^{-1} $year^{-1}$.

It is assumed that a fraction f_{nitrif} of this flux is lost to the environment as oxides of nitrogen. The resulting input flux of N to the environment is

$$I_{N,nitrif \to env} = f_{nitrif} O_{Namm \to nitrif} \cdot f_{nitrif} = 0.1. \tag{5.4g}$$

The input to nitrate pool N_{nit} is

$$I_{Namm \to Nnit} = O_{Namm \to nitrif} - I_{N,nitrif \to env}. \tag{5.4h}$$

The N flux to volatilization $O_{Namm \to vol}$ is

$$O_{Namm \to vol} = k_{vol} N_{amm}, \qquad k_{vol} = k_{vol20} f_{T,soil} f_{W,soil}, \\ k_{vol20} = 0.02 \text{ day}^{-1}. \tag{5.4i}$$

At 20°C and with N_{amm} = 0.001 kg N m^{-2}, the volatilization flux is 20 × 10^{-6} kg N m^{-2} day^{-1}, or 73 kg N ha^{-1} year^{-1}. The undoubted dependence of volatilization rate on windspeed is ignored (Whitehead, 1995, pp. 157–160). The generic soil temperature and water functions $f_{T,soil}$, $f_{W,soil}$ [Eqns (5.11a), (5.11b)] are assumed to apply. The influence of soil factors such as pH, clay content and cation exchange capacity is also ignored (Whitehead, 1995, pp. 160–161).

$O_{Namm \to rt}$ is the uptake of N by plant roots and associated mycorrhiza and is given by

$$O_{Namm \to rt} = u_{Namm} \tag{5.4j}$$

of Eqn (3.4i).

5.6 Nitrate N Pool, N_{nit}

5.6.1 Differential equation (kg N m^{-2} day^{-1})

This is

$$\frac{dN_{nit}}{dt} = I_{N,fert \to Nnit} + I_{N,env \to Nnit} + I_{Namm \to Nnit} \\ - (O_{Nnit \to imm} + O_{Nnit \to bioG} + O_{Nnit \to rt} + O_{Nnit \to denit} + O_{Nnit \to lch}), \\ N_{nit}(t = 0) = 0.0003 \text{ kg N m}^{-2}. \tag{5.5a}$$

5.6.2 Inputs

The nitrate N fertilizer application rate, $I_{N,fert \to Nnit}$, is defined in Eqn (7.6c). The atmospheric nitrate N input from deposition, $I_{N,env \to Nnit}$, is defined in Table 7.2. Both these fluxes are zero in the default scenario.

The major input to the N_{nit} pool is the nitrification flux given by Eqn (5.4h).

5.6.3 Outputs

The immobilizing flux of N out of the nitrate N pool is by definition the requirement for nitrate by the SOM transformations (Fig. 5.1). The flux is

$$O_{Nnit \to imm} = O_{Nnit,surf\text{-}li \to SOM} + O_{Nnit,soil\text{-}li \to SOM} \\ + O_{Nnit,uSOM \to sSOM} + O_{Nnit,pSOM \to sSOM}. \tag{5.5b}$$

See after Eqn (5.4e) for detailed explanation. The four fluxes are calculated in Eqns (5.1o), (5.2o), (5.7l) and (5.8l).

The $O_{Nnit \rightarrow bioG}$ flux is the nitrate N taken up for soil microbial biomass growth [Eqn (5.6g)].

$O_{Nnit \rightarrow rt}$ is the uptake of N by plant roots and associated mycorrhiza, given by

$$O_{Nnit \rightarrow rt} = u_{Nnit}, \tag{5.5c}$$

where root uptake of nitrate, u_{Nnit}, is given by Eqn (3.4i).

The denitrification flux, $O_{Nnit \rightarrow denit}$, is modelled similarly to the nitrification flux [Eqn (5.4f)], with an extra rate modifier, $f_{denit,W}$, to allow for the aerobicity of the soil as affected by soil relative water content [Eqn (6.2f)]. We use the simple expression [but see Parton *et al.* (1996), Almeida *et al.* (1997), for more detailed models of denitrification]:

$$O_{Nnit \rightarrow denit} = k_{denit} C_{bio} N_{nit}, \qquad k_{denit} = k_{denit20} f_{T,soil} f_{W,soil} f_{denit,W},$$

$$f_{denit,W} = \left(\frac{\theta_{so}}{\theta_{so,\max}} \right)^{q_{denit}}, \tag{5.5d}$$

$$k_{denit20} = 0.5\ (\text{kg biomass C m}^{-2})^{-1}\ \text{day}^{-1}, \quad q_{denit} = 10.$$

Denitrification is only active when the soil is anaerobic, and is quickly switched off when soil water content θ_{so} [Eqn (6.2f)] is below field capacity $\theta_{so,max}$ [Eqn (6.2a)] (see Whitehead, 1995, chapter 9 for a review). If $\theta_{so}/\theta_{so,max} = 0.95$, the $f_{denit,W}$ factor is 0.6. Whitehead (1995, p. 185) redraws data from Aulakh *et al.* (1992), showing that the denitrification rate decreases quickly to zero when the water-filled pore space is below 0.8. No dependence is included for soil type or porosity in Eqn (5.5d). Eqn (5.5d) relates denitrification rate to field capacity, whereas the relationship to water-filled pore space is probably more significant (Scholefield, personal communication). Note that soil texture affects which pores are water-filled at various fractions of water content at field capacity (e.g. Rowell, 1994, table 4.3, p. 63). The investigation of Arah and Smith (1989) suggests that aggregates of intermediate size are relatively efficient denitrifiers; their model uses parameters such as soil moisture, oxygen reduction potential and nitrate concentration. The parameter values above with C_{bio} = 0.2 kg C m^{-2} and N_{nit} = 0.002 kg N m^{-2} give an instantaneous denitrification flux at 20°C for a water-logged soil of 200×10^{-6} kg N m^{-2} day^{-1}, equivalent to 730 kg N ha^{-1} year^{-1}.

Whitehead (1995, p. 182) stresses that several conditions must be simultaneously satisfied for denitrification to occur: the nitrate concentration must be sufficiently high, conditions must be anaerobic so that nitrate is used as the terminal electron acceptor rather than oxygen when organic carbon is oxidized to obtain energy for biosynthesis (e.g. $4HNO_3 + 5\{CH_2O\} = 2N_2 + 7H_2O + 5CO_2$ + energy), there is an adequate supply of organic carbon, and the temperature must be such as to allow reactions to proceed. Ryden (1983) measured appreciable denitrification only when the air-filled porosity < 38%, nitrate > 5 μg N g^{-1} (equating to about 0.0015 kg N m^{-2} in present units), and the soil temperature is > 5 to 8°C. Eqn (5.5d) does not describe the full complexity of the process, and a more detailed soil submodel might take account of specialized groups of microbes with different energy costs associated with their activity and growth, and spatial heterogeneity in relation to pore sizes. However, Eqn (5.5d) has been found to give very satisfactory annual estimates of denitrification losses. See also the discussion in the immediately following leaching section (leaching and denitrification are closely linked).

The leaching flux is [cf. Eqn (5.3b) for C leaching]

$$O_{Nnit \rightarrow lch} = k_{leach} N_{nit}. \quad (5.5e)$$

The leaching constant k_{leach} (day^{-1}) is calculated in Eqn (6.2e). For a root depth of 0.1 m and a water drainage rate of 0.1 m of water $year^{-1}$, k_{leach} = 1 $year^{-1}$. The N lost to leaching assuming a nitrate N concentration of 0.001 kg N m^{-2} is 0.001 kg N m^{-2} $year^{-1}$ or 10 kg N ha^{-1} $year^{-1}$. The N concentration in the leachate is 0.001 kg N m^{-2} / 0.1 m^3 water m^{-2} = 0.01 kg N (m^3 water)$^{-1}$ = 10 mg N $litre^{-1}$ = 0.7 mmol $litre^{-1}$. This can be compared with a value of 0.86 mmol $litre^{-1}$ for an Oxfordshire soil solution quoted by Rowell (1994, p. 147).

Eqn (5.5e) calculates nitrate leaching. Any leaching of organic N is ignored, although there is evidence that leaching of organic N does occur (Jarvis, personal communication). To take this into account would require a soil organic N pool and other changes to the model.

Although Eqn (5.5e) is a simple linear equation, it is to be stressed that this may not give rise to a linear relationship between N inputs to the system (e.g. fertilizer) and leaching losses as there are many non-linear processes in the system [e.g. plant uptake in relation to N_{nit}; Eqn (3.4g, h)] which can affect leaching rate indirectly. Note that leaching is driven primarily by drainage [Eqn (6.2e)], which can depend strongly on pore-size distributions and soil heterogeneity, which are ignored here. Also, our standard weather scenario has a smooth rainfall function (Fig. 7.1c), whereas drainage is often very dependent on the intensity of particular rainfall events. Although rainfall events can be partially simulated by using the daily weather data option [Eqn (7.4a); Fig. 7.2], it is unrealistic to expect the present formulation to simulate seasonal leaching events at particular sites exactly. Nevertheless, as discussed above, the model predicts reasonable values of annual leaching losses and of N concentrations in leachate.

Some recent and successful leaching models are based on a division of soil water into two categories: a mobile water pool which can give leaching, and an immobile pool which may retain solutes (Russo *et al.*, 1989). Addiscott and Whitmore's (1991) SLIM model is of this type. The model requires site-specific calibration but then works better than models with only a single category of water (e.g. Gabrielle and Kengni, 1996).

The total immobilizing flux removed from the ammonium (N_{amm}) and nitrate (N_{nit}) pools combined is $O_{Nmin \rightarrow imm}$, given by

$$O_{Nmin \rightarrow imm} = O_{Namm \rightarrow imm} + O_{Nnit \rightarrow imm}. \quad (5.5f)$$

The individual immobilizing fluxes are given in Eqns (5.4e) and (5.5b). This immobilizing flux can be compared with the total mineralizing N flux of Eqn (5.4d).

The two combined mineral N pools are designated by N_{min} with

$$N_{min} = N_{amm} + N_{nit}. \quad (5.5g)$$

5.7 Soil Microbial Biomass Pool, C_{bio}

The metabolically important components of soil biomass are microbial, and it is envisaged that this pool is predominantly a microbial pool. The microbial biomass pool has a constant C:N ratio, $r_{C:N,bio}$ [Eqn (5.6e)] so the C content of the biomass is the only variable applying to this pool.

5.7.1 Differential equation (kg C m^{-2} day^{-1})

This is just the growth rate minus the death rate, giving

$$\frac{dC_{bio}}{dt} = G_{bio} - D_{bio}, \qquad C_{bio}(t=0) = 0.2 \text{ kg C m}^{-2}. \tag{5.6a}$$

5.7.2 Growth of microbial biomass

The growth of a microbial culture of biomass x is often represented by the *autocatalytic* growth equation, $dx/dt = \mu x$, where μ is the specific growth rate (e.g. Pirt, 1975, p. 4 *et seq.*). The dependence of specific growth rate μ on substrate s is then modelled using the Monod relation, $\mu = \mu_m / (1 + K/s)$, where μ_m is the maximum specific growth rate (saturating substrate) and K is sometimes called the saturation constant (Pirt, 1975, pp. 8–9). We shall modify this approach, using an equation similar to that proposed by Parnas (1975), namely:

$$G_{bio} = k_{bioG} C_{bio}^{q_{bio}}, \qquad k_{bioG} = k_{bioG20} f_{T,soil} f_{CN,bio} f_{bio,max},$$

$$f_{CN,bio} = \frac{1}{\left(1 + \frac{K_{C,bioG}}{C_{sol}}\right)\left(1 + \frac{K_{N,bioG}}{N_{eff,bio}}\right)}, \qquad N_{eff,bio} = N_{amm} + c_{N,bio} N_{nit},$$

$$f_{bio,max} = \text{maximum}\left(0, 1 - \frac{C_{bio}}{f_{SOM,bio} C_{SOM}}\right), \tag{5.6b}$$

$$q_{bio} = \frac{2}{3}, \qquad k_{bioG20} = 0.1 \text{ (kg C)}^{1-q_{bio}} \text{ day}^{-1},$$

$$K_{C,bioG} = 0.001 \text{ kg C m}^{-2}, \qquad K_{N,bioG} = 0.0002 \text{ kg N m}^{-2},$$

$$c_{N,bio} = 1 \text{ kg ammonium N (kg nitrate N)}^{-1},$$

$$f_{SOM,bio} = 0.05 \text{ kg C in biomass (kg C in SOM)}^{-1}.$$

Autocatalytic growth corresponds to a parameter value of $q_{bio} = 1$; a value less than unity as taken here gives a slower than exponential growth rate which might result from e.g. a surface limitation on the supply of some nutrient. With $q_{bio} = 1$, k_{bioG} is a true specific growth constant. Parnas (1975) used a growth rate constant of 0.01 day^{-1} at an undefined temperature. The soil temperature ($f_{T,soil}$) and soil water ($f_{W,soil}$) functions [Eqns (5.11a), (5.11b)] give the dependence on temperature and water potential. In the expression for $f_{CN,bio}$, a form of bisubstrate Michaelis–Menten kinetics (Dixon and Webb, 1964, pp. 70–75) is used to model the dependence of growth rate on the level of the two substrates C and N. $N_{eff,bio}$ (N effective as seen by the microbial biomass) is a weighted sum of the ammonium and nitrate N levels. The purpose of the $f_{bio,max}$ factor is to place a ceiling on the microbial population that is proportional to the total amount of soil organic matter in the three SOM pools [Eqn (5.10a)]. Without this limit, C_{bio} would sometimes rise to unreasonable values; many microbes need to adhere to a solid substrate such as a soil organic matter particle, and it may therefore be reasonable that microbial growth is limited in this way (e.g. Rowell, 1994, table 13.3). With $f_{SOM,bio}$ = 0.05, a C content of SOM of 10 kg C m^{-2} is able to support a maximum biomass of 0.5 kg biomass C m^{-2}. The *maximum* function is to ensure that this factor does not become negative for poorly chosen initial values of C_{bio} and C_{SOM}.

The C requirement for microbial biomass growth is obtained by dividing the growth rate G_{bio} by the conversion yield Y_{bio} of the growth process, and is an output from the soil C substrate pool, C_{sol} [Eqn (5.3a)], with

$$O_{Csol \to bioG} = \frac{G_{bio}}{Y_{bio}}, \quad Y_{bio} = 0.5. \tag{5.6c}$$

There is a component of respiration given by

$$R_{bioG} = (1 - Y_{bio}) O_{Csol \to bioG}. \tag{5.6d}$$

The N requirement for microbial biomass growth is

$$O_{Nmin \to bioG} = \frac{G_{bio}}{r_{C:N,bio}}, \quad r_{C:N,bio} = 8 \text{ kg C (kg N)}^{-1}. \tag{5.6e}$$

$r_{C:N,bio}$ is a fixed C:N ratio for the microbial biomass. This N flux is taken from the N_{amm} and N_{nit} pools according to fractions:

$$f_{Namm \to bio} = \frac{N_{amm}}{N_{eff,bio}}, \qquad f_{Nnit \to bio} = 1 - f_{Namm \to bio}. \tag{5.6f}$$

$N_{eff,bio}$ is given in Eqn (5.6b). This gives rise to outputs from the N_{amm} and N_{nit} pools [Eqns 5.4a), (5.5a)] of

$$\begin{aligned} O_{Namm \to bioG} &= f_{Namm \to bio} O_{Nmin \to bioG}, \\ O_{Nnit \to bioG} &= f_{Nnit \to bio} O_{Nmin \to bioG}. \end{aligned} \tag{5.6g}$$

5.7.3 Death of microbial biomass

The death rate of the microbial biomass population is assumed to be

$$\begin{aligned} &D_{bio} = k_{bioD} C_{bio}, \qquad k_{bioD} = k_{bioD20} f_{T,soil} f_{W,soil} f_{bio,min}, \\ &f_{bio,min} = \text{maximum}\left(0, 1 - \frac{C_{bio,min}}{C_{bio}}\right), \\ &k_{bioD20} = 0.01 \text{ day}^{-1}, \qquad C_{bio,min} = 0.001 \text{ kg C in biomass m}^{-2}. \end{aligned} \tag{5.6h}$$

This equation gives a decreasing proportional rate of loss with decreasing C_{bio} until a floor of $C_{bio,min}$ is approached. This prevents the microbial population being reduced to unrealistically low levels, and reflects the fact that many bacteria have refractile or dormant forms. The *maximum* function stops $f_{bio,min}$ becoming negative

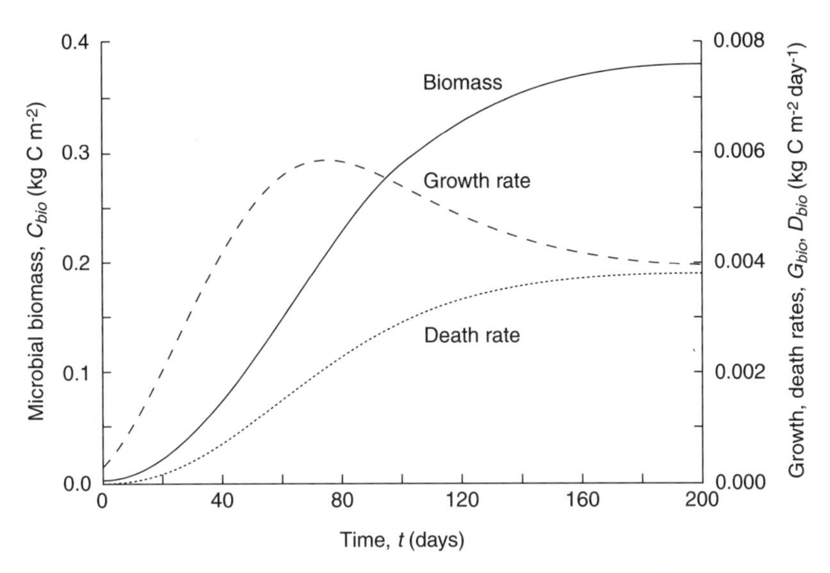

Fig. 5.2. Microbial biomass growth curve. This curve was generated using Eqns (5.6a), (5.6b) and (5.6h) with $C_{bio}(t = 0) = 0.001$ kg C m^{-2}, $f_{T,soil} = f_{W,soil} = 1$ (20°C, no water limitation). The substrate concentrations for biomass growth were constant, with $C_{sol} = 0.0005$ kg C m^{-2}, $N_{amm} = 0.0007$ kg N m^{-2}, $N_{nit} = 0.0003$ kg N m^{-2}. Also total soil organic matter C was constant at $C_{SOM} = 10.4$ kg C m^{-2}. Other parameters are as given.

for a very low initial value of C_{bio}. The functions $f_{T,soil}$ and $f_{W,soil}$ [Eqns (5.11a), (5.11b)] give the dependence on soil temperature and soil water potential.

Figure 5.2 shows a growth curve for the microbial biomass under constant conditions of environment (temperature = 20°C, no water limitation), substrate concentrations (C_{sol}, N_{amm}, N_{nit}) and soil organic matter C [C_{SOM}, Eqn (5.10a)]. Net growth is initially autocatalytic, slows down as a microbial biomass approaches the limit set by the $f_{bio,max}$ term in Eqn (5.6b), finally approaching an asymptote where growth and death rates are equal. The initial value of C_{bio} is the minimum value [$C_{bio,min}$ of Eqn (5.6h)] where the death rate is zero. The growth rates of Fig. 5.2 are similar to the observed rates of change of soil bacteria reported by Lundgren and Söderström (1983, Fig. 1).

5.7.4 Outputs associated with microbial biomass death

There are C and N fluxes associated with microbial biomass death:

$$O_{Cbio} = D_{bio}, \qquad O_{Nbio} = \frac{O_{Cbio}}{r_{C:N,bio}}. \tag{5.6i}$$

A fraction $f_{CbioD,R}$ of this C flux is respired and a fraction $f_{CbioD,Csol}$ is returned to the C substrate pool C_{sol}, giving

$$\begin{aligned} R_{bioD} &= f_{CbioD,R}O_{Cbio}, \qquad I_{C,bioD \to Csol} = f_{CbioD \to Csol}O_{Cbio}, \\ f_{CbioD,R} &= 0.5, \qquad f_{CbioD \to Csol} = 0.1. \end{aligned} \tag{5.6j}$$

The N associated with these two C fluxes is returned to N_{amm} pool:

$$I_{N,bioD \to Namm} = \frac{R_{bioD} + I_{C,bioD \to Csol}}{r_{C:N,bio}}. \tag{5.6k}$$

This leaves a N flux towards the unprotected and protected soil organic matter pools. With the second of Eqns (5.6i), this is

$$O_{Nbio \to SOM} = O_{Nbio} - I_{N,bioD \to Namm}. \tag{5.6l}$$

The remaining C flux [subtracting Eqns (5.6j) from Eqn (5.6i)] of

$$I_{Cbio \to SOM} = O_{Cbio} - R_{bioD} - I_{C,bioD \to Csol} \tag{5.6m}$$

enters the unprotected and protected SOM pools according to clay content [f_{clay}, Eqn (5.11)], with

$$\begin{aligned}
& I_{Cbio \to uSOM} = I_{Cbio \to SOM} f_{CbioD \to uSOM}, \qquad I_{Cbio \to pSOM} = I_{Cbio \to SOM} f_{CbioD \to pSOM}, \\
& f_{CbioD \to uSOM} = f_{CbioD \to uSOM,sandy} + c_{bioD,clay} f_{clay} \\
& \qquad \times \left(f_{CbioD \to uSOM,clay} - f_{CbioD \to uSOM,sandy} \right) \\
& f_{CbioD \to pSOM} = 1 - f_{CbioD \to uSOM}, \\
& c_{bioD,clay} = 3, \qquad f_{CbioD \to uSOM,sandy} = 0.7, \qquad f_{CbioD \to uSOM,clay} = 0.3.
\end{aligned} \tag{5.6n}$$

For a sandy soil ($f_{clay} = 0$), 70% of the C from microbial biomass death enters the unprotected SOM pool. As clay content increases, this percentage decreases, becoming 30% when $c_{bioD,clay} f_{clay} = 1$; that is, when $f_{clay} = \frac{1}{3}$.

To produce these fluxes of C to uSOM and pSOM with the specified C:N ratios [Eqns (5.7b), (5.8b)] requires an N flux of

$$I_{Nbio \to SOM} = \frac{I_{Cbio \to uSOM}}{r_{C:N,uSOM,new}} + \frac{I_{Cbio \to pSOM}}{r_{C:N,pSOM,new}}. \tag{5.6o}$$

This is usually less than the N flux of $O_{Nbio \to SOM}$ [Eqn (5.6l)] provided by the fraction of dying microbial biomass that is transformed into SOM, giving a flux into the ammonium N pool N_{amm} [Eqn (5.4a)] of

$$I_{N,bioD-SOM \to Namm} = O_{Nbio \to SOM} - I_{Nbio \to SOM}. \tag{5.6p}$$

The total N flux from microbial biomass into the ammonium N pool N_{amm} is

$$I_{Nbio \to Namm} = I_{N,bioD \to Namm} + I_{N,bioD-SOM \to Namm}, \tag{5.6q}$$

where we have added Eqn (5.6k). The first term on the right of this equation comprises two contributions: the N associated with the fraction of the dying microbial biomass that is broken down to CO_2 and ammonium N [the R_{bioD} term in Eqn (5.6k)], plus the N associated with the fraction of dying microbial biomass that is hydrolysed to C substrate (C_{sol}) and ammonium N [the $I_{C,bioD \to Csol}$ term in Eqn (5.6k)]. The second term on the right-hand side of Eqn (5.6q) is the surplus N arising from the transformation of a fraction of the C in dying microbial biomass into SOM with C:N ratios which are generally higher than the microbial biomass C:N ratio.

5.8 Unprotected Soil Organic Matter, C_{uSOM}, N_{uSOM}

This pool has a variable C:N ratio of $r_{C:N,uSOM}$ given by

$$r_{C:N,uSOM} = \frac{C_{uSOM}}{N_{uSOM}}. \tag{5.7a}$$

This can vary because *new* unprotected (and protected) SOM is given a variable C:N ratio $r_{C:N,uSOM,new}$, which depends on the current soil mineral N concentration N_{min} [Eqn (5.5g)], given by

$$\begin{aligned} r_{C:N,uSOM,new} &= r_{C:N,uSOM,max} - \left(r_{C:N,uSOM,max} - r_{C:N,uSOM,min}\right) \\ &\quad \times \frac{N_{min}}{N_{min} + K_{rC:N,uSOM}}, \\ r_{C:N,uSOM,max} &= 12, \quad r_{C:N,uSOM,min} = 6 \text{ kg C (kg N)}^{-1}, \\ K_{rC:N,uSOM} &= 0.002 \text{ kg N m}^{-2}. \end{aligned} \tag{5.7b}$$

At low soil mineral concentrations (N_{min}), the C:N ratio in new unprotected SOM is maximum (12), and decreases towards six as N_{min} increases. Outputs from the unprotected SOM pool have a C:N ratio given by the pool average [Eqns (5.7a)]. Inputs have the C:N ratio given by Eqns (5.7b).

5.8.1 Differential equations (kg C, N m^{-2} day^{-1})

For the C and N components of unprotected SOM these are

$$\begin{aligned} \frac{dC_{uSOM}}{dt} &= I_{Cbio \to uSOM} + I_{C,soil\text{-}li \to uSOM} + I_{C,surf\text{-}li \to uSOM} \\ &\quad - O_{C,uSOM \to min} - O_{C,uSOM \to sSOM}, \\ \frac{dN_{uSOM}}{dt} &= \frac{I_{Cbio \to uSOM} + I_{C,soil\text{-}li \to uSOM} + I_{C,surf\text{-}li \to uSOM}}{r_{C:N,uSOM,new}} \\ &\quad - \frac{O_{C,uSOM \to min} + O_{C,uSOM \to sSOM}}{r_{C:N,uSOM}}, \\ C_{uSOM}(t=0) &= 0.4 \text{ kg C m}^{-2}, \qquad N_{uSOM}(t=0) = 0.04 \text{ kg N m}^{-2}. \end{aligned} \tag{5.7c}$$

The initial C:N ratio is 10.

5.8.2 Inputs

The first C input ($I_{Cbio \to uSOM}$) is from microbial biomass death [Eqn (5.6n)]. The other two C inputs ($I_{C,soil\text{-}li \to uSOM}$, $I_{C,surf\text{-}li \to uSOM}$) are from the soil and surface litter lignin pools [Eqns (5.11), (5.21)]. The N inputs are obtained by dividing the C inputs by the C:N ratio of new unprotected SOM [Eqn (5.7b)].

5.8.3 Outputs

The output to mineralization is linear, with

$$O_{C,uSOM \rightarrow min} = k_{uSOM \rightarrow min} C_{uSOM},$$
$$k_{uSOM \rightarrow min} = k_{uSOM \rightarrow min20} f_{T,soil} f_{W,soil}, \quad (5.7d)$$
$$k_{uSOM \rightarrow min20} = 0.003 \text{ day}^{-1}.$$

The functions $f_{T,soil}$ and $f_{W,soil}$ [Eqns (5.11a), (5.11b)] give the dependence on soil temperature and soil water potential. At 20°C with no water limitation, the half-life of unprotected SOM is about 1 year. Note that mineralization is defined as the sum of the decomposition rates of the three SOM pools (Fig. 5.1). As currently formulated, mineralization is not turned down as the soil approaches field capacity and becomes anaerobic. This could lead to incorrect predictions in soils that are water-logged for much of the time.

The stabilization process converts unprotected SOM to stabilized SOM (C_{sSOM}), giving output C and N fluxes of

$$O_{C,uSOM \rightarrow sSOM} = k_{uSOM \rightarrow sSOM} C_{uSOM},$$
$$O_{N,uSOM \rightarrow sSOM} = k_{uSOM \rightarrow sSOM} N_{uSOM},$$
$$k_{uSOM \rightarrow sSOM} = k_{uSOM \rightarrow sSOM20} f_{T,soil} f_{W,soil}, \quad (5.7e)$$
$$k_{uSOM \rightarrow sSOM20} = 0.0002 \text{ day}^{-1}.$$

Temperature and water functions $f_{T,soil}$ and $f_{W,soil}$ are given in Eqns (5.11a), (5.11b). Both output fluxes of unprotected C are partially respired, giving fluxes of C to respiration of

$$R_{uSOM \rightarrow min} = f_{C,uSOM \rightarrow minR} O_{C,uSOM \rightarrow min},$$
$$R_{uSOM \rightarrow sSOM} = f_{C,uSOM \rightarrow sSOM,R} O_{C,uSOM \rightarrow sSOM}, \quad (5.7f)$$
$$f_{C,uSOM \rightarrow minR} = 0.4, \qquad f_{C,uSOM \rightarrow sSOM,R} = 0.2.$$

The fluxes of C into the soluble C (C_{sol}) and stabilized SOM (C_{sSOM}) pools are obtained by difference:

$$I_{C,uSOM \rightarrow Csol} = O_{C,uSOM \rightarrow min} - R_{uSOM \rightarrow min},$$
$$I_{C,uSOM \rightarrow sSOM} = O_{C,uSOM \rightarrow sSOM} - R_{uSOM \rightarrow sSOM}. \quad (5.7g)$$

The N outputs in the second of Eqns (5.7c) are obtained by dividing the C outputs [Eqns (5.7d), (5.7e)] by the average C:N ratio of unprotected SOM [Eqn (5.7a)]. The N flux associated with the C flux to mineralization [Eqn (5.7d)] flows into the ammonium N pool, N_{amm}. Also, the N flux associated with the fraction of the C flux to stabilized SOM which is respired [Eqn (5.7f)] is input to the N_{amm} pool, giving a total input to N_{amm} of

$$I_{N,uSOM \rightarrow Namm} = \frac{O_{C,uSOM \rightarrow min} + R_{uSOM \rightarrow sSOM}}{r_{C:N,uSOM}}. \quad (5.7h)$$

The C:N ratio is given in Eqn (5.7a).

Accompanying the C flux of $I_{C,uSOM \rightarrow sSOM}$ [Eqn (5.7g)] which actually flows into the stabilized SOM pool is an N flux of (dividing by the C:N ratio of the unprotected SOM pool)

$$\frac{I_{C,uSOM \to sSOM}}{r_{C:N,uSOM}}. \tag{5.7i}$$

To provide the N required for the C:N ratio of new stabilized SOM matter [Eqn (5.9b)] requires an N flux of (dividing by the C:N ratio of the new stabilized SOM)

$$\frac{I_{C,uSOM \to sSOM}}{r_{C:N,sSOM,new}}. \tag{5.7j}$$

The net N requirement for the process of transforming unprotected to protected SOM is the difference between the two quantities above, and is taken from the mineral N pools, that is, the ammonium N and nitrate N pools, which together are designated as N_{min}. This net N requirement is an output from N_{min} and is

$$O_{Nmin,uSOM \to sSOM} = I_{C,uSOM \to sSOM}\left(\frac{1}{r_{C:N,sSOM,new}} - \frac{1}{r_{C:N,uSOM}}\right). \tag{5.7k}$$

The N flux of Eqn (5.7k) is taken from the ammonium N (N_{amm}) and nitrate N (N_{nit}) pools in the same ratio in which microbial biomass takes up N from these two pools. With Eqns (5.6f), the outputs from the N_{amm} and N_{nit} pools are

$$O_{Namm,uSOM \to sSOM} = f_{Namm \to bio} O_{Nmin,uSOM \to sSOM},$$
$$O_{Nnit,uSOM \to sSOM} = f_{Nnit \to bio} O_{Nmin,uSOM \to sSOM}. \tag{5.7l}$$

This is written as an immobilizing flux, taking N from mineral N pools. Depending on the C:N ratios in the SOM pools, it may be negative, giving a mineralizing N flux which then flows into both the N_{amm} and N_{nit} pools. Normally the mineralizing N fluxes flow solely into the N_{amm} pool.

5.9 Protected Soil Organic Matter, C_{pSOM}, N_{pSOM}

This pool is treated similarly to the unprotected SOM pool of Section 5.8, except that rate constants and other parameters have different values. The pool has a average C:N ratio of $r_{C:N,pSOM}$ given by

$$r_{C:N,pSOM} = \frac{C_{pSOM}}{N_{pSOM}}. \tag{5.8a}$$

This can vary because *new* protected SOM is given a variable C:N ratio depending on soil mineral N concentration N_{min} [Eqn (5.5g)], with

$$r_{C:N,pSOM,new} = r_{C:N,pSOM,max} - \left(r_{C:N,pSOM,max} - r_{C:N,pSOM,min}\right) \times \frac{N_{min}}{N_{min} + K_{rC:N,pSOM}}, \tag{5.8b}$$

$$r_{C:N,pSOM,max} = 12, \quad r_{C:N,pSOM,min} = 6 \text{ kg C (kg N)}^{-1},$$

$$K_{rC:N,pSOM} = 0.002 \text{ kg N m}^{-2}.$$

At low soil mineral concentrations (N_{min}), the C:N ratios in new protected SOM is maximum (12), and decreases towards six as N_{min} increases. Outputs from the protected SOM pools have a C:N ratio given by the pool average [Eqn (5.8a)].

5.9.1 Differential equations (kg C, N m^{-2} day^{-1})

For the C and N components of the protected SOM pools, these are

$$\begin{aligned}\frac{dC_{pSOM}}{dt} &= I_{Cbio\to pSOM} + I_{C,soil\text{-}li\to pSOM} + I_{C,surf\text{-}li\to pSOM} \\ &\quad - O_{C,pSOM\to min} - O_{C,pSOM\to sSOM}, \\ \frac{dN_{pSOM}}{dt} &= \frac{I_{Cbio\to pSOM} + I_{C,soil\text{-}li\to pSOM} + I_{C,surf\text{-}li\to pSOM}}{r_{C:N,pSOM,new}} \\ &\quad - \frac{O_{C,pSOM\to min} + O_{C,pSOM\to sSOM}}{r_{C:N,pSOM}}, \\ C_{pSOM}(t=0) &= 4 \text{ kg C m}^{-2}, \qquad N_{pSOM}(t=0) = 0.4 \text{ kg N m}^{-2}.\end{aligned} \tag{5.8c}$$

5.9.2 Inputs

The first C input ($I_{Cbio\to pSOM}$) is from microbial biomass death and is given in Eqn (5.6n). The other two C inputs ($I_{C,soil\text{-}li\to pSOM}$, $I_{C,surf\text{-}li\to pSOM}$) are from the soil and surface litter lignin pools [Eqns (5.11), (5.21)]. The N inputs are obtained by dividing the C inputs by the C:N ratio of new protected SOM [Eqn (5.8b)].

5.9.3 Outputs

The mineralizing output is linear (Section 1.6):

$$\begin{aligned}O_{C,pSOM\to min} &= k_{pSOM\to min} C_{pSOM}, \\ k_{pSOM\to min} &= k_{pSOM\to min20} f_{T,soil} f_{W,soil}, \\ k_{pSOM\to min20} &= 0.00015 \text{ day}^{-1}.\end{aligned} \tag{5.8d}$$

Functions $f_{T,soil}$ and $f_{W,soil}$ [Eqns (5.11a), (5.11b)] give the dependence on temperature and water potential. At 20°C with no water limitation, the half-life of protected SOM is about 12 years.

The stabilization process converts protected SOM to stabilized SOM (C_{sSOM}), giving a output C flux of

$$\begin{aligned}O_{C,pSOM\to sSOM} &= k_{pSOM\to sSOM} C_{pSOM}, \\ k_{pSOM\to sSOM} &= k_{pSOM\to sSOM20} f_{T,soil} f_{W,soil}, \\ k_{pSOM\to sSOM20} &= 0.00002 \text{ day}^{-1}.\end{aligned} \tag{5.8e}$$

Temperature and water functions $f_{T,soil}$ and $f_{W,soil}$ are given in Eqns (5.11a), (5.11b). Both these output fluxes of C are partially respired, giving respiratory C fluxes of

$$\begin{aligned}R_{pSOM\to min} &= f_{C,pSOM\to minR} O_{C,pSOM\to min}, \\ R_{pSOM\to sSOM} &= f_{C,pSOM\to sSOM,R} O_{C,pSOM\to sSOM}, \\ f_{C,pSOM\to minR} &= 0.4, \qquad f_{C,pSOM\to sSOM,R} = 0.2.\end{aligned} \tag{5.8f}$$

The fluxes of C into the soluble C (C_{sol}) and stabilized SOM (C_{sSOM}) pools are obtained by difference:

$$I_{C,pSOM\to Csol} = O_{C,pSOM\to min} - R_{pSOM\to min},$$
$$I_{C,pSOM\to sSOM} = O_{C,pSOM\to sSOM} - R_{pSOM\to sSOM}. \quad (5.8g)$$

The N outputs in the second of Eqns (5.8c) are obtained by dividing the C outputs by the average C:N ratio in the protected SOM pool [Eqn (5.8a)]. The N flux is associated with the C flux to mineralization [Eqn (5.8d)] flows into the ammonium N pool, N_{amm}. See the remarks after Eqn (5.7d). Also, the N flux associated with the fraction of the C flux to stabilized SOM which is respired [Eqn (5.8f)] is input to the N_{amm} pool, giving a total input to N_{amm} of

$$I_{N,pSOM\to Namm} = \frac{O_{C,pSOM\to min} + R_{pSOM\to sSOM}}{r_{C:N,pSOM}}. \quad (5.8h)$$

The C:N ratio is given Eqn (5.8a).

Accompanying the C flux from the protected SOM pool of $I_{C,pSOM\to sSOM}$ [Eqn (5.8g)] which actually enters the stabilized SOM pool is an N flux of (dividing by the C:N ratio of the protected SOM pool)

$$\frac{I_{C,pSOM\to sSOM}}{r_{C:N,pSOM}}. \quad (5.8i)$$

To provide the N required for the C:N ratio $r_{C:N,sSOM,new}$ for new matter entering the stabilized SOM pool [Eqn (5.9b)] requires an N flux of (dividing by this C:N ratio)

$$\frac{I_{C,pSOM\to sSOM}}{r_{C:N,sSOM,new}}. \quad (5.8j)$$

The net N requirement is the difference between the two quantities above. It is taken from the mineral N pools: the ammonium N and nitrate N pools, together designated as N_{min}. This net N requirement is an output from N_{min} and is given by

$$O_{Nmin,pSOM\to sSOM} = I_{C,pSOM\to sSOM}\left(\frac{1}{r_{C:N,sSOM,new}} - \frac{1}{r_{C:N,pSOM}}\right). \quad (5.8k)$$

The N flux of Eqn (5.8k) is taken from the ammonium N (N_{amm}) and nitrate N (N_{nit}) pools in the same ratio in which microbial biomass takes up N from these two pools. With Eqns (5.6f), the outputs from the N_{amm} and N_{nit} pools are

$$O_{Namm,pSOM\to sSOM} = f_{Namm\to bio}O_{Nmin,pSOM\to sSOM},$$
$$O_{Nnit,pSOM\to sSOM} = f_{Nnit\to bio}O_{Nmin,pSOM\to sSOM}. \quad (5.8l)$$

See after Eqn (5.7l) for further explanation.

5.10 Stabilized Soil Organic Matter, C_{sSOM}, N_{sSOM}

The stabilized SOM pool has an average C:N ratio, $r_{C:N,sSOM}$, of

$$r_{C:N,sSOM} = \frac{C_{C,sSOM}}{N_{N,sSOM}}. \tag{5.9a}$$

This can vary because *new* stabilized SOM is given a variable C:N ratio depending on soil mineral N concentration N_{min} [Eqn (5.5g)], with

$$\begin{aligned} r_{C:N,sSOM,new} &= r_{C:N,sSOM,max} - \left(r_{C:N,sSOM,max} - r_{C:N,sSOM,min}\right) \\ &\quad \times \frac{N_{min}}{N_{min} + K_{rC:N,sSOM}}, \\ K_{rC:N,sSOM} &= 0.002 \text{ kg N m}^{-2}. \end{aligned} \tag{5.9b}$$

At low soil mineral concentrations (N_{min}), the C:N ratios in new protected SOM is maximum, and decreases as N_{min} increases. Outputs from the protected SOM pools have a C:N ratio given by the pool average [Eqns (5.9a)].

For stabilized SOM the range of C:N ratios depends on clay fraction, f_{clay} [Eqn (5.11)], according to

$$\begin{aligned} r_{C:N,sSOM,max} &= r_{C:N,sSOM,sandy\text{-}max} + c_{sSOM,clay} f_{clay} \\ &\quad \times \left(r_{C:N,sSOM,clay\text{-}max} - r_{C:N,sSOM,sandy\text{-}max}\right), \\ r_{C:N,sSOM,min} &= r_{C:N,sSOM,sandy\text{-}min} + c_{sSOM,clay} f_{clay} \\ &\quad \times \left(r_{C:N,sSOM,clay\text{-}min} - r_{C:N,sSOM,sandy\text{-}min}\right), \\ r_{C:N,sSOM,clay\text{-}max} &= 10,\ r_{C:N,sSOM,sandy\text{-}max} = 20 \text{ kg C (kg N)}^{-1}, \\ r_{C:N,sSOM,clay\text{-}min} &= 5,\ r_{C:N,sSOM,sandy\text{-}min} = 10 \text{ kg C (kg N)}^{-1}, \quad c_{sSOM,clay} = 3. \end{aligned} \tag{5.9c}$$

For a sandy soil with $f_{clay} = 0$, the C:N range is 10–20 kg C (kg N)$^{-1}$, depending on the soil mineral N concentration N_{min}, whereas a soil with a clay fraction of $\frac{1}{3}$ has a C:N range of 5–10 kg C (kg N)$^{-1}$ [Eqn (5.9b)]. With $f_{clay} = 0.1$ (our default value), the C:N range of stabilized SOM is 8.5–17 kg C (kg N)$^{-1}$.

5.10.1 Differential equations (kg C, N m^{-2} day^{-1})

For the C and N components of stabilized SOM, these are

$$\begin{aligned} \frac{dC_{sSOM}}{dt} &= I_{C,uSOM \to sSOM} + I_{C,pSOM \to sSOM} - O_{C,sSOM \to min}, \\ \frac{dN_{sSOM}}{dt} &= \frac{I_{C,uSOM \to sSOM} + I_{CpSOM \to pSOM}}{r_{C:N,sSOM,new}} - \frac{O_{C,sSOM \to min}}{r_{C:N,sSOM}}, \\ C_{sSOM}(t=0) &= 6 \text{ kg C m}^{-2},\ N_{sSOM}(t=0) = 0.5 \text{ kg N m}^{-2}. \end{aligned} \tag{5.9d}$$

The initial C:N ratio is 12 kg C (kg N)$^{-1}$.

5.10.2 Inputs

There are two C inputs: $I_{C,uSOM \to sSOM}$ from unprotected SOM is calculated as an output of the C_{uSOM} pool [Eqn (5.7g)]; similarly, $I_{C,pSOM \to sSOM}$ from protected SOM is an output of the C_{pSOM} pool [Eqn (5.8g)]. The N inputs are obtained by dividing the C inputs by the C:N ratio in new stabilized SOM [Eqn (5.9b)].

5.10.3 Output

There is a single C output to mineralization, given by

$$\begin{aligned} O_{C,sSOM\to min} &= k_{sSOM\to min}\, C_{sSOM}, \\ k_{sSOM\to min} &= k_{sSOM\to min20}\, f_{T,soil}\, f_{W,soil}, \\ k_{sSOM\to min20} &= 0.03\times 10^{-3}\ \text{day}^{-1}. \end{aligned} \tag{5.9e}$$

At 20°C with no water limitation, the half-life of stabilized SOM is about 100 years. See the remarks after Eqn (5.7d).

This output of C is partially respired, with respiratory C flux of

$$\begin{aligned} R_{sSOM\to min} &= f_{C,sSOM\to minR} O_{C,sSOM\to min}, \\ f_{C,sSOM\to minR} &= 0.4. \end{aligned} \tag{5.9f}$$

The flux of C into the soluble C pool (C_{sol}) is

$$I_{C,sSOM\to Csol} = O_{C,pSOM\to min} - R_{sSOM\to min}. \tag{5.9g}$$

The output of N is given by dividing the C output [Eqn (5.9e)] by the average C:N ratio of stabilized SOM [Eqn (5.9b)]. This provides a flux of N into the ammonium N pool (N_{amm}) of

$$I_{N,sSOM\to Namm} = \frac{O_{C,sSOM\to min}}{r_{C:N,sSOM}}, \tag{5.9h}$$

where the C:N ratio is given in Eqn (5.9a).

5.11 Totals for the Soil Submodel

The total C and N contents of the three soil organic matter pools together, C_{SOM} and N_{SOM}, are

$$\begin{aligned} C_{SOM} &= C_{uSOM} + C_{pSOM} + C_{sSOM}, \\ N_{SOM} &= N_{uSOM} + N_{pSOM} + N_{sSOM}. \end{aligned} \tag{5.10a}$$

The C:N ratio of soil organic matter as a whole is

$$r_{C:N,SOM} = \frac{C_{SOM}}{N_{SOM}}. \tag{5.10b}$$

The C:N ratio of the total SOM can change as the ratios for the separate pools change [Eqns (5.7a), (5.8a), (5.9a)], and as the relative proportions of the three pools change.

Total soil respiration is

$$R_{soil} = R_{surf\text{-}li} + R_{soil\text{-}li} + R_{urine} + R_{bioG} + R_{bioD} + R_{uSOM \to min} + R_{uSOM \to sSOM} + R_{pSOM \to min} + R_{pSOM \to sSOM} + R_{sSOM \to min}. \quad (5.10c)$$

The terms in this equation can be found in Eqns (5.1i), (5.2i), (5.4b), (5.6d), (5.6j), (5.7f), (5.8f) and (5.9f). Note that root respiration [Eqn (3.3l)] is counted in with plant respiration.

5.12 Effects of Temperature and Water on Soil and Litter Processes

It is assumed that soil processes have the same basic temperature dependence as plant processes [Fig. 3.6; Eqn (3.11a)], but calculated at the appropriate temperature. For the surface processes [e.g. Eqns (5.1e)], it is assumed that the surface temperature is midway between air temperature T_{air}, and soil temperature T_{soil}. The soil and surface temperature functions, $f_{T,soil}$ and $f_{T,surf}$, are given by

$$f_{T,soil} = f(T_{soil}), \qquad f_{T,surf} = f(T_{surf}), \qquad T_{surf} = \tfrac{1}{2}(T_{air} + T_{soil}). \quad (5.11a)$$

where the function f is given in Eqn (3.11a). Further discussion of the temperature function can be found in Section 3.11.

The water submodel (Chapter 6) calculates the chemical activity of water in soil, $a_{W,so}$ [Eqn (6.7a)], which depends on the soil water potential [ψ_{so}, Eqn (6.2g)]. For processes taking place on the soil surface, the function $f_{W,surf}$ used to modify the rates of the surface processes is calculated in Eqn (6.7c), where shoot water activity is used as a surrogate for surface water activity. Thus

$$f_{W,soil} = a_{W,so}^{q_{W,soil}}, \qquad f_{W,surf} = a_{W,sh}^{q_{W,surf}}, \qquad q_{W,soil} = 20, \qquad q_{W,surf} = 30. \quad (5.11b)$$

Denitrification is modified by soil relative water content as in Eqn (5.5d).

5.13 Simulations

In this section some simulations of the decoupled soil and litter submodel are given, to show its main dynamic features. For this purpose, diurnal and seasonal weather is switched off, and air and soil temperatures are set to 20°C. The water submodel (Chapter 6) is switched off. There is no leaching of C and N [Eqns (5.3b), (5.5e)], and no denitrification of N [Eqn (5.5d)]. Plant dynamics are switched off so that the plant state variables remain at their initial values. Root and shoot substrate concentrations [Eqn (3.1h)] are set equal. The turnover rates of the shoot and root [Eqns (3.5a), (3.5c)] are made equal, so that the plant inputs to surface litter and soil litter are identical. There is continuous grazing with 7 sheep ha^{-1} [Eqn (7.6f)], giving an animal input to the surface litter pools.

The consequence of these assumptions is that all the driving variables of the soil and litter submodel are constant, so that the predicted dynamic behaviour of the model is intrinsic to the soil and litter submodel alone. Note that these simulations are performed at 20°C; this is considerably warmer than our standard environment (Fig. 7.1b) with a annual average temperature of 10°C, where reactions proceed at one-third of the 20°C rate (Fig. 3.6, cubic).

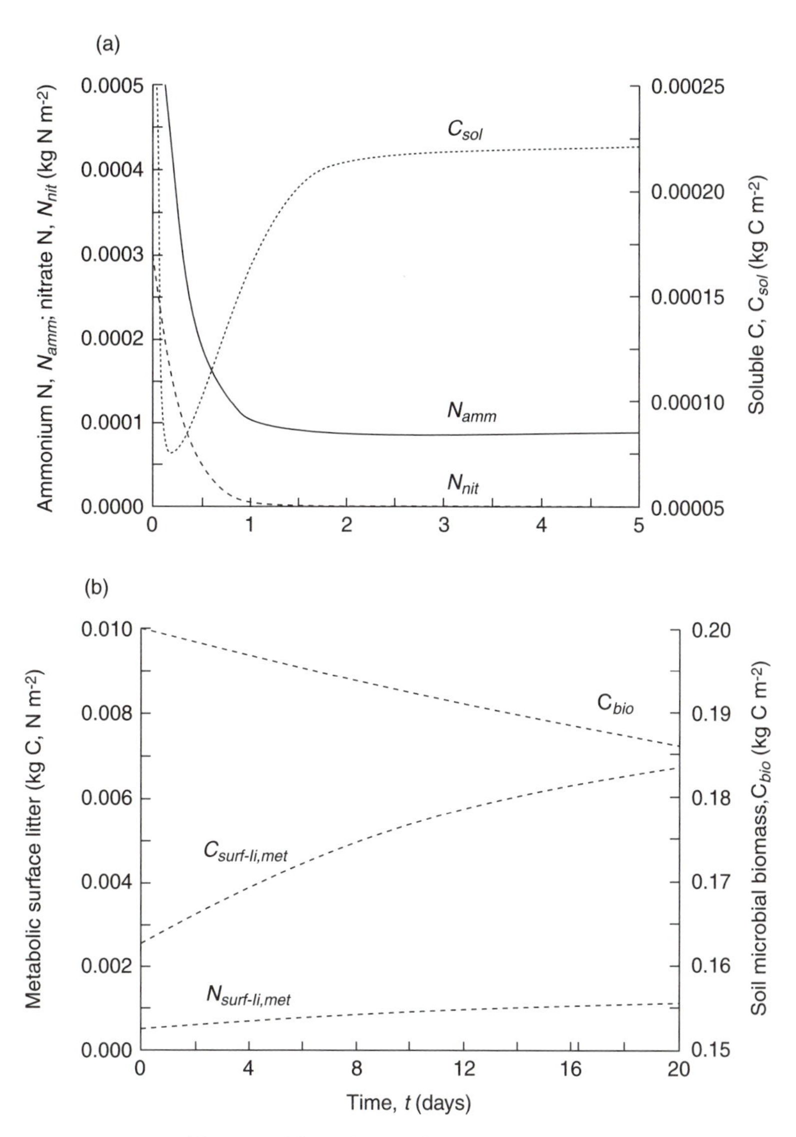

Fig. 5.3. Dynamics of fast variables of the soil and litter submodel up to 20 days. Conditions are: a constant environment; air and soil temperatures of 20°C; no water limitations, no leaching of C or N [Eqns (5.3b), (5.5e)], and no denitrification. Plant state variables remain at their initial values adjusted so that the root and shoot substrate concentrations [Eqn (3.1h)] are equal. The inputs to surface litter and soil litter pools are equal. There is continuous grazing at 7 sheep ha^{-1}.
(a) Mineral N pools: ammonium N and nitrate N, N_{amm} and N_{nit} [Eqns (5.4a), (5.5a)]; soluble C pool, C_{sol} [Eqn (5.3a)]. (b) The soil microbial biomass C pool, C_{bio} [Eqn (5.6a)]; surface litter metabolic C and N pools, $C_{surf\text{-}li,met}$ and $N_{surf\text{-}li,met}$ [Eqns (5.1a)] (soil litter metabolic pools are similar).

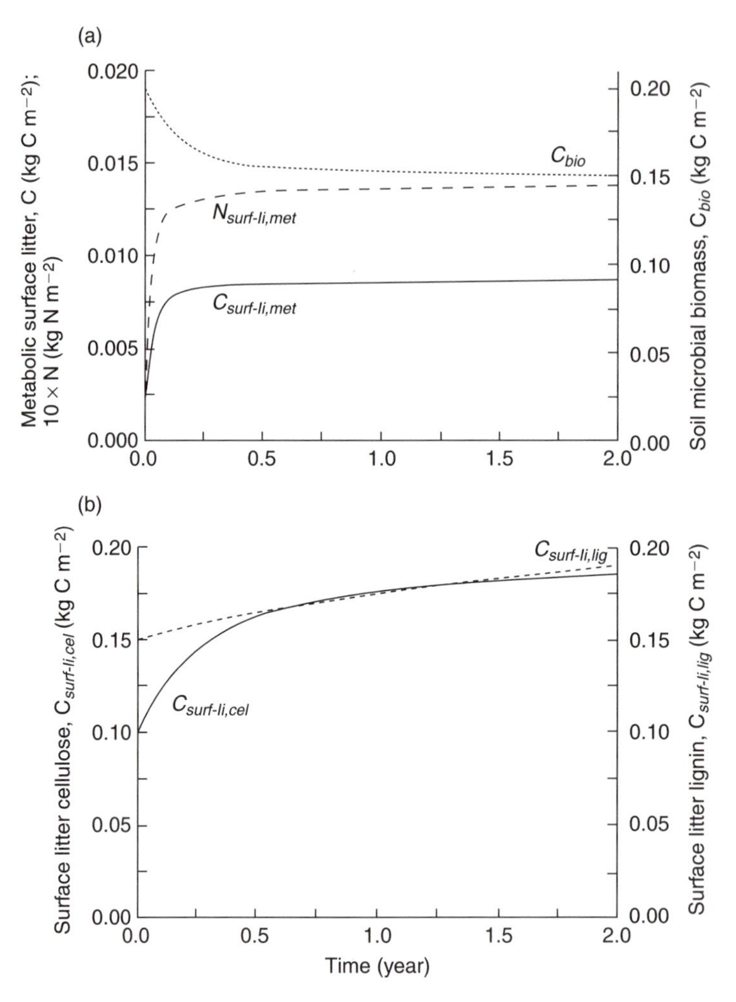

Fig. 5.4. Two-year dynamics of soil and litter submodel. See Fig. 5.3 legend for general conditions. (a) The soil microbial biomass C pool, C_{bio} [Eqn (5.6a)]; surface litter metabolic C and N pools, $C_{surf\text{-}li,met}$ and $N_{surf\text{-}li,met}$ [Eqns (5.1a)]. (b) Surface litter cellulose and lignin C pools, $C_{surf\text{-}li,cel}$ and $C_{surf\text{-}li,lig}$ [Eqns (5.1a)] (All surface and soil litter pools are similar).

Figure 5.3a shows the fastest pools in the soil and litter subsystem: the mineral N pools and the soluble C pool. The soluble C pool C_{sol} [Eqn (5.3a)] falls sharply within 6–12 h to a value from which a slower increase occurs. The initial fall arises from the large imbalance under these circumstances between microbial biomass utilization of C_{sol} and the inputs to the pool, relative to the size of the pool. After these initial changes these three pools (Fig. 5.3a) change at similar rates as they are quite tightly coupled. The slow increase in C_{sol} is mostly due to the metabolic C pool increasing (Fig. 5.3b) giving a greater input to C_{sol}. Microbial biomass decreases slowly on a 20-day time scale (Fig. 5.3b); the dynamics of the biomass pool in isolation are shown in Fig. 5.2. The metabolic C and N pools [Eqns (5.1a),

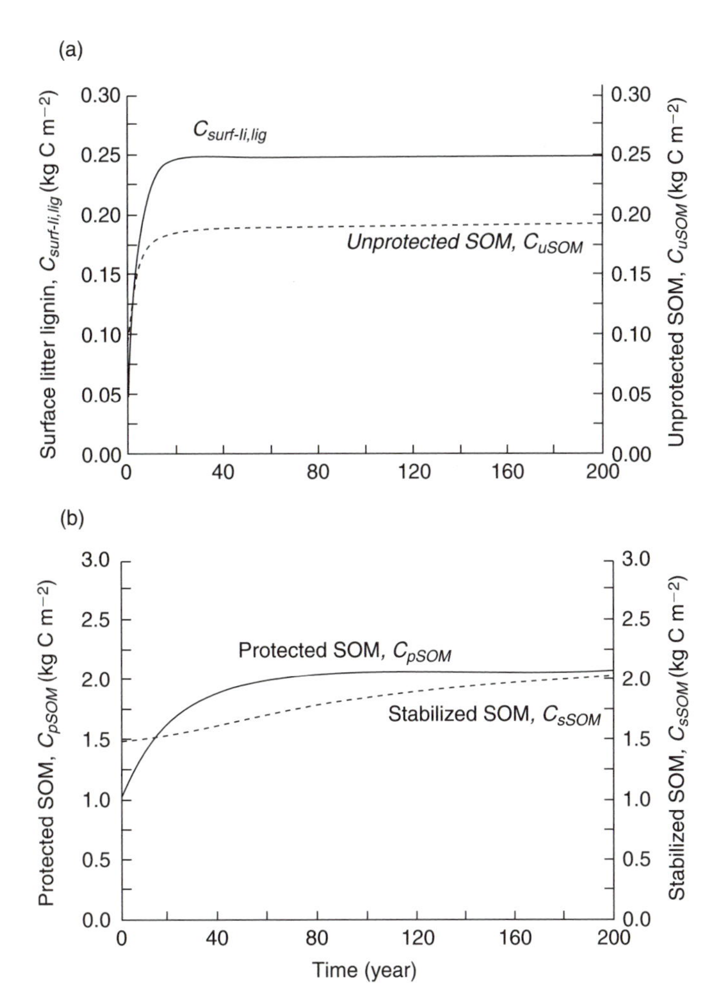

Fig. 5.5. Two-hundred-year dynamics of soil and litter submodel. See Fig. 5.3 legend for general conditions. The initial values of the lignin and SOM pools are set to low values. (a) Surface litter lignin C pool, $C_{surf\text{-}li,lig}$ (soil litter lignin pool is similar) and unprotected soil organic matter pool (SOM), C_{uSOM} [Eqns (5.1a), (5.7c)]. (b) Protected and stabilized SOM pools, C_{pSOM} and C_{sSOM} [Eqns (5.8c), (5.9d)].

(5.2a)] are the slowest in the group shown in Fig. 5.3: they move asymptotically towards their steady-state values.

Figure 5.4 shows a 2-year simulation. On this time scale fast pools such as the mineral N pools and soluble C (see Fig. 5.3a) move at a rate determined by slower pools, which are shown in Fig. 5.4a. The metabolic litter pools move towards steady-state values with a time constant of a few weeks. The microbial biomass settles down at a similar rate, but note that the asymptote of of the microbial biomass growth equation [Eqn (5.6b)] depends on total SOM carbon which changes slowly (Fig. 5.6). In Fig. 5.4b the cellulose pool moves towards its steady state value with a time constant of months, whereas the lignin pool is much slower.

A 200-year simulation is shown in Fig. 5.5. In Fig. 5.5a it is seen that unprotected soil organic matter (C_{uSOM}) and lignin ($C_{surf\text{-}li,lig}$) have similar kinetics, responding over a few years. In Fig. 5.5b the protected SOM fraction (C_{pSOM}) responds over about 20 years, whereas the stabilized SOM fraction (C_{sSOM}) is demonstrating sigmoidal behaviour with respect to time and approaching a steady state over several hundreds of years.

A 1000-year run is shown in Fig. 5.6. While the protected SOM pool takes about 100 years to settle (see also Fig. 5b), the stabilized SOM pool takes about 500 years to reach within 1% of its steady-state value. Remember that this is at a uniform 20°C and the time scale is three times as long when the average mean temperature is 10°C as in the standard environment (Table 7.3; Fig. 3.6, cubic). It may be noted that soil mineral N concentration [N_{min}, Eqn (5.5g)] is not greatly affected by the levels of the slower pools such as protected and stabilized SOM (see next section). N_{min} represents the two mineral N pools [Eqn (5.5g)], which are the only variables directly affecting plant growth through N uptake by the root [Eqn (3.4h)].

5.14 Representation of Very Slow Pools in SOM Models

Very slow pools such as the stabilized SOM pools in this model (C_{sSOM}, N_{sSOM}, Fig. 5.6) can present substantial problems to experimentalists and to modellers. For the experimentalist it is difficult to define what these pools are and to measure very small fluxes. The modeller needs to be able to work with steady-state solutions to investigate and understand the model because these give results that are usually

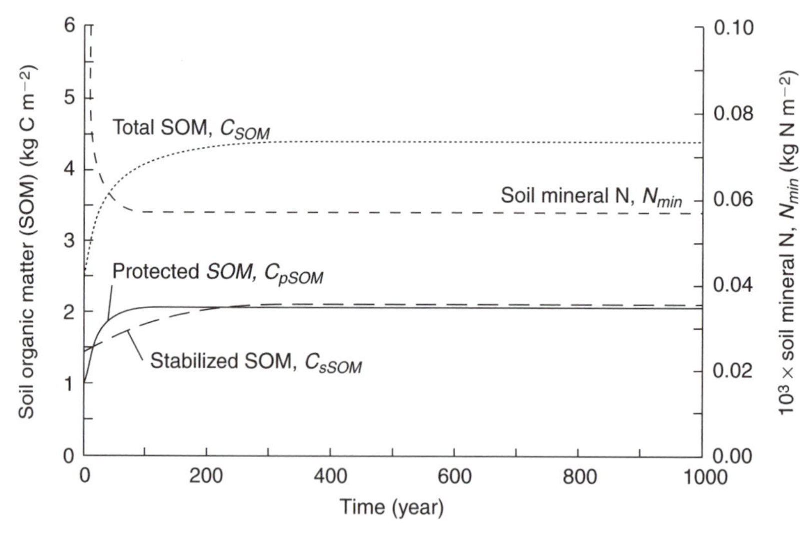

Fig. 5.6. One-thousand-year dynamics of soil and litter submodel. See Fig. 5.3 legend for general conditions. The initial values of the lignin and SOM pools are set to low values. Protected, stabilized and total SOM pools, C_{pSOM}, C_{sSOM} and C_{SOM} [Eqns (5.8c), (5.9d), (5.10a)], and the labile soil mineral N pool, N_{min} [Eqn (5.5g), see Fig. 5.3], are shown.

independent of possibly arbitrary initial conditions. Also, evaluating the consequences of climate change requires long-term projections to be made. Obtaining steady-state solutions can be very tedious if a very slow pool is involved. Jenkinson's model (1990) has an inert pool with neither inputs or outputs; essentially, this appears to be a way of introducing a constant so that observed data can be better fitted, but biologically, such a concept seems to be of questionable value. Parton's model (Schimel *et al.*, 1994) has a passive pool with a turnover time of over 1000 years at 20°C.

A very slow pool is necessarily connected to the rest of the system with small fluxes. At a certain point on the scale of increasing slowness such a pool ceases to affect the system appreciably and can be ignored for many purposes. The fast pools are generally small pools with relatively large input and output fluxes, and short turnover times. The influence of the levels of the slow pools with their small fluxes on the fast pool concentrations may be quite small (e.g. N_{min} in Fig. 5.6).

To determine the impact of the stabilized SOM pool, which is the slowest pool in the present model, on various aspects of model performance, we have done an experiment in which the stabilized pool is switched off – that is, all inputs and outputs to the stabilized pool are zero. This is achieved by setting the three rate constants which determine the inputs and outputs to the pool equal to zero. The results are shown in Table 5.2, where the equilibrium soil state variables are given.

Eliminating the stabilized SOM pool from the model greatly affects the total C (and N) sequestered by the soil, because, with our numerical assumptions, the stabilized SOM pool is both the slowest and the largest pool in the system. However, most other variables are little changed. It can be seen in Table 5.2 that the mineral N pools, which affect plant growth, are virtually the same, as is the soil microbial biomass pool. The soluble C pool, C_{sol}, shows the largest change, an increase of about 40%. The other two SOM pools (unprotected SOM, C_{uSOM} and protected SOM, C_{pSOM}) increase by 5% and 10% respectively.

Very slow pools in SOM models will affect the long-term C and N sequestering ability of the model greatly. They may however have little effect on plant growth

Table 5.2. Steady state values after 500 years with and without the stabilized soil organic matter pools [C_{sSOM}, N_{sSOM}, Eqn (5.9d)]. A constant environment is applied; air and soil temperatures are 20°C; no water limitations, no leaching of C or N [Eqns (5.3b), (5.5e)] and no denitrification [Eqn (5.5d)] apply. Plant state variables remain at their initial values adjusted so the root and shoot substrate concentrations [Eqn (3.1h)] are equal. The inputs to surface litter and soil litter pools are equal. There is constant grazing with 7 sheep ha^{-1} [Eqn (7.6f)]. The C_{sSOM}, N_{sSOM}, pools are eliminated by putting the three rate constants $k_{pSOM \to sSOM20}$, $k_{uSOM \to sSOM20}$, $k_{sSOM \to min20}$ [Eqns (5.8e), (5.7e), (5.9e)] equal to zero. The litter pools are unchanged by elimination of the C_{sSOM} pool.

N_{amm} (10^{-6} kg N m^{-2})	N_{nit} (10^{-6} kg N m^{-2})	C_{sol} (10^{-3} kg C m^{-2})	C_{bio} (kg C m^{-2})	C_{uSOM} (kg C m^{-2})	C_{pSOM} (kg C m^{-2})	C_{sSOM} (kg C m^{-2})
56	0.37	1.3	0.13	0.19	2.1	2.1
Without the stabilized SOM pool:						
56	0.37	1.8	0.13	0.20	2.3	arbitrary constant

and primary productivity. Note that the present model does not include any effect of soil organic matter on the soil water characteristic; this could be an important influence of these large relatively inert pools on long-term responses in particular environments.

6 Water Submodel

6.1 Introduction

The water submodel described in this chapter is constructed for use with plant growth simulators that represent internal plant substrates and variable shoot:root allocation, as in Chapter 3. The model calculates water flow from soil to root, root to shoot, and shoot to the atmosphere, for a closed-canopy grassland.

The scheme used is drawn in Fig. 6.1. It has three state variables: the masses of water in the soil, root and shoot, and represents the processes of rainfall, rainfall interception and evaporation from the canopy, drainage, movement of water from soil to root, root to shoot, and evapotranspiration. Direct evaporation from the soil is ignored in this application: it is likely to be negligible for a closed grass canopy; it is also difficult to calculate direct soil evaporation realistically with a single soil water compartment as in Fig. 6.1. The fluxes of water from soil to root, and root to shoot, are driven by water potential differences. This approach worked satisfactorily, although it should be noted that this widely used formalism may not always be appropriate, e.g. for irreversible processes operating far from equilibrium (Nobel, 1991, pp. 165–176). Tissue water potential and its components are calculated from tissue water content and other plant variables and parameters. The Penman–Monteith equation is used for plant evapotranspiration. The model is able to simulate osmoregulation and describes a variable relationship between tissue water potential, its pressure and osmsotic components and relative water content. The model, essentially as presented here, has been described by Thornley (1996). An analogous version has also been used in the ITE Edinburgh forest model (Thornley, 1991b; Thornley and Cannell, 1992, 1996) with satisfactory results.

Modelling plant water status in relation to soil water availability is an area where there are still disagreements about many details (e.g. see Passioura, 1984; Ludlow, 1987), but there is largely a consensus about the general approach and the qualitative responses both observed and predicted (e.g. Jones, 1992). The level of complexity in a water submodel is partly determined by the type of plant growth model with which the water submodel is to be used. The grassland growth model

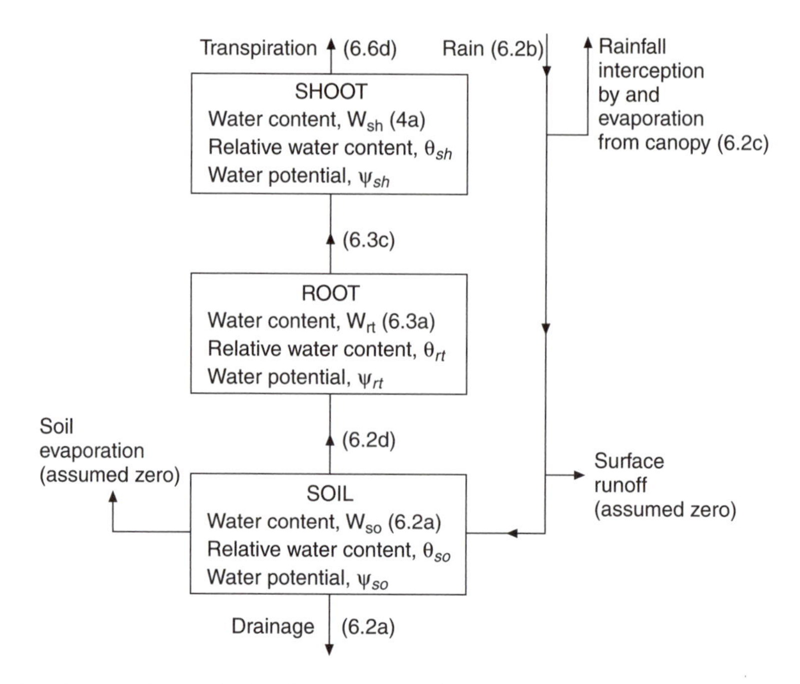

Fig. 6.1. Water submodel. The three state variables of the submodel are the masses of water in the soil, root and shoot (W_{so}, W_{rt}, W_{sh}) and are shown in the boxes. Equation numbers point to flux equations or differential equations. Other important variables of the model are the relative water contents, θ_{so}, θ_{rt}, θ_{sh}, and the water potentials, ψ_{so}, ψ_{rt}, ψ_{sh}. These latter variables are derived from the three state variables [Eqns (6.2f), (6.2g), (6.5d), (6.5c)]. Direct soil evaporation and surface runoff are ignored in the present application.

represents plant internal substrate pools as well as plant structural dry matter. This permits a rather detailed treatment of the relations between tissue water content, the components of tissue water potential, and relative tissue water content. Arguably, it is important to get this part of the submodel reasonably correct as it leads to osmoregulatory effects. Some other aspects of the model, such as rainfall interception and evaporation, and soil-to-root water flow, are treated quite simply.

Jones (1988) has recently reviewed water relations in grassland. He emphasizes that, frequently, shortage of water is an important factor limiting yield. While fertilizers are commonly used on agricultural grassland, irrigation is seldom applied, although Doyle (1981) has shown that it can be economically viable, even in temperate regions.

Johnson *et al.* (1991) describe a water submodel that includes root/shoot message control of stomatal conductance. Preliminary work with their extended model showed that the results were not much changed by this additional complexity, and program execution was far slower. The grassland water model of Coughenour (1984) has many similarities with the present approach: the model has a plant water pool, substrate pools, water movement driven by a water potential gradient with resistances depending on hydraulic conductivities. His model does not represent the relationships between osmotica and structural dry mass in the plant, water content and water potential. He uses a multi-layered soil model for treating the flows of heat and of water, but does not present a detailed description of the model. The grassland model of Verberne (1992) treats the soil as a multi-layered system. There are no plant water pools and there is no treatment of plant water relations. The potential evapotranspiration flux is distributed over the different soil layers; the rate of removal from a given soil layer depends on the moisture content in that layer. Soil surface evaporation depends on the moisture content in the upper soil layer but is extracted in a distributed manner from all layers. Capillary rise between layers is not taken into account. Neither Coughenour (1984) nor Verberne (1992) appear to consider the numerical impact of layer thickness, number of soil layers and root depth on the predicted results from their models. Jensen *et al.*'s model (1993) treats in detail many of the areas which are simplified here: e.g. crop interception of rain and evaporation from foliage, radiation interception with sunlit and shade components, and the root contact concept is used in the calculation of root water uptake. These authors use 12 soil layers and apply their model to barley growing on a sandy soil.

Units are a perennial problem. Although the use of traditional units has advantages for many readers, at some point the nettle of making an arguably more rational choice of units must be grasped. I have chosen to use energy units per unit mass [J $(\text{kg water})^{-1}$] for water potential, rather than pressure units which measure energy per unit volume (pascal, Pa; Pa = N m^{-2} = J m^{-3}); the latter depend on the variable density of water ρ_W (kg m^{-3}) (Thornley and Johnson, 1990, p. 50). The logic of this choice is that the mass of water is conserved, whereas volumes of water are not conserved. However, the equivalent in pressure units is generally given. Energy units and pressure units are related by (also see the Appendix)

$$\text{water potential}\left[J(\text{kg water})^{-1}\right] = \frac{\text{water potential}\left(\text{Pa} = \text{J m}^{-3}\right)}{\rho_w\left(\text{kg water m}^{-3}\right)}. \quad (6.1a)$$

Assuming that the density of water ρ_W = 1000 kg water m^{-3}, therefore

$$1\ \text{J kg}^{-1} \equiv 1000\ \text{Pa} = 1\ \text{kPa}. \quad (6.1b)$$

Conveniently, this leads to

$$100\ \text{J kg}^{-1} \equiv 0.1\ \text{MPa} = c.\ 1\ \text{atmosphere}, \quad (6.1c)$$

with 100 J kg^{-1} corresponding to approximately 1 atmosphere.

It is regrettable that the SI unit for the mole was defined so that 1 SI mole of water has a mass of 0.018 kg, rather than 18 kg, which would be consistent with the base mass unit of SI, the kilogram (Thornley and Johnson, 1991). I have used a 'kg mole' of $10^3 \times$ SI mole, so that the molar mass of water has the familiar numerical value of 18, with units of kg $(\text{kg mole})^{-1}$, and the molar mass is equal to the dimensionless relative molecular mass of 18. Without this choice, the molar mass of water would be 0.018 kg $(\text{SI mole})^{-1}$, and that of sucrose (say) would be 0.342 kg (SI $\text{mole})^{-1}$. This should not cause the reader any difficulties, since molar masses are always divided by the gas constant, which is also 10^3 times its usual value [e.g. see Eqn (6.5a) for the osmotic potential].

For grassland, the gravitational component of water potential can be ignored (see Nobel, 1991, for a general discussion of water potential). The water potential ψ is the sum of the osmotic and pressure components, ψ_{os} and ψ_{pr}, with

$$\psi = \psi_{os} + \psi_{pr}. \tag{6.1d}$$

Water potential is measured relative to a pool of pure water with a free surface. In plants the water potential is generally negative, with a large negative osmotic component (ψ_{os}) arising from solutes and a positive pressure component (ψ_{pr}) giving positive turgor which is of lesser magnitude than ψ_{os} (e.g. Fig. 6.4a; Jones, 1992, Fig. 4.3, p. 78). Negative values of the pressure component are sometimes referred to as 'suction'.

Some simulations are given in this chapter to illustrate the particular mechanisms assumed in this submodel. Other interesting aspects of the water submodel are concerned with the responses of the grassland ecosystem as a whole. These responses can involve quite subtle interactions between water, nitrogen, leaching, temperature and CO_2, and are discussed in Chapters 8 and 9. The water submodel is also an important component of the biennial oscillations which the whole-system model predicts under certain conditions (Section 8.14). This suggests that the water relations may play a central role in the stability, or otherwise, of grassland ecosystems.

State variables, other variables and parameters, with values where appropriate, are listed in Table 6.1.

6.2 Soil Water Pool, W_{so}

The soil water state variable W_{so} has units of kg water (m^2 ground)$^{-1}$; this is the mass of water in a notional soil depth, d_{soil} (m). Assuming a water density of 1000 kg water m^{-3}, 1 kg water (m^2 ground)$^{-1}$ has a depth of 1 mm.

6.2.1 Differential equation

This is a conditional equation depending upon whether there is a net input to the soil water pool, and whether the soil is already at field capacity. It is given by

$$\begin{aligned}
&\text{If } \left[\left(I_{W,so} > 0\right) \text{ and } \left(\theta_{so} \geq \theta_{so,\max}\right)\right] \\
&\frac{dW_{so}}{dt} = 0, \quad O_{W,so\rightarrow drain} = I_{W,so}; \\
&\text{else } \frac{dW_{so}}{dt} = I_{W,so}, \quad O_{W,so\rightarrow drain} = 0. \\
&\theta_{so,\max} = 0.4 \text{ m}^3 \text{ water (m}^3 \text{ soil)}^{-1}, \\
&W_{so}(t=0) = d_{soil}\rho_W \theta_{so,\max} = 400 \text{ kg water m}^{-2}.
\end{aligned} \tag{6.2a}$$

Here $I_{W,so}$ (kg water m^{-2} day^{-1}) is the net flux into the soil water pool [Eqn (6.2b)], θ_{so} is the soil relative water content [Eqn (6.2f)], and $\theta_{so,max}$ is the soil relative water content at field capacity. $O_{W,so\rightarrow drain}$ (kg water m^{-2} day^{-1}) is the output to drainage. The initial value of W_{so} is calculated from the soil depth [d_{soil}, Eqn (6.2f)], the density of water [ρ_W, Eqn (6.2b)] and $\theta_{so,max}$; this puts the soil initially at field capacity.

Table 6.1. Symbols of water submodel. Equation references are: (i) for state variables, to their differential equation; (ii) for parameters, to the equation where the parameter is introduced and explained; (iii) for other water submodel variables, to the equation where the variable is defined; (iv) for variables from other submodels, to the equation in this chapter where the variable is used, and to the equation where the variable is defined. DM = dry matter.

State variables	Description	Initial value
W_{rt}, W_{sh}, W_{so}	Water contents of root, shoot, soil [(6.3a), (6.4a), (6.2a)]	1, 1, 400 kg water m^{-2}
Parameters		**Values**
$c_{r,leaf}$	Leaf reflection coefficient (6.6a)	0.15
$c_{\psi,pr}$	Parameter affecting pressure component of plant water potential (6.5b)	0.2 kg structural DM (kg water)$^{-1}$
$c_{W,rs\text{-}rt}$, $c_{W,so\text{-}rs}$	Constants affecting resistance between soil and root (6.2i)	0.5×10^6 m s^{-1} 80 m^2
$c_{WT,pl}$	Constant for water transport in plant (6.3c)	0.005 kg water (kg structural DM)$^{-1}$ [J (kg water)$^{-1}$]$^{-1}$ day^{-1}
d_{soil}	Soil depth (6.2f)	1 m
$f_{rain,i,e,m}$	Fraction of rain intercepted that evaporates each calendar month (m = 1–12) (6.2c)	0.1, 0.15, 0.2, 0.3, 0.35, 0.4, 0.4, 0.35, 0.3, 0.2, 0.15, 0.1
$f_{S,os\text{-}ac}$	Fraction of substrate (storage) mass that is osmotically active(6.5a)	2
h_{ref}	Reference height for meteorological measurements [6.6c, Table 7.2]	2 m
$K_{so,max}$	Maximum value of soil hydraulic conductivity(6.2h)	0.1 kg m^{-3} s
$K_{W,rs\text{-}rt}$	Soil–root resistance constant (6.2i)	1 kg structural DM m^{-2}
k_{can}	Canopy extinction/rainfall interception coefficient [(6.2c), (3.2b)]	0.5 m^2 ground (m^2 leaf)$^{-1}$
k_{vK}	Von Karman constant (6.6c)	0.4
$q_{W,pl}$, $q_{W,surf}$, $q_{W,soil}$, $q_{W,ph}$, $q_{W,uN}$	Parameters for effects of water stress on plant biochemistry (6.7b), surface litter biochemistry (6.7c), soil biochemistry (6.7b), photosynthesis (6.7d), N uptake (6.7e)	20, 30, 20, 2, 3
$q_{\psi,soW}$	Exponent of soil water characteristic (6.2g)	5 (2 for sand, 18 for clay)
R_{gas}	Gas constant (6.5a)	8314 J (kg mole)$^{-1}$ K^{-1}
w_{ind}	Wind speed (6.6c)	1 m s^{-1}
ε	Cell wall rigidity parameter (6.5b)	2×10^6 Pa
$\theta_{so,max}$	Maximum value of soil relative water content (field capacity) (6.2a)	0.4 m^3 water (m^3 soil)$^{-1}$
μ_S, μ_W	Molar masses of substrate (storage) (6.5a) and water (6.7a)	20, 18 kg (kg mole)$^{-1}$
ρ_W	Density of water (6.2b)	1000 kg m^{-3}
χ_{leaf}	Leaf transmittance [(6.6a), (3.2b)]	0.1
$\Psi_{so,max}$	Water potential of soil at field capacity (6.2g)	−10 J kg^{-1}
Other variables		**Units**
$a_{W,i}$, i = rt, sh, so	Chemical activities of water in root, shoot, soil (6.7a)	
$f_{rain,i}$	Fraction of rain intercepted by foliage (6.2c)	
$f_{W,i}$, i = rt, sh, $soil$, $surf$, ph, u_N	Effect of water on root, shoot, soil, ground surface, photosynthetic, N uptake processes [(6.7b), (6.7c), (6.7d), (6.7e)]	
g_{blcon}	Boundary layer conductance (6.6c)	m s^{-1}
$g_{W,rt\text{-}sh}$	Conductivity for water flow between root and shoot (6.3c)	kg m^{-4} day
$I_{W,so}$	Input of water to soil (6.2b)	kg water m^{-2} day^{-1}

Continued over

Table 6.1. *Continued*

Other variables		Units
$I_{W,rt \to sh}$	Input of water from root to shoot (6.4b)	kg water m^{-2} day^{-1}
$I_{W,so \to rt}$	Input of water from soil to root (6.3b)	kg water m^{-2} day^{-1}
$j_{NetR,abs\text{-}can}$	Net radiation absorbed by canopy (6.6b)	J m^{-2} day^{-1}
K_{so}	Soil hydraulic conductivity (6.2h)	kg m^{-3} s
k_{leach}	Leaching rate (6.2e)	day^{-1}
$O_{W,rain\text{-}int \to atm}$	Output of water, rain intercepted by foliage, to atmosphere (6.2c)	kg water m^{-2} day^{-1}
$O_{W,rt \to sh}$	Output of water, root to shoot (6.3c)	kg water m^{-2} day^{-1}
$O_{W,sh \to an}$	Output of water, shoot to grazing animals (6.4c)	kg water m^{-2} day^{-1}
$O_{W,sh \to atm}$	Output of water, shoot to atmosphere (6.6d)	kg water m^{-2} day^{-1}
$O_{W,sh \to hv}$	Output of water, shoot to harvesting (6.4d)	kg water m^{-2} day^{-1}
$O_{W,so \to drain}$	Output of water, soil to drainage (6.2a)	kg water m^{-2} day^{-1}
$O_{W,so \to rt}$	Output of water, soil to root (6.2d)	kg water m^{-2} day^{-1}
$r_{W,so\text{-}rt}$	Resistance to water flow from soil to root (6.2i)	kg^{-1} m^4 s^{-1}
t	Time, the independent variable	day
$W_i(\psi_i = 0)$	Water content of root, shoot at zero water potential (full turgor)[(6.5d), (6.5e), $i = rt, sh$]	kg water m^{-2}
$\theta_{rt}, \theta_{sh}; \theta_{so}$	Relative water contents in root, shoot; soil [(6.5d), $i = rt, sh$; (6.2f)]	
$\psi_i, \psi_{i,os}, \psi_{i,pr}$ $i = rt, sh$	Water potential: total, osmotic and pressure components in root and shoot [(6.5c), (6.5a), (6.5b)]	J (kg water)$^{-1}$
ψ_{so}	Soil water potential (6.2g)	J (kg water)$^{-1}$
Animal variables		
$I_{DM,pl \to an,gnd}$	Plant DM input to animal [(6.4c), (4.2b)]	kg total DM m^{-2} day^{-1}
Environmental variables		
j_{NetR}	Instantaneous net radiation flux incident on canopy [(6.6b), (7.5b)]	J m^{-2} day^{-1}
r_{ain}	Rainfall [(6.2b), Section 7.5]	m day^{-1}
$r_{el\text{-}hum}$	Relative humidity [(6.6d); (7.5g)]	
T_{air}, T_{soil}	Air, soil temperatures [(6.5a), (6.7a), (7.5e) and Section 7.5]	°C
w_{ind}	Wind speed [(6.6c), Table 7.2]	m s^{-1}
Plant variables		
d_{rt}	Root depth [(6.2e); (3.9c)]	m
f_{harv}	Fraction of shoot material removed in a harvest [(6.4d), (3.8b)]	
g_s	Stomatal conductance [(6.6e), (3.2u)]	m s^{-1}
h_{can}	Canopy height [(6.6c), (3.8a)]	m
L_{AI}	Leaf area index [(6.2c), (6.6e); (3.1b)]	m^2 leaf (m^2 ground)$^{-1}$
$M_{S,rt}, M_{S,sh}$	Substrate DM of root, shoot [(6.5a), (3.1e)]	kg substrate DM m^{-2}
M_{sh}	Total shoot DM [(6.4c), (3.1f)]	kg total DM m^{-2}
$M_{X,rt}, M_{X,sh}$	Structural DMs of root, shoot [(6.2i), (6.5b), (3.1a)]	kg structural DM m^{-2}
$p_{ulse,harv}$	Harvesting pulse function [(6.4d), (7.6d)]	day^{-1}
$r_{DM:FM}$	Ratio of dry mass to fresh mass (6.5f)	
ρ_{rt}	Root density [(6.2i), (3.9a)]	kg structural DM (m^3 soil)$^{-1}$

The net flux into the soil water pool is

$$I_{W,so} = \rho_W r_{ain} - O_{W,rain\text{-}int \to atm} - O_{W,so \to rt},$$
$$\rho_W = 1000 \text{ kg water m}^{-3}. \tag{6.2b}$$

ρ_W is the density of water. r_{ain} (m day^{-1}) is the daily rainfall (Section 7.5 and Table 7.3). The first output term, $O_{W,rain\text{-}int \to atm}$ [Eqn (6.2c)], is rainfall intercepted by the foliage and evaporating to the atmosphere. The second output term, $O_{W,so \to rt}$ [Eqn (6.2d)], is root uptake of water.

6.2.2 Outputs

There are mechanistic treatments of rainfall interception and evaporation (e.g. Cienciala *et al.*, 1994; Leuning *et al.*, 1994; Hörmann *et al.*, 1996). However, these can only be applied to rainfall events and are not suited to the present context. We use a simple expression for this flux, namely

$$O_{W,rain\text{-}int \to atm} = \rho_W \, r_{ain} f_{rain,i} f_{rain,i,e,m},$$
$$f_{rain,i} = 1 - e^{-k_{can} L_{AI}},$$
$$f_{rain,i,e,m = 1,\ldots,12} = 0.1,\ 0.15,\ 0.2,\ 0.3,\ 0.35,\ 0.4,\ 0.4,\ 0.35,\ 0.3,\ 0.2,\ 0.15,\ 0.1, \tag{6.2c}$$
$$k_{can} = 0.5 \text{ m}^2 \text{ ground (m}^2 \text{ leaf)}^{-1}.$$

ρ_W is the density of water [Eqn (6.2b)]. r_{ain} (m day^{-1}) is the daily rainfall (Section 7.5). $f_{rain,i}$ is the fraction of rain intercepted by the canopy; this is assumed to depend on the canopy extinction coefficient k_{can} [Eqn (3.2b)] and leaf area index L_{AI} [Eqn (3.1b)]. The fraction of the intercepted rainfall that evaporates is given a numerical value for each month, $f_{rain,i,e,m}$, m = 1 (Jan), ... , 12. These numbers depend on the intensity and duration of rainfall events as well as the evaporative demand in each month.

The flux of water from soil to root, $O_{W,so \to rt}$ (kg water m^{-2} day^{-1}), is assumed to be proportional to a water potential difference divided by a resistance, with

$$O_{W,so \to rt} = \frac{86{,}400 \left(\psi_{so} - \psi_{rt} \right)}{r_{W,so \to rt}}. \tag{6.2d}$$

86,400 converts units of s^{-1} to day^{-1}. The soil and root water potentials, ψ_{so} and ψ_{rt}, are calculated in Eqns (6.2g) and (6.5c) (with $i = rt$). The resistance for transport from soil to root $r_{W,so \to rt}$ is given in Eqn (6.2i).

Direct evaporation of water from the soil is ignored. For crops with closed canopies and abundant litter such as grassland and forest, this term is mostly small. It would also be difficult to calculate in a model that treats soil water as a single compartment and ignores rainfall events.

6.2.3 Leaching

The rate constant for leaching, k_{leach} (day^{-1}), is obtained by dividing the depth of water which drains per day (m day^{-1}) by the root depth, d_{rt} [m, Eqn (3.9c)]. The depth of water draining per day equals the mass output of water to drainage, $O_{W,so\to drain}$ [kg m^{-2} day^{-1}, Eqn (6.2a)], divided by the density of water ρ_W [kg m^{-3}, Eqn (6.2b)]. Therefore

$$k_{leach} = \frac{O_{W,so\to drain}}{\rho_W}\frac{1}{d_{rt}}. \tag{6.2e}$$

This expression probably overestimates the leaching rate constant, which is used to calculate C and N leaching in Eqns (5.3b) and (5.5e).

There is an unresolved problem concerning the calculation of leaching, which seems not to be satisfactorily soluble using a single compartment soil water scheme. In Eqn (6.2e), root depth d_{rt} appears in the denominator. This is typically 0.2 m. An alternative would be to use soil depth d_{soil} [Eqn (6.2f)] which is 1 m, rather than the root depth of 0.2 m. This would make the rate constant for leaching five times smaller. In the first case it is assumed that the soil nitrate (say) of 0.002 kg N m^{-2} is distributed over the root depth of 0.2 m; in the second case it is distributed over the notional soil depth of 1 m, leading to a lower volume concentration of N and less leaching. Notwithstanding this difficulty, the model simulates leaching fluxes satisfactorily. Surface runoff is ignored.

6.3 Soil Water Relations and Soil–Plant Resistance

The soil relative water content θ_{so} [m^3 water (m^3 soil)$^{-1}$] depends on the state variable W_{so} [kg water (m^2 ground)$^{-1}$] according to

$$\theta_{so} = \frac{W_{so}}{\rho_W d_{soil}}, \qquad d_{soil} = 1\ \text{m}, \tag{6.2f}$$

ρ_W is the density of water [Eqn (6.2b)]. d_{soil} is a fixed soil depth. θ_{so} has a maximum value of $\theta_{so,max}$ [field capacity; see Eqn (6.2a)].

The soil water potential ψ_{so} [J (kg water)$^{-1}$] is assumed to depend on soil relative water content θ_{so} according to a three-parameter equation proposed by Gregson *et al.* (1987)

$$\psi_{so} = \psi_{so,max}\left(\frac{\theta_{so,max}}{\theta_{so}}\right)^{q_{\psi,soW}},$$
$$\psi_{so,max} = -10\ \text{J (kg water)}^{-1}, \qquad q_{\psi,soW} = 5\ (2\ \text{for sand, 18 for clay}). \tag{6.2g}$$

$q_{\psi,soW}$ is a dimensionless constant characteristic of soil type: at the soil type extremes, $b = 2$ for a sandy soil, and $b = 18$ for a clay soil. $\psi_{so,max}$ is the water potential when soil relative water content θ_{so} has its maximum value of $\theta_{so,max}$ [field capacity; Eqn (6.2a)]. A value of -10 J kg^{-1} corresponds to a soil suction of 10 kPa (multiply by -1000).

Eqn (6.2g) is illustrated in Fig. 6.2a. These responses can be compared with Kramer (1983, p. 95) and Rowell (1994, pp. 82–83). Eqn (6.2g) can be 'derived' assuming identical circular pores with radius changing with axial distance (Thornley and Johnson, 1990, pp. 425–429). Gregson *et al.*'s equation is simple and easy to apply, but it does not contain sufficient parameters to describe some soils

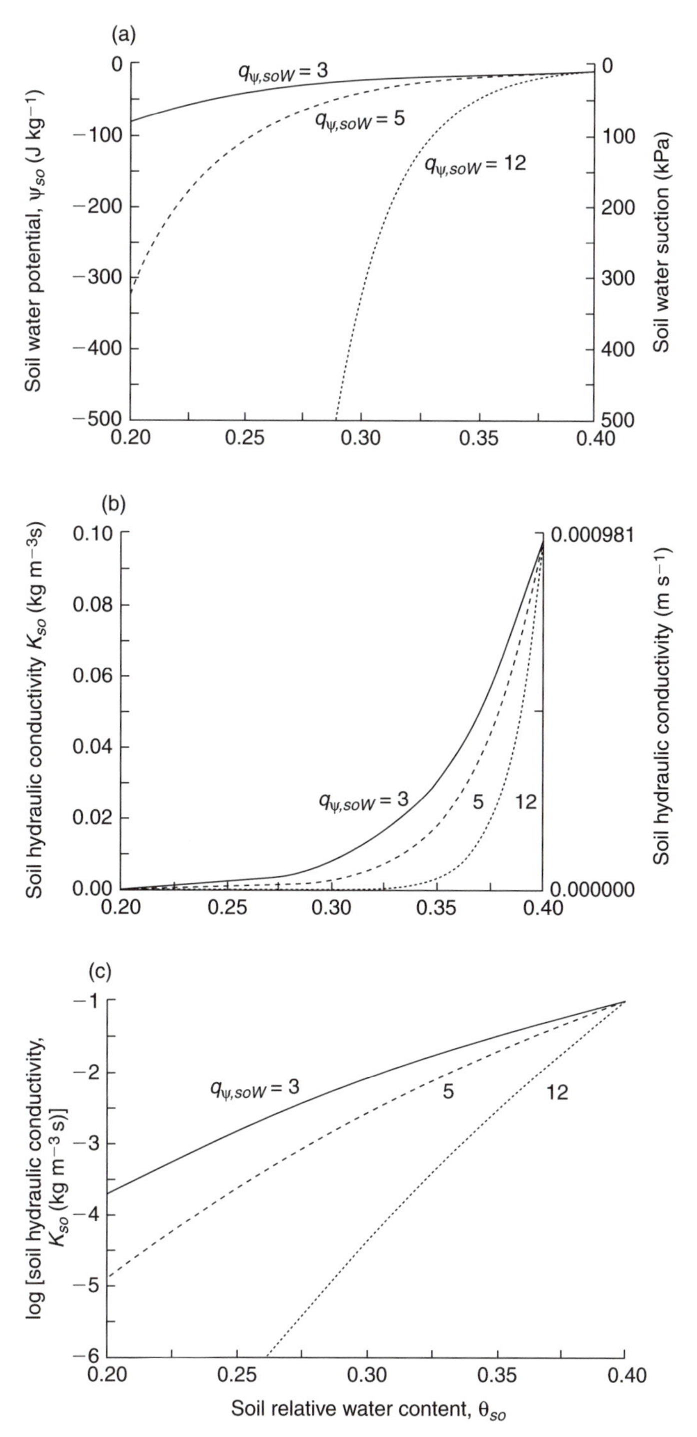

Fig. 6.2. Soil water potential and soil hydraulic conductivity. Responses to soil water content are illustrated, using Eqns (6.2g) and (6.2h) with the parameters given or as defined on the graph.

realistically, and it fails when the soil is close to saturation. More complex approaches have been developed, e.g. as described by van Genuchten (1980) with a four-parameter model, and Wilson *et al.* (1992) who consider a three-component pore-size distribution with macro-, meso- and micro-pores. I am indebted to Jon Arah (personal communication) for his thoughts on this topic.

The hydraulic conductivity of the soil is required for the calculation of the soil to root water transport resistance $r_{W,so\to rt}$ used in Eqn (6.2d). The hydraulic conductivity is denoted by K_{so} with units of kg m^{-3} s. These units describe the mass of water (kg) moved per unit time (s^{-1}) per unit area (m^{-2}) per unit gradient in water potential (J kg^{-1} $m^{-1})^{-1}$. These units of kg m^{-3} s are related to units of m s^{-1} which are sometimes used for hydraulic conductivity by a factor of g/ρ_W [m s^{-2}/(kg m^{-3}) = kg^{-1} m^4 s^{-2}], where g is the acceleration due to gravity and ρ_W is density of water. Thus 1 kg m^{-3} s^{-1} × 9.81 m s^{-2} / (1000 kg m^{-3}) = 0.00981 m s^{-1}, approximately 1 cm s^{-1}. These latter units have the possible disadvantage that they depend on the acceleration due to gravity and the density of water. It is assumed that hydraulic conductivity depends on soil relative water content according to

$$K_{so} = K_{so,max}\left(\frac{\theta_{so}}{\theta_{so,max}}\right)^{(2q_{\varphi,soW}+3)},$$
$$K_{so,max} = 0.1 \text{ kg m}^{-3} \text{ s.} \qquad (6.2h)$$

This equation is due to Campbell (1974), and is similar in form to Eqn (6.2g) where the exponent $q_{\psi,soW}$ is defined. The saturating value of K_{so} when θ_{so} has its maximum value of $\theta_{so,max}$ [Eqn (6.2a)] is $K_{so,max}$. Eqn (6.2h) describes a very strong dependence of hydraulic conductivity on soil relative water content. For a single compartment soil model that ignores heterogeneity, the dependence given by Campbell's equation may be too severe. Eqn (6.2h) is drawn in Figs 6.2b, c. These curves can be compared with Kramer (1983, p. 78) and Rowell (1994, pp. 82–83). Eqn (6.2h) is acceptable in a generic study but is not usually sufficiently realistic for a particular application. Jensen *et al.* (1993) found Eqn (6.2h) to be satifactory for a coarse textured soil where the exponent $q_{\psi,soW}$ is between three and four. A more elaborate approach to soil hydraulic conductivity is used by Hillel and van Bavel (1976), who employ a formula from Jackson (1972), based on the work of Childs and Collis-George (1950). This approach sums contributions to hydraulic conductivity arising from the water content increments into which the actual water content can be broken down. Talsma (1985) reports the application of Eqn (6.2h) to five different soils, using laboratory and field measurements. In the field studies, the equation performed poorly for sandy loams and light clays.

We assume that the resistance to water flow between soil and plant, $r_{W,so\text{-}rt}$ (kg^{-1} m^4 s^{-1}), used in Eqn (6.2d) for calculating the water flux from soil to root, is

$$r_{W,so\text{-}rt} = \frac{c_{W,so\text{-}rs}\rho_{rt}}{K_{so}M_{X,rt}} + \frac{c_{W,rs\text{-}rt}}{\rho_{rt}}\left(\frac{M_{X,rt} + K_{W,rs\text{-}rt}}{M_{X,rt}}\right),$$
$$c_{W,so\text{-}rs} = 80 \text{ m}^2, \quad c_{W,rs\text{-}rt} = 0.5\times10^6 \text{ m s}^{-1}, \qquad (6.2i)$$
$$K_{W,rs\text{-}rt} = 1 \text{ kg structural DM m}^{-2}.$$

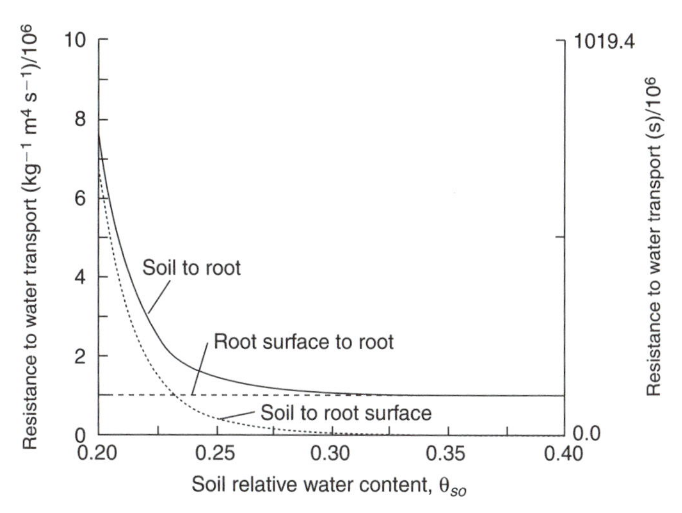

Fig. 6.3. Resistance between soil and root. Eqn (6.2i) is plotted with the parameters given in Eqns (6.2g), (6.2h) and (6.2i). Root density and root structural mass are constant with $\rho_{rt} = 1$ kg structural DM m^{-3} and $M_{X,rt} = 1$ kg structural DM m^{-2}. The resistance scale on the right is obtained by multiplying by ρ_W/g where $\rho_W = 1000$ kg m^{-3} is the density of water and $g = 9.81$ m s^{-2} is the acceleration due to gravity [see discussion before Eqn (6.2h)]. With the right-hand-scale units, the volume flux of water (m^3 m^{-2} s^{-1}) equals the head of water (m) divided by the resistance (s).

The first term represents the resistance between bulk soil (*so*) and root surface (*rs*). Increasing root structural mass $M_{X,rt}$ [Eqn (3.1a)] at a constant root density ρ_{rt} [Eqn (3.9a)], or decreasing root density ρ_{rt} at a constant root structural mass $M_{X,rt}$, both give a more extensive root system, which reduces the resistance between bulk soil and root surface. Increasing soil hydraulic conductivity [K_{so}, Eqn (6.2h)] also reduces this resistance. Parameter $c_{W,so\text{-}rs}$ scales the first term. The second term represents the resistance between the root surface and the root as a whole. It is independent of soil relative water content, and is scaled by $c_{W,rs\text{-}rt}$. Increasing root mass gives an asymptote. Decreasing root density at constant root mass gives longer pathways and higher resistance. Eqn (6.2i) with constant root density ρ_{rt} and root structural mass $M_{X,rt}$ is illustrated in Fig. 6.3. In a simulation of the whole system, root density and structural mass will change as the soil dries down. Note the rapid increase in soil to root surface resistance as soil relative water content decreases, corresponding to changing soil hydraulic conductivity (Fig. 6.2c with $q_{\psi soW} = 5$). Eqn (6.2i) is unlikely to be valid for low values of root mass which are not important in the present application. The equation is constructed to make use of the variables available from the plant submodel. It is discussed further by Thornley and Johnson (1990, pp. 432–437).

Taylor and Klepper (1978) have discussed the difficulties in calculating the water flux from soil to plant, most of which are still unresolved. Gardner (1991), considering the typical current model of water uptake, states 'Such models have sufficient number of parameters that they can be fitted to data reasonably well. Most water uptake patterns … reveal, remarkable similarities … . This similarity is not predictable with current models.' I have found that Eqn (6.2i) works quite well as an empirical expression, and it provides a satisfactory interface with the plant submodel variables.

6.4 Root Water Pool, W_{rt}

The differential equation for this pool is [kg water (m^2 ground)$^{-1}$ day^{-1}]

$$\frac{dW_{rt}}{dt} = I_{W,so \to rt} - O_{W,rt \to sh}, \qquad W_{rt}(t=0) = 1 \text{ kg water m}^{-2}. \tag{6.3a}$$

The initial value of the root water pool depends on the size of the root system; for the root initial values of Table 3.1, the value of 1 kg water m^{-2} gives an initial total dry mass to fresh mass ratio of 0.19 (this is sometimes called the dry matter or DM concentration). The DM concentration varies greatly diurnally, seasonally, and with position in the sward (e.g. McDonald, 1981, p. 116; see also Fig. 6.4b). However, it should be said that a model such as this one, which does not take account of cytoplasmic and non-cytoplasmic water, or the compartmentation of water and solutes in organelles, is unlikely to make realistic predictions of dry matter concentration.

The input to this pool is equal to the output [Eqn (6.2d)] from the soil water pool:

$$I_{W,so \to rt} = O_{W,so \to rt}. \tag{6.3b}$$

The flux of water from root to shoot, $O_{W,rt \to sh}$, is

$$O_{W,rt \to sh} = g_{W,rt-sh}\left(\psi_{rt} - \psi_{sh}\right), \quad g_{W,rt-sh} = c_{WT,pl}\frac{M_{X,rt}\, M_{X,sh}}{M_{X,rt} + M_{X,sh}},$$
$$c_{WT,pl} = 0.005 \text{ kg water (kg structural DM)}^{-1} \text{ [J (kg water)}^{-1}]^{-1} \text{ day}^{-1}. \tag{6.3c}$$

The flux is driven by the water potential difference, the root and shoot water potentials, ψ_{rt} and ψ_{sh}, are calculated below [Eqn (6.5c)]. The conductivity between root and shoot, $g_{W,rt\text{-}sh}$, scales with the structural dry mass (DM) of the plant, but in such a way that the smaller of the root or shoot DMs has the major effect. We have found that this expression is more satisfactory than the simpler equation $g_{W,rt \to sh} = c_{WT,pl}(M_{X,rt} + M_{X,sh})$ (see Section 3.3.3 for an interpretation of this equation). It is assumed that neither temperature nor plant water status affects this conductivity. $c_{TW,pl}$ is a constant.

6.5 Shoot Water Pool, W_{sh}

The differential equation for the shoot water pool is

$$\frac{dW_{sh}}{dt} = I_{W,rt \to sh} - O_{W,sh \to atm} - O_{W,sh \to an} - O_{W,sh \to hv},$$
$$W_{sh}(t=0) = 1 \text{ kg water m}^{-2}. \tag{6.4a}$$

Units are kg water (m^2 ground)$^{-1}$ day^{-1}. The initial value of W_{sh} depends on the size of the shoot system; for the shoot initial values of Table 3.1, the value of 1 kg water m^{-2} gives a total dry mass to fresh mass ratio of 0.19 [see discussion after Eqn (6.3a)].

The input to the shoot water pool is equal to the output [Eqn (6.3c)] from the root water pool:

$$I_{W,rt \to sh} = O_{W,rt \to sh}. \tag{6.4b}$$

The first output term $O_{W,sh \to atm}$ is plant evapotranspiration. This is calculated in Eqn (6.6d).

The output second term $O_{W,sh \to an}$ is the flux of water to the grazing animals. This is

$$O_{W,sh \to an} = \frac{I_{DM,pl \to an,gnd}}{M_{sh}} W_{sh}. \tag{6.4c}$$

$I_{DM,pl \to am,gnd}$ is the plant dry matter (DM) input to the animal per unit area of ground, calculated in Eqn (4.2b). Total shoot DM, M_{sh}, is given in Eqn (3.1f). It is assumed that shoot water, W_{sh}, is uniformly distributed over the plant shoot.

The last output term in Eqn (6.4a) is $O_{W,sh \to hv}$, the loss of water when shoot material is harvested. This is given by

$$O_{W,sh \to hv} = p_{ulse,harv} f_{harv} W_{sh}. \tag{6.4d}$$

The pulse harvesting function $p_{ulse,harv}$ is determined by management and is given in Eqn (7.6d). It is assumed that harvesting occurs over a period of 1 day, so that $p_{ulse,harv}$ is unity on the day that harvesting occurs. The dimensions of $p_{ulse,harv}$ are day^{-1}. f_{harv} is the fraction of the shoot material removed in a harvest; it is defined in Eqn (3.8b), and is determined at the beginning of the day on which harvesting occurs.

6.6 Plant Water Relations

The plant water state variables are W_{rt}, the mass of water in the root, and W_{sh}, the mass of water in the shoot, and are denoted by W_i, i = *rt*, *sh*. In this section the osmotic potential $\psi_{i,os}$, the pressure potential $\psi_{i,pr}$, total tissue water potential ψ_i, and the relative water content, θ_i for i = *rt* (root), *sh* (shoot), are derived. Energy units of J kg^{-1} are used for water potential and its components (1 J $kg^{-1} \equiv$ 1 kPa $\equiv$ 0.01 bar; see Section 6.1 and Appendix). Historically, it appears that Tang and Wang (1941) first presented an analysis of plant water relationships of the type outlined below.

It is assumed that tissue water is all cytoplasmic: in herbaceous species cytoplasmic water is about 85% of total water; in xerophytes this could be 30% (Acock and Grange, 1981; also Grange, R.I., personal communication). The osmotic potential is given by

$$\psi_{i,os} = -\frac{R_{gas}(T_i + 273) f_{S,os-ac} M_{S,i}}{\mu_S W_i}, \tag{6.5a}$$

$$R_{gas} = 8314 \text{ J (kg mole)}^{-1} \text{K}^{-1}, \quad f_{S,os-ac} = 2, \mu_S = 20 \text{ kg (kg mole)}^{-1}.$$

This is the Van't Hoff equation (e.g. Maron and Prutton, 1965, p. 329; Nobel, 1991, p. 72). The parameters of Eqn (6.5a) are: R_{gas}, the gas constant ($10^3 \times$ usual value; Section 6.1); $f_{S,os\text{-}ac}$, the fraction of the substrate material $M_{S,i}$ which is osmotically active; and μ_S, the mean molar mass of this substrate material [= 20 g (SI mole)$^{-1}$; Section 6.1]. T_i denotes the temperature of the root ($i = rt$) assumed equal to soil temperature T_{soil}, or that of the shoot ($i = sh$), assumed equal to air temperature T_{air}. Air temperature T_{air} is calculated in Eqn (7.5e). For soil temperature T_{soil} see Section 7.5. $M_{S,i}$ ($i = rt, sh$) is the mass of substrates in the root, shoot, given in Eqns (3.1e).

The pressure potential is obtained from the mass of water in the tissue (W_i) and the structural dry mass ($M_{X,i}$) [Eqns (3.1a)], with

$$\psi_{i,pr} = \varepsilon\left(\frac{c_{\psi,pr}W_i}{M_{X,i}} - 1\right)\Big/\rho_W, \qquad \varepsilon = 2\times10^6 \text{ Pa}, \quad c_{\psi,pr} = 0.2 \text{ kg structural DM (kg water)}^{-1}. \tag{6.5b}$$

Here ε is a rigidity parameter similar to a Young's modulus (Jones, 1992, p. 79); the value assumed here is consistent with the extensibility measurements on *Phaseolus vulgaris* leaves reported by van Volkenburgh *et al.* (1983). The constant $c_{\psi,pr}$ relates structural content to water content at zero cell wall extension, and is regarded as an adjustable parameter. The density of water is ρ_W [Eqn (6.2b)]. This equation assumes an instantaneous linear relationship between stress and strain, without hysteresis (Passioura, 1994). Nobel (1991, pp. 102–103), discusses the *kinetics* of cell volume change, a topic outside the scope of the current study. If the shoot pressure potential $\psi_{sh,pr}$ falls to or below -500 J kg^{-1} (-0.5 MPa), it is assumed that the plant has wilted and the simulation is stopped.

The total water potential, ψ_i, is the sum of the osmotic and pressure components, with

$$\psi_i = \psi_{i,os} + \psi_{i,pr}. \tag{6.5c}$$

Gravity is ignored for a grass crop, but not when this water submodel is applied to trees.

The tissue relative water content, θ_i, is defined by

$$\theta_i = \frac{W_i}{W_i(\psi_i = 0)}. \tag{6.5d}$$

The numerator is the actual water content of the tissue at the prevailing value of the tissue water potential [Eqn (6.5c)], and the denominator is the water content when the tissue water potential is zero. Using Eqns (6.5a) and (6.5b) to substitute for $\psi_{i,os}$ and $\psi_{i,pr}$ in Eqn (6.5c), and re-arranging, gives a quadratic equation in W_i:

$$0 = \left(\frac{\varepsilon c_{\psi,pr}}{\rho_W M_{X,i}}\right)W_i^2 - \left(\psi_i + \frac{\varepsilon}{\rho_W}\right)W_i - \frac{R_{gas}(T_i + 273)f_{S,os\text{-}ac}M_{S,i}}{\mu_S}. \tag{6.5e}$$

This equation relates water content of the tissue (W_i) to water potential (ψ_i). The variables for the tissue structural and substrate (storage) dry masses ($M_{X,i}$ and $M_{S,i}$)

appear in the coefficients. It can be solved for water content W_i in terms of water potential ψ_i. In particular, by setting $\psi_i = 0$, it can be solved for $W_i(\psi_i = 0)$, the water content at zero water potential. This is then substituted into Eqn (6.5d) to give relative water content θ_i. These equations lead to a relationship between tissue water potential and tissue relative water content that depends on plant variables such as the ratio of substrate to structure.

The dry mass concentration, which is the ratio of dry mass to total tissue mass, is

$$r_{DM:FM} = \frac{M_{X,i} + M_{S,i}}{M_{X,i} + M_{S,i} + W_i}. \tag{6.5f}$$

The relationship between tissue water potential and tissue relative water content is shown in Fig. 6.4 for two values of plant tissue substrate concentration. It can be seen that for a given tissue relative water content θ_i, higher substrate levels (dashed lines) decrease the total water potential (ψ_i) and its osmotic component ($\psi_{i,os}$), but increase the pressure component of water potential ($\psi_{i,pr}$). Figure 6.4a can be compared with a typical measured response given by Jones (1992, Fig. 4.3, p. 78). At 5% osmoticum the pressure potential is zero at $\theta_i = 0.58$, whereas at 10% osmoticum this occurs at $\theta_i = 0.47$. The dry mass to fresh mass fraction of the tissue is also decreased by higher substrate concentrations as more water enters the tissue. This is illustrated in Fig. 6.4b. The water contents of vegetative plant tissues reviewed by Kramer (1983, p. 6, pp. 349–352) vary widely between about 60% and 95% with substantial diurnal changes.

The plant wilts when the shoot pressure potential is -500 J kg^{-1} (-0.5 MPa) (by assumption). Increased osmotic effects (e.g. $f_{S,os\text{-}ac}$, $M_{S,sh}$ larger; μ_S smaller) cause the mass of water in the shoot at full turgor [$W_{sh}(\psi_{sh} = 0)$] to increase [Eqn (6.5e); Fig. 6.4b]; wilting then occurs at lower values of relative water content [Fig. 6.4a, Eqn (6.5d)]. Thus, our formulation of plant water relations includes a degree of osmotic adjustment, enabling the plant to adapt partially to drought conditions by increasing the magnitude of the osmotic component of the water potential (Jones, 1988, pp. 222–225, Fig. 6.8). However, Kramer defines osmotic adjustment as 'a decrease in osmotic potential greater than can be explained by solute concentration', and points out that the wide differences in tolerance of dehydration are not understood (Kramer, 1983, pp. 403, 404). In the present pasture model one of the responses of the plant to increasing atmospheric CO_2 is to increase substrate levels which makes the plant better able to withstand water stress. Although increased atmospheric CO_2 decreases fully open stomatal conductance, the higher values of shoot relative water content cause the stomata to be more open [Eqn (3.2u)], so that actual conductance and transpiration can be increased.

6.7 Plant Evapotranspiration, $O_{W,sh \rightarrow atm}$

This instantaneous flux has units of kg water (m^2 ground)$^{-1}$ day^{-1}, and is calculated by means of the Penman–Monteith equation (Penman, 1948; Monteith, 1965).

The net radiation flux absorbed by the canopy drives plant evapotranspiration.

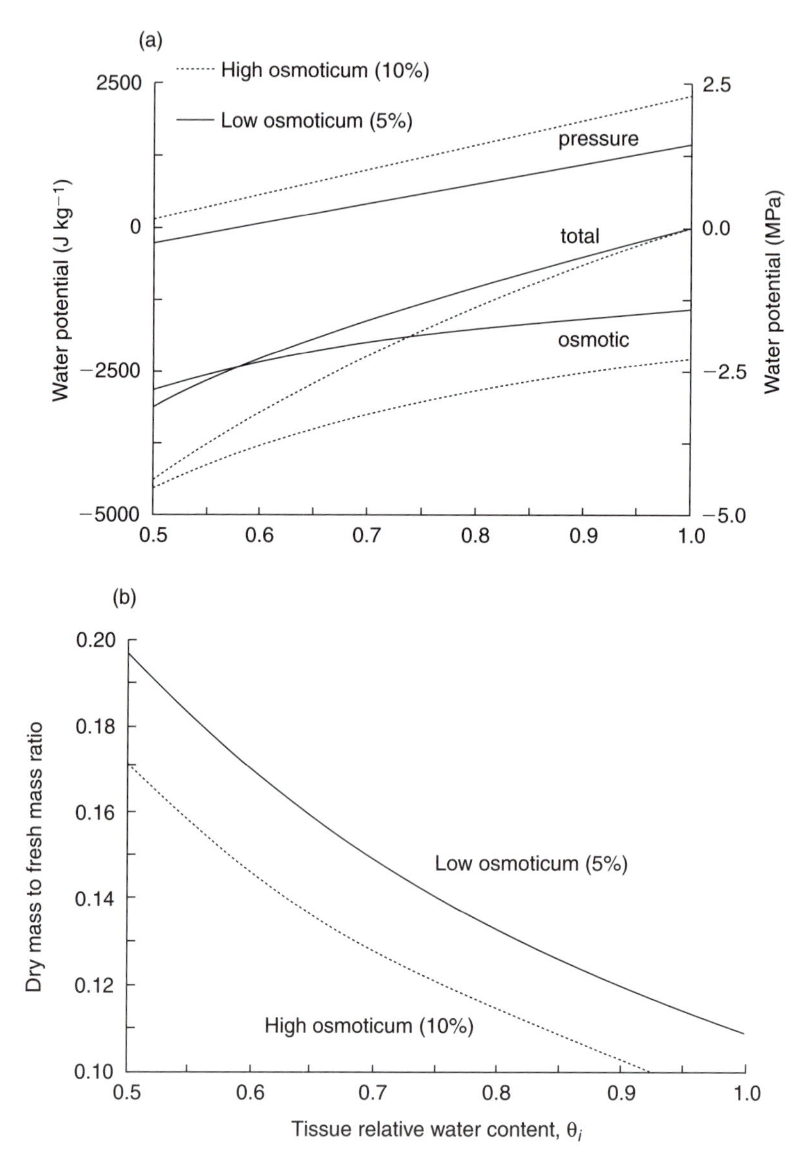

Fig. 6.4. Plant tissue water relations. (a) Components of tissue water potential [Eqns (6.5a), (6.5b), (6.5c)]. (b) Dry mass to fresh mass fraction [Eqn (6.5f)]. Tissue water content at full turgor $W_i(\psi_i = 0)$ is determined by solving Eqn (6.5e) with $\psi_i = 0$, and $M_{X,i} = 1$ kg structural DM m^{-2} and $M_{S,i} = 0.05$ (low osmoticum) or 0.1 (high osmoticum) kg substrate matter m^{-2}. Other parameters are as given in Eqns (6.5a), (6.5b) and (6.2b). Tissue relative water content θ_i is varied by taking values of tissue water W_i less than the full turgor value $W_i(\psi_i = 0)$, and solving Eqns (6.5a), (6.5b) and (6.5c) for the components of water potential and total water potential, and applying Eqn (6.5f) to give the dry mass concentration. 1 J $kg^{-1} \equiv 1$ kPa $\equiv 0.01$ bar.

It is assumed that the fraction of the incident net radiation absorbed by the canopy is the same as the fraction of PAR absorbed, $f_{PAR,abscan}$, given by

$$f_{PAR,abs-can} = \left(\frac{1 - c_{r,leaf} - \chi_{leaf}}{1 - \chi_{leaf}}\right)\left(1 - e^{-k_{can} L_{AI}}\right), \tag{6.6a}$$

$$c_{r,leaf} = 0.15, \quad \chi_{leaf} = 0.1.$$

$c_{r,leaf}$ is the leaf reflection coefficient. χ_{leaf} and k_{can} are the coefficients of leaf transmission and canopy extinction [Eqn (6.2c), also Eqn (3.2b)]. L_{AI} is leaf area index [Eqn (3.1b)]. This equation is a more accurate form of the equation $f_{PAR,abs-can} = 1 - \exp(-k_{can}L_{AI})$, although the numerical change is small. It is derived by Thornley and Johnson (1990, p. 204).

The net radiation flux absorbed by the canopy, $j_{NetR,abs-can}$ (J m^{-2} day^{-1}), is

$$j_{NetR,abs-can} = f_{PAR,abs-can} j_{NetR}. \tag{6.6b}$$

j_{NetR} (J m^{-2} day^{-1}) is the instantaneous net radiation in per day units [Eqn (7.5b)].

The boundary layer conductance, g_{blcon} (m s^{-1}), is (after Campbell, 1977, p. 138)

$$g_{blcon} = \frac{k_{vK}{}^2 w_{ind}}{\ln\left[\left(h_{ref} + \zeta_{heat,vap} - d\right)/\zeta_{heat,vap}\right] \ln\left[\left(h_{ref} + \zeta_{mntm} - d\right)/\zeta_{mntmp}\right]}, \tag{6.6c}$$

$$k_{vK} = 0.4, \quad \zeta_{heat,vap} = 0.026\,h_{can}, \quad \zeta_{mntm} = 0.13\,h_{can}, \quad d = 0.77\,h_{can}.$$

h_{ref} (m) is the reference height at which the wind speed, w_{ind} (m s^{-1}), is measured (Table 7.2). h_{can} (m) is canopy height [Eqn (3.8a)]. k_{vK} (dimensionless) is von Karman's constant; $\zeta_{heat,vap}$ (m) and ζ_{mntm} (m) are roughness parameters for heat and vapour exchange, and turbulent momentum transport; and d (m) is the zero plane displacement.

Neglecting the soil heat flux, the instantaneous water vapour flux from transpiration, $O_{W,sh\to atm}$ [kg water (m^2 ground)$^{-1}$ day^{-1}], is given by

$$O_{W,sh\to atm} = \frac{s j_{NetR,abs-can} + 86{,}400 \lambda g_{blcon}\left(\rho_{swv} - \rho_{wv}\right)}{\lambda\left[s + \gamma\left(1 + g_{blcon}/g_{can}\right)\right]}, \tag{6.6d}$$

$$\rho_{wv} = r_{el\text{-}hum}\rho_{swv}.$$

The parameters of this equation are as follows. $s = d\rho_{swv}/dT$ (kg m^{-3} K^{-1}) is the gradient of the saturated water vapour density, ρ_{swv} (kg m^{-3}), with respect to temperature (Table 6.2); $j_{NetR,abs-can}$ is the net radiation absorbed [Eqn (6.6b)]; 86,400 s day^{-1} is to convert boundary layer conductance [Eqn (6.6c)], g_{blcon} (m s^{-1}), to per day units; λ (J kg^{-1}) is the latent heat of evaporation of water (Table 6.2); ρ_{wv} (kg m^{-3}) is the actual water vapour density, obtained by multiplying relative humidity [$r_{el\text{-}hum}$, Eqn (7.5g)] by the saturated vapour density (ρ_{swv}) at air temperature T_{air} as given (Table 6.2); γ (kg m^{-3} K^{-1}) is the psychrometric parameter (Table 6.2); g_{blcon} (m s^{-1}) is the boundary layer conductance [Eqn (6.6c)]; and g_{can} (m s^{-1}) is the canopy conductance [Eqn (6.6e)]. The quantities, s, ρ_{swv}, λ and γ all depend on temperature. They are widely tabulated, and in the present units, they are given in Table 6.2. Their values are calculated at the current instantaneous air temperature (T_{air}) by interpolation.

Table 6.2. Temperature-dependent quantities needed for the calculation of evapotranspiration [Eqn (6.6d)]. T is temperature. $s = d\rho_{swv}/dT$. λ is the latent heat of vaporization of water. ρ_{swv} is the density of saturated water vapour. γ is the psychrometric parameter for standard atmospheric pressure. After Campbell (1977, tables A1, A3) and Jones (1992, appendices 3 and 4).

T (°C)	s (10^{-3} kg m^{-3} K^{-1})	λ(MJ kg^{-1})	ρ_{swv} (10^{-3} kg m^{-3})	γ(10^{-3} kg m^{-3} K^{-1})
−5	0.24	2.51	3.7	0.527
0	0.33	2.50	4.9	0.521
5	0.45	2.49	6.8	0.515
10	0.60	2.48	9.4	0.509
15	0.78	2.47	12.8	0.503
20	1.01	2.45	17.3	0.495
25	1.30	2.44	23.1	0.488
30	1.65	2.43	30.4	0.482
35	2.07	2.42	39.7	0.478
40	2.57	2.41	51.2	0.474

The first term in Eqn (6.6d) is driven by radiation, and the second term by vapour density gradient and wind.

Canopy conductance g_{can} used in Eqn (6.6d) is calculated by

$$g_{can} = L_{AI} g_s, \tag{6.6e}$$

where stomatal conductance g_s is computed with Eqn (3.2u) and leaf area index L_{AI} with Eqn (3.1b).

6.8 Effects of Water Status on Plant and Soil Processes

The effects of water status on chemical and biochemical processes are only partially known (Dixon and Webb, 1964). Some relevant work can be found in studies of food preservation (e.g. Sinell, 1980) and silage (e.g. McDonald, 1981). In both these areas the degree of dryness has important effects on biochemical and microbiological activity. The theoretical work of Hearon (1952) on multi-enzyme systems points at the likely simplicity of overall behaviour. This is reinforced by data shown by Pirt (1975, Fig. 15.1): he gives the response of specific growth rate to water activity for three microbial species – the curves are qualitatively similar although quantitively very different. We assume that most biochemical plant and soil processes can be treated similarly. This excludes diffusion-like processes which are assumed to be independent of both water status and temperature (but see Thornley and Johnson, 1990, p. 134; Nobel, 1991, p. 145).

The concept of chemical activity is discussed in many textbooks (e.g. Maron and Prutton, 1965; Nobel, 1991). The rationale for using water activity is, apart from dependence on substrate concentration: biochemical rates $\propto$ enzyme activity $\propto$ enzyme conformation $\propto$ degree of hydration $\propto$ water activity. The chemical activities of water in the root, shoot and soil, $a_{W,i}$, $i = rt, sh, so$, are calculated from the water potential (ψ) and temperature (T) with

$$a_{W,sh} = \exp\left(\frac{\mu_W \Psi_{sh}}{R_{gas}\left(T_{air} + 273.15\right)}\right), \quad a_{W,rt} = \exp\left(\frac{\mu_W \Psi_{rt}}{R_{gas}\left(T_{soil} + 273.15\right)}\right),$$
$$a_{W,so} = \exp\left(\frac{\mu_W \Psi_{so}}{R_{gas}\left(T_{soil} + 273.15\right)}\right),$$
$$\mu_W = 18 \text{ kg (kg mole)}^{-1}. \tag{6.7a}$$

R_{gas} is the gas constant [Eqn (6.5a)]. μ_W is the molar mass of water.

The rate constants for root, shoot and soil processes are multiplied by empirical functions, $f_{W,i}$, $i = rt, sh, so$, which depend on water activity according to

$$f_{W,sh} = a_{W,sh}^{q_{W,pl}}, \quad f_{W,rt} = a_{W,rt}^{q_{W,pl}}, \quad f_{W,soil} = a_{W,soil}^{q_{W,soil}}.$$
$$q_{W,pl} = q_{W,soil} = 20. \tag{6.7b}$$

The same exponent q is applied to plant and soil processes.

Eqns (6.7a) and (6.7b) are plotted in Fig. 6.5, which can be compared with Fig. 15.1 of Pirt (1975), where a water activity of 0.95 multiplies the specific growth rate at water activity of 1.0 by 0 (*Salmonella*), 0.5 (*Staphyloccoccus*) and 1.4 (*Aspergillus*).

The water submodel does not simulate surface properties, and the surface is likely to be much drier than the bulk soil. Shoot water activity is assumed to be the best surrogate for surface water activity. The surface multiplier $f_{W,surf}$ is

$$f_{W,surf} = a_{W,sh}^{q_{W,surf}}, \qquad q_{W,surf} = 30. \tag{6.7c}$$

The surface processes are assigned a high exponent, because surface water activity seems likely to be lower than shoot water activity.

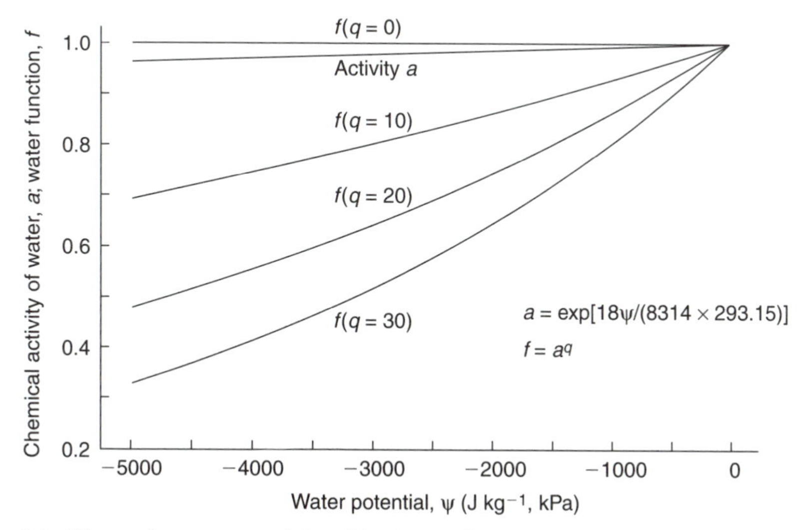

Fig. 6.5. Effects of water potential on biochemical rates. Eqn (6.7a) for water activity and Eqn (6.7b) for the effect on biochemical rates are plotted for different values of the exponent q.

There is still uncertainty concerning the effects of water stress on photosynthesis. There is evidence that electron transport may be little affected by water potential (Jones, 1992, p. 200) but ATP synthesis may be inhibited (Lawlor, 1987, p. 156). The leaf photosynthesis model of Chapter 3 has two principal parameters, the initial slope α and asymptote P_{max} of the light response curve [Eqns (3.2c), (3.2s), (3.2t)]. Water stress can close the stomata (g_s), as described in Eqn (3.2u); this reduces P_{max} as given by the factor $f_{gs,Pmax}$ in Eqn (3.2t). Jones (1988, p. 230) presents some unpublished data of Parsons and Woledge, in which increasing water deficit depresses the photosynthetic asymptote, but has a smaller effect on the initial slope. There are many indirect effects of water stress on photosynthesis, e.g. via specific leaf area, allocation. It is often argued that direct effects are not usually observed until the water stress is greater than that required to close the stomata (Jones, 1988, p. 232). In field-grown ryegrass most of the decrease in leaf photosynthesis in stressed swards was attributable to stomatal closure rather than a decrease in internal conductances (Jones *et al.*, 1980). We assume that there is an additional small effect of water stress on α and P_{max}, given by a multiplier $f_{W,ph}$ with

$$f_{W,ph} = a_{W,sh}^{q_{W,ph}}, \quad q_{W,ph} = 2. \tag{6.7d}$$

A much lower exponent is assumed for photosynthesis than for plant processes in general [Eqn (6.7b)].

The incremental specific leaf area, v_{sla} [m^2 leaf (kg structural dry matter)$^{-1}$], of Eqn (3.6c), is assumed to be proportional to shoot relative water content, θ_{sh} [Eqn (6.5d)], with parameter $c_{SLA,W} = 1$. This effect of water stress on leaf area expansion is in addition to the indirect effects of water status on shoot and root structural dry mass growth rates [Eqns (3.3c)] and shoot:root allocation [see also Zur and Jones (1981), Hsiao and Bradford (1983) and Ludlow (1987)]. However, Hughes (1974, p. 32) reports that Thomas and Norris observed that in ryegrass (S. 23), leaf extension is insensitive to ambient water stress under field conditions. Jones (1988, p. 220) presents data showing a strong relationship between leaf extension rate and leaf water potential.

It is assumed that the mineral N uptake by plant roots is decreased by the factor $f_{W,uN}$ where

$$f_{W,uN} = f_{W,rt}\left(\frac{\theta_{so}}{\theta_{so,max}}\right)^{q_{W,uN}}, \tag{6.7e}$$

$$q_{W,uN} = 3.$$

This equation allows for the effect of soil relative water content θ_{so} on N uptake; $\theta_{so,max}$ is the maximum value of θ_{so} [Eqns (6.2a), (6.2f)]. $q_{W,uN}$ is a parameter. This decrease is additional to the decrease of $f_{W,rt}$ [Eqn (6.7b)] which applies to the root biochemical machinery responsible for N uptake [Eqn (3.4h)].

Experiments on the effects of water supply on fertilizer responses in grassland have given variable results (e.g. Whitehead, 1995, pp. 243–245). This is probably due to the many interactions that can occur, through plant growth, leaching and denitrification.

See Eqn (5.5d) for the effect of soil relative water content on denitrification.

7 Environment and Management

7.1 Introduction

The model is driven by daily environmental and management data. There are three options for the daily data (Section 7.5): they can be generated internally assuming a seasonal sinusoidal variation for radiation, temperature, rainfall and relative humidity; or they can be read from an external file; or daily values may be calculated internally by interpolation from monthly means which are read in.

There are no stochastic weather generators in the program. Nonhebel (1994) examined the use of average instead of actual weather on the predicted yields of crop growth simulators, finding that overpredictions of 5–15% were obtained by using average data, although in water-limited conditions larger errors could occur. Arguably, the use of stochastic elements in plant ecosystem models at the present time will obscure the connections between environment, mechanisms and response which we are attempting to disentangle. The options provided (Section 7.5) of using average weather, actual weather, or some synthetic weather pattern, seem likely to be sufficient for most requirements.

A within-day time step of Δt = 1/64 day = 22.5 minutes is generally used in the model. Occasionally the simulation conditions require that Δt is halved to 1/128 day in order to give stable integration. Account is taken of the diurnal variation in air temperature (T_{air}), relative humidity ($r_{el\text{-}hum}$) and radiation (photosynthetically active radiation and net radiation) by calculating a diurnal time course from daily variables (Section 7.6).

Management processes, such as fertilizer application (Fig. 7.5), harvesting (Fig. 7.6) and stocking (Fig. 7.7) may be applied in any pattern during the year.

The default values of the environment given here and in the program are typical for central southern Britain. Sometimes these conditions are compared with a northern Britain upland grassland (e.g. Fig. 9.11), for which the appropriate values are given in {..} in Tables 7.2, 7.3 and 7.4.

7.2 Time

The independent variable denoting time is t (day). Many essential quantities are derived below from the independent variable t. For example, date and time of year are needed for keeping a record of events, but also because they define day length, sun angle etc., which affect the environment, e.g. the energy incident on the pasture. t is a continuous variable with $t \geq 0$ day. $t = 0$ day is at the beginning (0 h) of 1 January. Leap years are ignored. A simulation normally starts at $t = 0$ on 1 January. By setting a program parameter t_{ic} to 59 (say; see Table 7.1) at runtime (Section 10.2), the simulation can be made to begin on 1 March or on any date. The default value of $t_{ic} = 0$ for a 1 January start to the simulation.

The integer and decimal parts of t are

$$t_{int} = \text{integer part}(t), \qquad t_{dec} = \text{decimal part}(t). \tag{7.1a}$$

The hour of the day is

$$h_{our} = 24t_{dec}. \tag{7.1b}$$

Hence, if $t = 4.375$ day, then $t_{int} = 4$ day and $t_{dec} = 0.375$ day $\equiv$ 09.00 hour. $t = 4.375$ day occurs on the 5th day, so that t_{int} is the day number minus one.

Expressing the time variable in years, y_{ear} is

$$y_{ear} = \frac{t}{365}. \tag{7.1c}$$

Numbering the years, beginning with 1, then the year number, $y_{ear,i}$, is

$$y_{ear,i} = \text{integer part}\left(y_{ear}\right) + 1. \tag{7.1d}$$

Next number the days of the year, starting with 1 on 1 January, to give the Julian day number, i_{Julian} ($1 \leq i_{Julian} \leq 365$), with

$$i_{Julian} = t_{int} + 1 - 365\left(y_{ear,i} - 1\right). \tag{7.1e}$$

We use a recipe derived from a procedure proposed by Stuff and Dale (1973) to convert the Julian day number i_{Julian}, to the day of the month, $i_{day,mo}$, and the month, i_{month}. Note that $1 \leq i_{day,mo} \leq 31$ and $1 \leq i_{month} \leq 12$. Using x as a working variable, the conversion is achieved by

$$\begin{aligned} & x = i_{Julian} + 306 \\ & \text{if } x > 365 \text{ then } x = x - 365 \\ & j_{month} = \text{integer part } [(x + 91.3)/30.6] \\ & i_{day,mo} = \text{integer part } (x - 30.6 j_{month} + 92.3) \\ & \text{if } j_{month} \geq 13 \text{ then } i_{month} = j_{month} - 12 \\ & \qquad\qquad \text{else } i_{month} = j_{month}. \end{aligned} \tag{7.1f}$$

When using the model it is helpful to be able to convert easily from calendar date to the time variable t. Table 7.1 provides the means of doing this.

7.3 Day Length

Day length [τ_{day} (day)] is calculated from latitude [ϕ_{lat} (rad)] and the Julian day number [Eqn (7.1e)]. Thus, if it is wished to start a simulation on other than

Table 7.1. Correspondence between the (continuous) time variable t (day) and 0 h on the given calendar date. $t = 0$ day at 0 h on 1 January. Any integer × 365 can be added to the values of t. To obtain the Julian day number [Eqn (7.1e)] add 1 to the values of t. To start the program on 1 Mar, rather than the default of 1 Jan, set program parameter $t_{ic} = 59$ at runtime (Section 10.2).

t (day)	Date	t (day)	Date	t (day)	Date
0	1 Jan	150	31 May	272	30 Sep
30	31 Jan	151	1 Jun	273	1 Oct
31	1 Feb	171	21 Jun	303	31 Oct
58	28 Feb	180	30 Jun	304	1 Nov
59	1 Mar	181	1 Jul	333	30 Nov
79	21 Mar	211	31 Jul	334	1 Dec
89	31 Mar	212	1 Aug	354	21 Dec
90	1 Apr	242	31 Aug	364	31 Dec
119	30 Apr	243	1 Sep	365	1 Jan
120	1 May	263	21 Sep		

1 January, then it is important to tell the program the starting date, by setting t_{ic} to the appropriate value, as mentioned in Section 7.2.

The year angle [ϕ_{year} (rad)] is obtained from the Julian day number, i_{Julian}, by

$$\phi_{year} = \frac{(i_{Julian} - 80)}{365} 2\pi. \tag{7.2a}$$

The year angle is zero on 21 March ($i_{Julian} = 80$, see Table 7.1).

The solar declination [$\phi_{sol\text{-}dec}$ (rad)] is the angle between the earth's equatorial plane and the line joining the sun and the earth. At the equinoxes ($i_{Julian} = 80$, ≈ 263 day), when the year angle $\phi_{year} = 0$ and π rad (0 and 180°), $\phi_{sol\text{-}dec} = 0$ rad (0°). On 21 June ($i_{Julian} = 172$), $\phi_{year} = \frac{1}{2}\pi$ rad (90°), $\phi_{sol\text{-}dec} = +0.4093$ rad (23.45°), and on 21 December ($i_{Julian} = 355$), $\phi_{year} = (3/2)\pi$ rad (270°), $\phi_{sol\text{-}dec} = -0.4093$ rad (−23.45°). An empirical equation has been given by Usher (1970), for calculating solar declination from year angle:

$$\begin{aligned}\phi_{sol-dec} = \frac{2\pi}{360}\big(&0.38092 - 0.76996\cos\phi_{year} + 23.26500\sin\phi_{year}\\ &+0.36958\cos 2\phi_{year} + 0.10868\sin 2\phi_{year} + 0.01834\cos 3\phi_{year}\\ &-0.16650\sin 3\phi_{year} - 0.00392\cos 4\phi_{year} + 0.00072\sin 4\phi_{year}\\ &-0.00051\cos 5\phi_{year} + 0.00250\sin 5\phi_{year} + 0.00442\cos 6\phi_{year}\big).\end{aligned} \tag{7.2b}$$

An approximate equation for solar declination (not used here) is

$$\phi_{sol-dec} = \frac{2\pi}{360}\left\{23.4\sin\left[\frac{2\pi}{365}(i_{Julian} - 80)\right]\right\}. \tag{7.2c}$$

The solar elevation, $\phi_{sol\text{-}elev}$ (rad), for latitude ϕ_{lat} (rad) is given by

$$\sin\phi_{sol\text{-}elev} = \sin\phi_{lat}\sin\phi_{sol\text{-}dec} + \cos\phi_{lat}\cos\phi_{sol\text{-}dec}\cos\phi_{hour}. \tag{7.2d}$$

Here ϕ_{hour} (rad) is the hour angle of the sun: the angle on a horizontal plane between the sun and the local meridian. It is obtained from

$$\phi_{hour} = 2\pi(t_{dec} - t_{noon}), \tag{7.2e}$$

where t_{dec} is given by Eqn (7.1a). t_{noon} (day) is the local time of solar noon; t_{noon} = 0.5 day when this is midday. The difference between local solar time and mean solar time with time of year (the 'equation of time') does not concern us: it is discussed, for example, by France and Thornley (1984, p. 97) and Jones (1992, p. 362).

The sun is treated as a point, and refraction by the atmosphere is ignored. The day length, τ_{day} (day), is calculated by [solve Eqn (7.2d) for ϕ_{hour} with $\phi_{sol\text{-}elev} = 0$]

$$\tau_{day} = \frac{2\cos^{-1}(-\tan\phi_{lat}\tan\phi_{sol\text{-}dec})}{2\pi}. \tag{7.2f}$$

τ_{day} (day) is a fraction between 0 and 1: thus, at the equinoxes, $\phi_{sol\text{-}dec} = 0$ rad, $\cos^{-1}(0) = \frac{1}{2}\pi$ rad (90°), and $\tau_{day} = 0.5$ day. To obtain the day length in hours and seconds, we use

$$\tau_{day,hr} = 24 \times \tau_{day} \quad \text{and} \quad \tau_{day,\sec} = 3600 \times \tau_{day,hr}. \tag{7.2g}$$

The solar elevation is the angle between the sun and a horizontal plane; at noon this is given by Eqn (7.2d) with $\phi_{hour} = 0$ rad [Eqn (7.2e)]:

$$\sin\phi_{sol\text{-}elev,noon} = \sin\phi_{lat}\sin\phi_{sol\text{-}dec} + \cos\phi_{lat}\cos\phi_{sol\text{-}dec}. \tag{7.2h}$$

The solar elevation is maximum at $\pi/2$ rad if $\phi_{sol\text{-}dec} = \phi_{lat}$; however, for latitudes above 23.45° it is maximum when $\phi_{sol\text{-}dec}$ is maximum at a year angle of $\pi/2$ rad [$\phi_{year} = \frac{1}{2}\pi$, midsummer, Eqns (7.2a), (7.2b)].

7.4 Constant Environmental Variables

Quantities usually regarded as diurnally and seasonally constant using the default program settings are listed in Table 7.2. These are as follows.

A flux of N equivalent to 50 kg N ha^{-1} $year^{-1}$ (default value) is put into the soil ammonia pool (N_{amm}, Fig. 5.1), denoted by $I_{N,env\rightarrow Namm}$. This flux is due to atmospheric deposition. A value of 50 kg N ha^{-1} $year^{-1}$ is typical for a moderately polluted environment in Europe or the USA. A heavily polluted environment could have an input as high as 100 kg N ha^{-1} $year^{-1}$. An unpolluted preindustrial environment might have an input of 5 kg N ha^{-1} $year^{-1}$ (e.g. see Fig. 7.8, which is for northern Britain). Some part of the N input in the parameter $I_{N,env\rightarrow Namm}$ could be regarded as arising from nitrogen fixation by the nitrogen-fixing component of the sward. Non-symbiotic N fixation is calculated independently in the soil submodel (Section 5.5.2). The default flux of N deposition into the soil nitrate pool (N_{nit}) is zero.

The wind speed, w_{ind}, is 1 m s^{-1}, with a reference height of 2 m. France and Thornley (1984, p. 110) reported a seasonal variation of about 12% based on 30-year records taken at Littlehampton, West Sussex, on the south coast of the UK, where the mean wind speed was 2.25 m s^{-1}. Ephrath *et al.* (1996) report large diurnal variation in wind speed for selected sites on certain days in the summer. Wind at

Table 7.2. Constant environmental quantities using program default settings. These values are for central southern Britain (latitude 51.54°N, elevation = 50 m). Note that N deposition of 13.7 $\times 10^{-6}$ kg N m^{-2} day^{-1} is equivalent to (multiply by 365 × 10000) 50 kg N ha^{-1} year^{-1}. Values for a northern Britain environment (latitude 55.27°N, elevation = 242 m) are given in {..}. CO_2 concentration may be subject to climate change (Section 7.8).

Symbol	Description	Default value
h_{ref}	Reference height for meteorological measurements	2 m
$I_{N,env \rightarrow Namm}$	Environmental N input to soil ammonium pool	13.7 {5.48} $\times 10^{-6}$ kg N m^{-2} day^{-1} (50 {20} kg N ha^{-1} year^{-1})
$I_{N,env \rightarrow Nnit}$	Environmental N input to soil nitrate pool	0 kg N m^{-2} day^{-1} year^{-1}
w_{ind}	Wind speed at height of 2 m	1 {2} m s^{-1}
$C_{O_2,vpm}$	Atmospheric CO_2 concentration	350 vpm (μmol mol^{-1})

night is unimportant in the model as the stomata are closed. It may be a better approximation to assume that wind only occurs during the daylight hours, and calculate a mean wind speed accordingly.

Apart from possible climate change (Section 7.8), atmospheric CO_2 concentration is regarded as constant. CO_2 concentration in vpm is converted into a true concentration (kg CO_2 m^{-3}) by means of

$$C_{O_2,air} = \frac{C_{O_2,vpm}}{10^6} \frac{273.15}{(T_{air} + 273.15)} \frac{P_{air}}{101{,}325} 1.9636. \tag{7.3a}$$

In this equation −273.15°C is absolute zero, 101,325 Pa is standard atmospheric pressure, and 1.9636 kg CO_2 m^{-3} is the concentration of pure CO_2 at standard temperature and pressure (T_{air} = 0°C, P_{air} = 101,325 Pa). At 350 vpm, 20°C and standard atmospheric pressure, $C_{O_2,air} = 0.6404 \times 10^{-3}$ kg CO_2 m^{-3} (see the Appendix).

Although the first three environmental quantities in Table 7.2 are constant in the program default settings, there is the provision for changing them on a daily basis if daily environmental data are provided (Section 7.5.2, Fig. 7.2). Also, the climate change option (Section 7.8, Fig. 7.8) allows the second and the last of the environmental variables in Table 7.2 to change on an annual basis.

7.5 Environmental Variables on a Daily Schedule

Three options are provided. First, daily values are generated assuming the seasonal variation is sinusoidal; this is the default regime of the program; however, constant seasonal values can be obtained by making the amplitude of the seasonal variation zero [Eqn (7.4a)]. Second, daily values can be read from a data file. Third, daily values are obtained by interpolation from monthly values which are read in. With any of these three options, the default diurnal variation in radiation, air temperature and humidity can be switched off (Section 7.6.4).

7.5.1 Sinusoidal generation of daily values

The daily environmental variables on a seasonal sinusoidal schedule are listed in Table 7.3. In some environments an assumption that the seasonal variation of daily values is sinusoidal gives an acceptable and convenient approximation to long-term average weather. This is so for the UK. If the quantity x is regarded as sinusoidal, its specification requires an annual mean value (x_{mean}), an annual variation (x_{var}) [= ($\frac{1}{2}$(annual maximum – annual minimum)], and the phase [$t_{ph,x}$ (day)] which determines the date when the maximum is attained. The equation used for the sinusoid generation of daily values is

$$\begin{aligned} x = {} & \left(1 - \sigma_{dy\text{-}data}\right)\left(1 - \sigma_{mo\text{-}data}\right)\left\{x_{mean} + \sigma_{seasonal}\, x_{\mathrm{var}} \sin\left[2\pi\left(t - t_{ph,x}\right)\right]/365\right\} \\ & + \sigma_{dy\text{-}data}\left\{\textit{read in daily data}\right\} \qquad (7.4a) \\ & + \sigma_{mo\text{-}data}\left\{\textit{use interpolations of monthly data}\right\}, \\ & \sigma_{dy\text{-}data} = 0, \quad \sigma_{mo\text{-}data} = 0, \quad \sigma_{seasonal} = 1 \text{(default values)}. \end{aligned}$$

The default values of the switches $\sigma_{dy\text{-}data}$ (daily data) and $\sigma_{mo\text{-}data}$ (monthly environmental data) give the sinusoidal seasonal variation. Setting either of the switches to unity uses daily or interpolated monthly data (see below). Setting the switch $\sigma_{seasonal} = 0$ eliminates seasonal variation.

Equation (7.4a) makes use of the parameters in Table 7.3 to give the seasonal variation in radiation, air temperature (daily maximum and minimum), soil temperature, rainfall and relative humidity (daily maximum and minimum). These quantities are illustrated in Fig. 7.1 for southern Britain.

7.5.2 Input of daily values from a file

The second weather data option is to read daily values from a data file, as in Fig. 7.2. To do this the switch $\sigma_{dy\text{-}data}$ in Eqn (7.4a) must be set to unity. Daily environmental values generated by the sinusoid assumption of Eqn (7.4a) can be output to the file dailyout.dat (set a program switch, s_dy_data_out = 1, at runtime, Section 10.2). This output file has the same format as the input file shown in Fig. 7.2.

7.5.3 Interpolation of monthly values

Monthly values can be specified as in Fig. 7.3 for interpolation to give daily values. For this the switch $\sigma_{mo\text{-}data}$ in Eqn (7.4a) must be set to unity. Interpolation is done in the program INITIAL section (Fig. 10.1) as explained below to give arrays of daily weather data for use through the year.

First define arrays for the number of days in each month (d_{pm}), the Julian number of the midpoint of each month [J_{mpm}, not necessarily an integer as is Julian day number, Eqn (7.1e)], and the Julian day number for the first day of each month (J_{1dm}):

$$\begin{aligned} & d_{pm}(1..12) = 31, 28, 31, 30, 31, 30, 31, 31, 30, 31, 30, 31 \text{ day}, \\ & J_{mpm}(1..12) = 16, 45.5, 75, 105.5, 136, 166.5, 197, 228, 258.5, 289, 319.5, 350, \qquad (7.4b) \\ & J_{1dm}(1..12) = 1, 32, 60, 91, 121, 152, 182, 213, 244, 274, 305, 335. \end{aligned}$$

Table 7.3. Environmental variables on a daily schedule with sinusoidal seasonal variation. The values given for the annual mean, annual variation, and phase are substituted for x_{mean}, x_{var} and $t_{ph,x}$ in the sinusoidal weather generator [Eqn (7.4a)]; they are based on 30-year means for central southern Britain (latitude 51.54°N, elevation = 50 m) (the year begins on 1 January). Values for upland northern Britain (latitude 55.27°N, elevation = 242 m) are given in {..}.

Symbol	Description	Default values: annual mean (x_{mean}) ± annual variation (x_{var})	Phase ($t_{ph,x}$) (date of maximum)
$j_{PAR,dy}$	Daily light receipt of photosynthetically active radiation (PAR)	$4.8 \pm 4.1 \times 10^6$ J m^{-2} day^{-1} {$3.8 \pm 3.3 \times 10^6$}	81 day (21 June)
$T_{air,max}$	Air temperature, daily maximum value (15 h)	14 ± 7°C {11 ± 7}	115 day (25 July)
$T_{air,min}$	Air temperature, daily minimum value (dawn)	6.5 ± 5.5°C {4 ± 5}	115 day (25 July)
T_{soil}	Soil temperature, daily mean (at 10 cm depth)	10 ± 6°C {8.5 ± 6}	116 day (26 July)
r_{ain}	Rainfall	0.00181 ± 0.000602 m day^{-1} 0.66 ± 0.22 m year^{-1} {0.00418 ± 0.00139 m day^{-1}} {1.53 ± 0.51 m year^{-1}}	233 day (20 November)
$r_{el\text{-}hum,1}$	Relative humidity (dawn)	0.8 ± 0.1 {0.84 ± 0.07}	273 day (30 December)
$r_{el\text{-}hum,2}$	Relative humidity (15.00 h)	0.7 ± 0.1 {0.75 ± 0.13}	273 day (30 December)

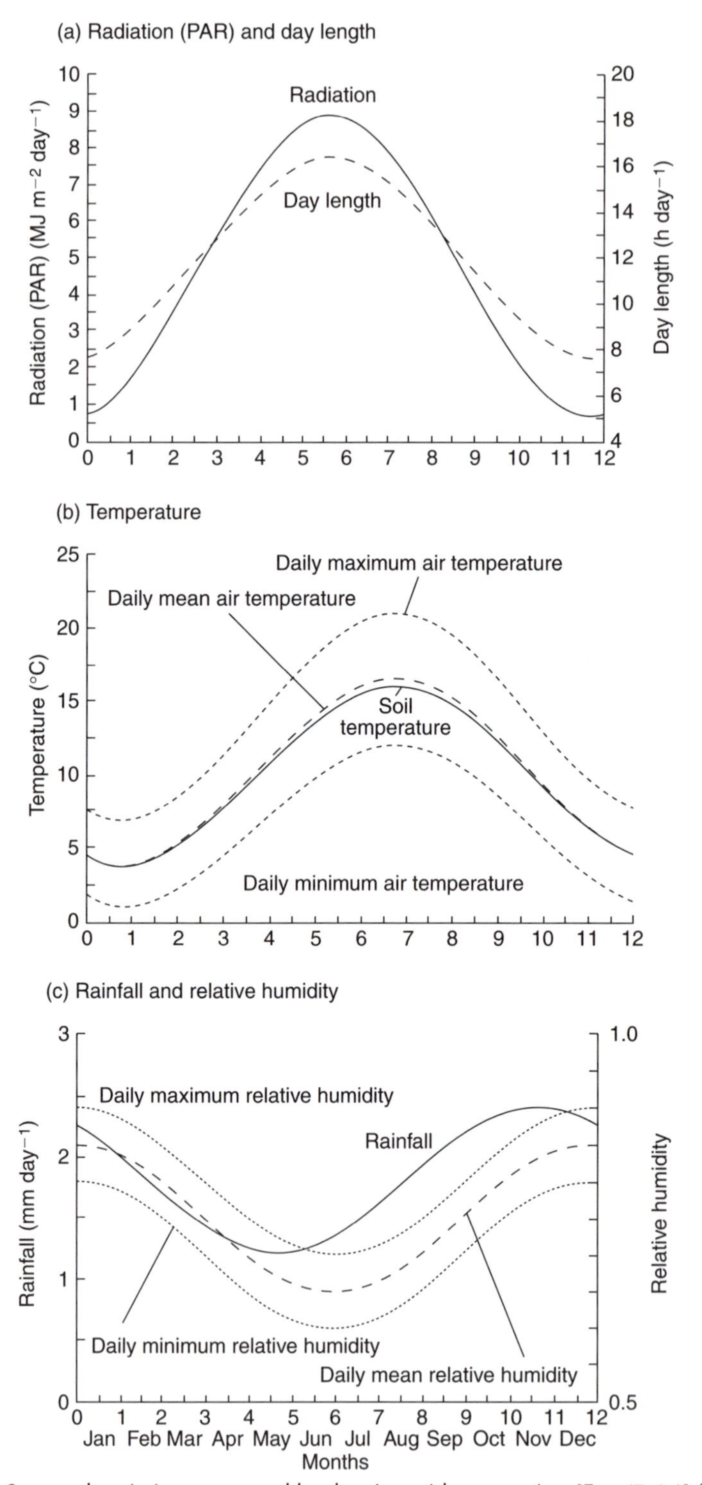

Fig. 7.1. Seasonal variation generated by the sinusoid assumption [Eqn (7.4a)] for southern Britain (Table 7.3). (a) Photosynthetically active radiation and day length. (b) Air and soil temperatures. (c) Rainfall and relative humidity.

'Title'
51.54 = latitude in decimal degrees.

'idaymo	imonth	iJulian	iyear	jPARdy	Tairmax	Tairmin	Tsoil	rain	wind	RH1	RH2	lNenv_Namm	lNenv_Nnit
1	1	1	1	7.77E+05	7.5	1.4	4.5	.00226	1.0	.90	.80	1.37E−05	0.00E+00
2	1	2	1	7.91E+05	7.5	1.4	4.5	.00225	1.0	.90	.80	1.37E−05	0.00E+00
3	1	3	1	8.06E+05	7.4	1.3	4.4	.00225	1.0	.90	.80	1.37E−05	0.00E+00
4	1	4	1	8.23E+05	7.4	1.3	4.4	.00224	1.0	.90	.80	1.37E−05	0.00E+00
5	1	5	1	8.40E+05	7.4	1.3	4.3	.00223	1.0	.90	.80	1.37E−05	0.00E+00
6	1	6	1	8.59E+05	7.3	1.3	4.3	.00223	1.0	.90	.80	1.37E−05	0.00E+00
7	1	7	1	8.79E+05	7.3	1.2	4.3	.00222	1.0	.90	.80	1.37E−05	0.00E+00
8	1	8	1	9.01E+05	7.3	1.2	4.2	.00221	1.0	.90	.80	1.37E−05	0.00E+00
9	1	9	1	9.23E+05	7.2	1.2	4.2	.00220	1.0	.90	.80	1.37E−05	0.00E+00
10	1	10	1	9.47E+05	7.2	1.2	4.2	.00219	1.0	.90	.80	1.37E−05	0.00E+00

Fig. 7.2. Daily environmental data file, daily.dat, for input of daily environmental data to the ACSL program (set a program switch, s_dy_data = 1, at runtime, Section 10.2). The program reads and dumps the first line of characters. In the second line latitude is read in. The next line is read and dumped. The program then reads one line at the beginning of each day. The first four integer columns are unused by the program; they are: day of the month (idaymo), month (imonth), Julian day (iJulian), and year number (iyear) [See Eqns (7.1f), (7.1e) and (7.1d)]. The next ten columns correspond to the seven entries in Table 7.3 and three entries in Table 7.2, and are respectively: (1) daily receipt of photosynthetically active radiation (jPARdy, J m^{-2} day^{-1}), (2) air temperature daily maximum (Tairmax, °C), (3) air temperature daily minimum (Tairmin, °C), (4) soil temperature (Tsoil, °C, daily mean), (5) daily rainfall (rain, m day^{-1}), (6) wind speed (wind, m s^{-1}, daily mean), (7) relative humidity (RH1, dawn, a fraction), (8) relative humidity (RH2, 15.00 h, a fraction), (9) ammonium deposition (lNenv_Namm, kg N m^{-2} day^{-1}), (10) nitrate deposition (lNenv_Nnit, kg N m^{-2} day^{-1}).

'Kew 51 28 N, 0 19 W. 5 m. England'
51.47 This is the latitude in decimal degrees.

'Tmax,	Tmin,	RH1,	RH2,	Precipitation (mm/month),	Bright_sunshine (h/month)'
6.30	2.20	86.00	77.00	54.00	46.00
6.90	2.20	85.00	72.00	40.00	64.00
10.10	3.30	81.00	64.00	37.00	113.00
13.30	5.50	71.00	56.00	37.00	160.00
16.70	8.20	70.00	57.00	46.00	199.00
20.30	11.60	70.00	58.00	45.00	213.00
21.80	13.50	71.00	59.00	57.00	198.00
21.40	13.20	76.00	62.00	59.00	188.00
18.50	11.30	80.00	65.00	49.00	142.00
14.20	7.90	85.00	70.00	57.00	98.00
10.10	5.30	88.00	78.00	64.00	53.00
7.30	3.50	87.00	81.00	48.00	40.00

Fig. 7.3. Monthly environmental data file, monthly.dat, used for interpolation to give daily data. These data are 30-year averages from the UK Meteorological Office tables (Meteorological Office, 1982, p. 152). To use monthly data in the ACSL program (set a program switch, s_mo_data = 1, at runtime, Section 10.2). The program reads and dumps the first line of characters. In the second line latitude is read in. The third line is read and dumped. The next 12 lines are monthly means of maximum and minimum air temperature (°C), morning and afternoon values of relative humidity (%), and monthly totals of precipitation and bright sunshine hours. The first month is January.

Next convert the columns in Fig. 7.3 into the present units with

$$RH1 := 0.01\,RH1, \quad RH2 := 0.01\,RH2,$$

$$\text{Precipitation}\left(\text{m day}^{-1}\right) := \frac{0.001}{d_{pm}}\,\textit{Precipitation}, \tag{7.4c}$$

$$\textit{Bright\_sunshine}\left(\text{h day}^{-1}\right) := \frac{\textit{Bright\_sunshine}}{d_{pm}}.$$

Here ':=' denotes 'is replaced by'.

The recipe for calculating the interpolated daily values $y(1..365)$ from the monthly values $x(1..12)$ is as follows. For Julian day number $i_{Julian} = 1, \ldots, 365$, day of the month $i_{day,mo}$ and month i_{month} are obtained with Eqns (7.1f). Then
First half of January $[i_{Julian} < J_{mpm}(1)]$:

$$y(i_{Julian}) = x(12) + [x(1) - x(12)]\frac{i_{Julian} + 365 - J_{mpm}(12)}{J_{mpm}(1) + 365 - J_{mpm}(12)}. \tag{7.4d}$$

Second half of every month from January to November $\{(i_{month} \le 11)$ AND $[J_{mpm}(i_{month}) \le i_{Julian}]$ AND $[i_{Julian} < J_{1dm}(i_{month} + 1)]\}$ (AND denotes logical AND):

$$y(i_{Julian}) = x(i_{month}) + [x(i_{month} + 1) - x(i_{month})] \times \frac{i_{Julian} - J_{mpm}(i_{month})}{J_{mpm}(i_{month} + 1) - J_{mpm}(i_{month})}. \tag{7.4e}$$

First half of every month from February to December $\{(i_{month} \ge 2)$ AND $[J_{1dm}(i_{month}) \le i_{Julian}]$ AND $[i_{Julian} < J_{mpm}(i_{month})]\}$:

$$y(i_{Julian}) = x(i_{month}) - [x(i_{month}) - x((i_{month} - 1)] \times \frac{J_{mpm}(i_{month}) - i_{Julian}}{J_{mpm}(i_{month}) - J_{mpm}(i_{month} - 1)}. \tag{7.4f}$$

Last half of December $[J_{mpm}(12) \le i_{Julian}]$:

$$y(i_{Julian}) = x(12) + [x(1) - x(12)]\frac{i_{Julian} - J_{mpm}(12)}{J_{mpm}(1) + 365 - J_{mpm}(12)}. \tag{7.4g}$$

Bright sunshine hours, $h_{b\text{-}s}$ (h day^{-1}), for each day of the year, are given by using Eqns (7.4d), (7.4e), (7.4f), (7.4g) to interpolate the last column in Fig. 7.3. These values must be converted to daily light receipt [J (PAR) m^{-2} day^{-1}].

The relative hours of bright sunshine, $R_{h,b\text{-}s}$ (dimensionless) is obtained by dividing by the day length in hours, $\tau_{day,hr}$ of Eqn (7.2g), with

$$R_{h,b\text{-}s} = \frac{h_{b\text{-}s}}{\tau_{day,hr}}. \tag{7.4h}$$

We use our own version of the Ångström formula (see Martínez-Lozano *et al.*, 1984, for a review) to provide a value of $j_{PAR,dy,b\text{-}s}$ (J PAR m^{-2} day^{-1} from bright sunshine), the daily receipt of photosynthetically active radiation (PAR) from the relative hours of bright sunshine:

$$j_{PAR,dy,b\text{-}s} = f_{Ang,PAR} c_{solar} \sin(\phi_{sol\text{-}elev,noon}) \times \tau_{day,\sec}(a_{Ang} + b_{Ang} R_{h,b\text{-}s}),$$

$$f_{Ang,PAR} = 0.28, \quad c_{solar} = 1370 \text{ J m}^{-2}\text{s}^{-1}, \quad a_{Ang} = 0.28, \quad b_{Ang} = 0.54. \quad (7.4i)$$

c_{solar} is the solar constant, the total radiation flux density at the top of the atmosphere on a surface normal to the sun's direction. The solar elevation term [$\phi_{sol\text{-}elev,noon}$, Eqn (7.2h)] corrects this to a horizontal surface. A simple adjustment for day length [$\tau_{day,sec}$, Eqn (7.2g)] is applied. a_{Ang} and b_{Ang} are dimensionless constants used in Ångström's formula; see Tables III and IV of Martínez-Lozano *et al.* (1984) to see *a* and *b* values for many sites. The dimensionless factor $f_{Ang,PAR}$ allows for atmospheric absorption, converts total radiation to PAR, and allows empirically for other required corrections; this factor was adjusted so that for Kew (UK) (Fig. 7.3), the values of daily PAR estimated from bright sunshine hours by Eqn (7.4i) are acceptably accurate, especially during the summer months.

7.6 Diurnally Varying Environmental Variables

There are three diurnally varying variables in the model. These are irradiance (including net radiation), air temperature and relative humidity. Diurnal values during the day are calculated from daily data, and, as discussed above (Section 7.5), there are several options for the daily data. Ephrath *et al.* (1996) discuss the calculation of diurnal variation from daily variables. In addition to irradiance, temperature and relative humidity, they include the diurnal variation of wind speed which, although it can be considerable, is ignored here; they give formulae, and compare their equations with measurements made in Israel, California and the Netherlands; their contribution is valuable, but, as they point out, the methods they present cannot be considered as independent of site or season. Our method is based mostly on sinusoidal interpolation.

7.6.1 Radiation

The instantaneous photosynthetically active radiation, $j_{PAR,sc}$ [J (PAR) m^{-2} s^{-1}], can be treated either as a full sine wave during the light period (by default), or as a step function with height equal to {daily light receipt [J (PAR) m^{-2} day^{-1}]} / [day length (s day^{-1})] (set program switch, s_PAR_sin = 0, at runtime, Section 10.2). $j_{PAR,sc}$ is calculated from the daily PAR receipt, $j_{PAR,dy}$ [J (PAR) m^{-2} day^{-1}], by means of

$$x = \text{PULSE}(t_{dawn}, 1, \tau_{day}), \qquad y = \frac{j_{PAR,dy}}{\tau_{day,\sec}},$$

$$\text{if } \sigma_{diurnal} = 0, \qquad j_{PAR,sc} = y,$$

$$\text{else if } \sigma_{PAR,\sin} = 1, \qquad j_{PAR,sc} = xy\left\{1 + \cos\left[\frac{2\pi(t_{dec} - 0.5)}{\tau_{day}}\right]\right\}, \quad (7.5a)$$

$$\text{else } j_{PAR,sc} = xy.$$

$$\sigma_{diurnal} = 1, \qquad \sigma_{PAR,\sin} = 1 \text{(default values)}.$$

The time of dawn, t_{dawn}, is given by Eqn (7.5c); day length in days, τ_{day}, is given by Eqn (7.2f). The PULSE function is an ACSL function which is explained in detail after Eqn (7.6b); it has the value 0 during the night and 1 during the daylight hours. x and y are working variables. Daily radiation receipt, $j_{PAR,dy}$, is provided as input data (Table 7.3, or Fig. 7.2). Day length in seconds, $\tau_{day,sec}$, is given by Eqn (7.2g). If diurnal variation is switched off with $\sigma_{diurnal} = 0$, radiation is assigned its average value during the daylight hours over the whole 24 hour period. In this case photosynthesis and transpiration which are driven by radiation are scaled down by the day length (τ_{day}). This allows for the non-linearity of the photosynthetic response to light (Fig. 3.3). See also Section 7.6.4. If diurnal variation is switched on ($\sigma_{diurnal} = 1$), then there is a choice for the radiation variation during the daylight hours. With $\sigma_{PAR,sin} = 1$, a full sine wave is used. This approximation for diurnal radiation was suggested by Charles-Edwards and Acock (1977, equation 7), and is discussed in France and Thornley (1984, pp. 101–102). Alternatively, a step function for daily radiation can be applied by putting the switch $\sigma_{PAR,sin} = 0$. These two options are shown in Fig. 7.4a.

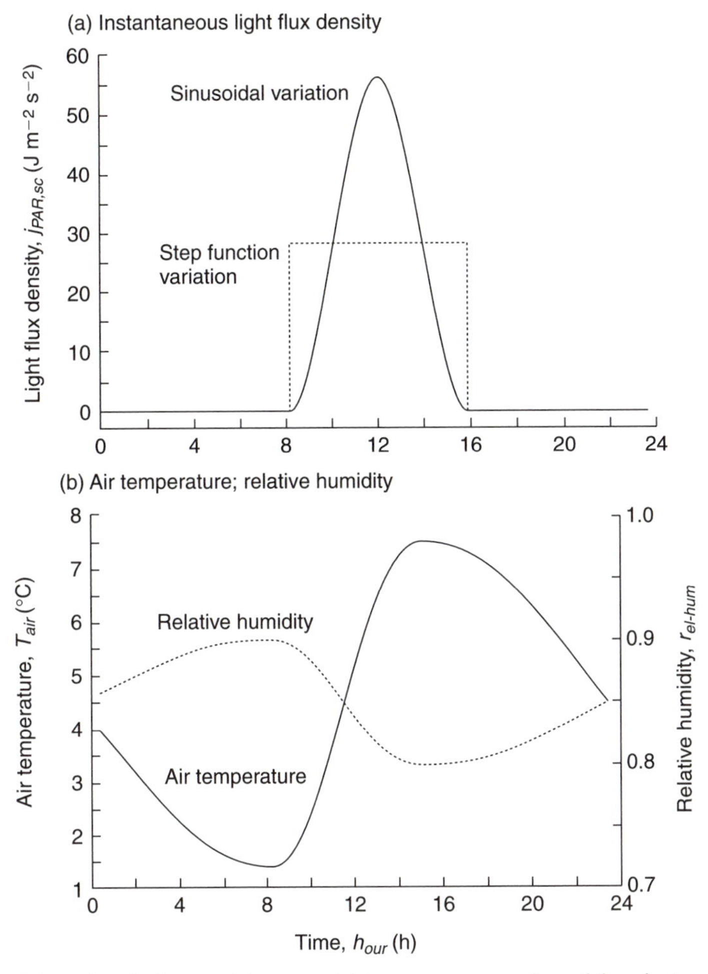

Fig. 7.4. Diurnal variation on 1 January. (a) Instantaneous value of the photosynthetic light flux density, $j_{PAR,sc}$, given by the sinusoid and square wave options given in Eqns (7.5a). (b) Air temperature, T_{air}, [Eqn (7.5e)], and relative humidity, $r_{el\text{-}hum}$ [Eqn (7.5g)].

Instantaneous net radiation, j_{NetR} (J m^{-2} s^{-1}), is calculated by assuming a constant ratio between net radiation and PAR, $r_{NetR,PAR}$:

$$j_{NetR} = r_{NetR,PAR} j_{PAR,sc},$$
$$r_{NetR,PAR} = 1.4 \text{ J (net radiation)[J (PAR)]}^{-1}. \tag{7.5b}$$

This follows a prescription used by Jensen *et al.* (1993, p. 6 and Fig. 3). Jensen *et al.* (1993) assumed that net radiation is 0.68 × global radiation and that PAR is 0.5 × global radiation, leading to net radiation being 1.36 × PAR.

7.6.2 Air temperature

Air temperature, T_{air} (°C) is treated as a sinusoid, whose minimum value is at dawn, and whose maximum is 3 hours after solar noon (15.00 h; decimal fraction of day = 0.625 day). Dawn is given by t_{dawn} (day), with

$$t_{dawn} = \tfrac{1}{2}(1 - \tau_{day}), \tag{7.5c}$$

where τ_{day} is the day length [Eqn (7.2f)].

The maximum and minimum values of daily air temperature, $T_{air,max}$ and $T_{air,min}$ are obtained by either seasonal sinusoidal interpolation (Section 7.5.1), or by input of daily values (Section 7.5.2), or by interpolation of monthly values (Section 7.5.3). The daily mean temperature and its variation are assumed to be

$$T_{mean} = \tfrac{1}{2}(T_{air,max} + T_{air,min}), \quad T_{\mathrm{var}} = \frac{\sigma_{diurnal}}{2}(T_{air,max} - T_{air,min}). \tag{7.5d}$$

The diurnal variation can be switched off with $\sigma_{diurnal} = 0$ (program switch, s_diurnal).

For the three periods of the day [0 h to dawn, t_{dawn}; dawn to 15.00 h (0.625 day); and 15.00 h to 24 h], the air temperature, T_{air}, at decimal time of day t_{dec} [Eqn (7.1a)] is

$$\begin{aligned}
&\text{if } \left(0.0 \le t_{dec} \le t_{dawn}\right) \\
&\qquad T_{air} = T_{mean} - T_{\mathrm{var}} \sin\left(2\pi \frac{\left[t_{dec} + 1 - \tfrac{1}{2}\left(1 + t_{dawn} + 0.625\right)\right]}{2\left(1 + t_{dawn} - 0.625\right)}\right), \\
&\text{if } \left(t_{dawn} < t_{dec} < 0.625\right) \\
&\qquad T_{air} = T_{mean} + T_{\mathrm{var}} \sin\left(2\pi \frac{\left[t_{dec} - \tfrac{1}{2}\left(t_{dawn} + 0.625\right)\right]}{2\left(0.625 - t_{dawn}\right)}\right), \\
&\text{if } \left(0.625 \le t_{dec} < 1\right) \\
&\qquad T_{air} = T_{mean} - T_{\mathrm{var}} \sin\left(2\pi \frac{\left[t_{dec} - \tfrac{1}{2}\left(0.625 + 1 + t_{dawn}\right)\right]}{2\left(1 + t_{dawn} - 0.625\right)}\right).
\end{aligned} \tag{7.5e}$$

This procedure can give rise to a small temperature discontinuity at midnight due to the seasonal (daily) change in the maximum and minimum temperatures. The diurnal air temperature wave for 1 January is shown in Fig. 7.4b. It is asymmetric with respect to daylight and midday.

7.6.3 Relative humidity

The diurnal variation in relative humidity $r_{el\text{-}hum}$ is calculated similarly to the daily variation of air temperature (Section 7.6.2), except that it is assumed that maximum humidity occurs when air temperature is minimum (dawn) and minimum humidity when the air temperature is maximum (15.00 h). The maximum (early morning) and minimum (early afternoon) values of relative humidity, $r_{el\text{-}hum,1}$ and $r_{el\text{-}hum,2}$, are obtained either by seasonal sinusoidal interpolation (Section 7.5.1), or by input of daily values (Section 7.5.2), or by interpolation of monthly values (Section 7.5.3). The daily mean relative humidity and its variation are assumed to be

$$\begin{aligned} r_{el\text{-}hum,mn} &= \tfrac{1}{2}\left(r_{el\text{-}hum,1} + r_{el\text{-}hum,2}\right), \\ r_{el\text{-}hum,var} &= \frac{\sigma_{diurnal}}{2}\left(r_{el\text{-}hum,1} - r_{el\text{-}hum,2}\right). \end{aligned} \tag{7.5f}$$

The diurnal variation can be switched off with $\sigma_{diurnal} = 0$ (program switch, s_diurnal, Section 10.2).

For the three periods of the day [0 h to dawn, t_{dawn}; dawn to 15.00 h (0.625 day); and 15.00 h to 24 h], the relative humidity $r_{el\text{-}hum}$ at decimal time of day t_{dec} [Eqn (7.1a)] is

$$\begin{aligned} &\text{if } \left(0.0 \le t_{dec} \le t_{dawn}\right) \\ &\quad r_{el\text{-}hum} = r_{el\text{-}hum,mn} + r_{el\text{-}hum,\mathrm{var}} \sin\left(2\pi \frac{\left[t_{dec} + 1 - \frac{1}{2}\left(1 + t_{dawn} + 0.625\right)\right]}{2\left(1 + t_{dawn} - 0.625\right)}\right), \\ &\text{if } \left(t_{dawn} < t_{dec} < 0.625\right) \\ &\quad r_{el\text{-}hum} = r_{el\text{-}hum,mn} - r_{el\text{-}hum,\mathrm{var}} \sin\left(2\pi \frac{\left[t_{dec} - \frac{1}{2}\left(t_{dawn} + 0.625\right)\right]}{2\left(0.625 - t_{dawn}\right)}\right), \\ &\text{if } \left(0.625 \le t_{dec} < 1\right) \\ &\quad r_{el\text{-}hum} = r_{el\text{-}hum,mn} + r_{el\text{-}hum,\mathrm{var}} \sin\left(2\pi \frac{\left[t_{dec} - \frac{1}{2}\left(0.625 + 1 + t_{dawn}\right)\right]}{2\left(1 + t_{dawn} - 0.625\right)}\right). \end{aligned} \tag{7.5g}$$

A small discontinuity in the relative humidity at midnight may occur, arising from the daily change in the maximum and minimum relative humidity values. The diurnal relative humidity wave for 1 January is illustrated in Fig. 7.4b.

Castellví *et al.* (1996) compare methods for calculating vapour pressure deficit and relative humidity on various time scales. They discuss some new estimation methods which not site-specific, and which they found to give good results. Their equations can be used to estimate evapotranspiration from measurements of temperature, precipitation and insolation.

7.6.4 Diurnally constant environment

It can be useful to examine the time-course of the system in a constant environment. As has already been mentioned, this can be achieved by setting a switch $\sigma_{diurnal}$ to 0 [Eqns (7.5a), (7.5d) and (7.5f)] (its default value in the program is 1). For air temperature and relative humidity, the daily variations T_{var} of Eqn (7.5d) and $r_{el-hum,var}$ of Eqn (7.5f) are simply multiplied by $\sigma_{diurnal}$. For photosynthetic radiation [Eqn (7.5a)] and net radiation [Eqn (7.5b)], the radiation fluxes are made equal to their mean values during the light period (i.e. $j_{PAR,dy}/\tau_{day,sec}$). The fluxes arising from these radiation fluxes (photosynthesis and transpiration) are applied at a constant rate through the 24-hour day but are multiplied by τ_{day} (day)/(1 day) [see Eqn (7.2f)], so that the 24-hour flux is the same as the flux would have been during the light period. The stomata, which only affect photosynthesis and transpiration, also appear to be continually open when diurnal variation is switched off [Eqn (3.2u)].

7.7 Management Parameters

These comprise fertilizer application, harvesting, and stocking. In the default environment there is no fertilizer application or harvesting, but animals graze for 7 months at a moderate stocking density. A summary of management parameters is in Table 7.4.

Table 7.4. Management parameters. Default values are given. In default there is no fertilizer application or harvesting. Under Stocking, {..} denotes values for northern Britain. [1..40] etc. indicates an array of 40 elements. The notation 40 * 0.0 indicates that the value 0.0 is repeated 40 times.

Symbol	Description	Value
Fertilizer		
f_{ertN} [1..40]	Quantity of N applied at each fertilizer application (7.6a)	40 * 0.0 kg N m^{-2}
$f_{N,fert \to Namm}$, $f_{N,fert \to Nnit}$	Fractions of N fertilizer entering soil ammonium and nitrate pools (7.6c)	0.9314, 0.0686
n_{fert}	Number of fertilizer applications (7.6a)	40
t_{fert} [1..40]	Times of fertilizer applications (7.6b)	82, 89, … , 355 day
σ_{fert}	Fertilizer switch (7.6a)	0 (off)
Harvesting		
h_{harv}	Harvest height (7.6e)	0.03 m
t_{harv}[1..30]	Times of harvesting (7.6d)	66, 73, … , 262, 269 day
σ_{harv}	Switch for all harvests (7.6d)	0 (off)
$\sigma_{harv[1..30]}$	Switch for separate harvests (7.6d)	30 * 0 (all off)
Stocking		
$n_{animals,c}$	Constant stocking density through year (7.6f)	0.0007 {0.0005} sheep m^{-2}
n_{stock}[1..26]	Stocking densities for 26 periods of stocking (7.6f)	0.0012 {0.0008}, 25 * 0 sheep m^{-2}
t_{stock}[1..26]	Times when stock are introduced (7.6f)	90 (1 Apr), 25 * 10^{20} day
σ_{seasnl}	Seasonality switch (0 for no seasonality) (7.6f)	1
σ_{stock}	Stocking switch; switches stocking on/off (7.6f)	0 (off), 1 (on, default)
τ_{stock}[1..26]	Duration of stocking periods (7.6f)	214 (7 months), 25 * 0.0 day

7.7.1 Fertilizer application

The 'standard' or default treatment is no fertilizer application. However, the program allows fertilizer to be applied up to 40 times a year on any day of the year in a pattern which repeats each year if the simulation continues for several years. Constant inputs of N can be applied to the soil ammonium and/or nitrate pools using the parameters $I_{N,env \rightarrow Namm}$ and $I_{N,env \rightarrow Nnit}$ of Table 7.2 for environmental N inputs.

First define a fertilizer application function, $p_{ulse,fert}$ (kg N m^{-2} day^{-1}), for the total N fertilizer input, by

$$P_{ulse,fert} = \sigma_{fert} \sum_{i_{fert}=1}^{n_{fert}} \left\{ f_{fertN}\left[i_{fert}\right] \mathrm{PULSE}\left(t_{fert}\left[i_{fert}\right], 365, 1.0\right) \right\}, \tag{7.6a}$$

$$\sigma_{fert} = 0, \quad n_{fert} = 40 \text{ (maximum)}, \qquad f_{fertN}[1..40] = 40 \times 0.0 \text{ kg N m}^{-2}.$$

In this equation, σ_{fert} (s_fert in the program) is a switch (0, 1) used to switch fertilizer application off/on; n_{fert} is the number of fertilizer applications in the year; i_{fert} is an index running from 1 to n_{fert}; $f_{ertN}[i_{fert}]$ (kg N m^{-2}) is the quantity of N applied at the i_{fert} application, zero by default (note that 0.01 kg N m^{-2} equals 100 kg N ha^{-1}). The default values for the 40 fertilizer application times $t_{fert}[i_{fert}]$ are weekly beginning on 24 March, with

t_{fert} [1..40] =
82, 89, 96, 103, 110, 117, 124, 131, 138, 145,
152, 159, 166, 173, 180, 187, 194, 201, 208, 215,
222, 229, 236, 243, 250, 257, 264, 271, 278, 285,
292, 299, 306, 313, 320, 327, 334, 341, 348, 355. (7.6b)
(24, 31 Mar; 7, 14, 21, 28 Apr; 5, 12, 19, 26 May;
2, 9, 16, 23, 30 Jun; 7, 14, 21, 28 Jul; 4,
11, 18, 25 Aug; 1, 8, 15, 22, 29 Sep; 6, 13,
20, 27 Oct; 3, 10, 17, 24 Nov; 1, 8, 15, 22 Dec.)

The PULSE function, which we can write as PULSE (t_{begin}, p_{eriod}, w_{idth}), is a very useful ACSL function: it has the value 0 or 1; it becomes 1 when the time variable $t \geq t_{begin}$; it then remains 1 over a time period of w_{idth} after which it becomes 0 again; and this behaviour repeats with period p_{eriod} [every year in Eqn (7.6a)]. In Eqn (7.6a), w_{idth} = 1 day so that the fertilizer is applied at a steady rate over 1 day, which avoids extreme transients that might be caused by adding it instantaneously. The units of the PULSE function are time^{-1}, or here, day^{-1}.

The N inputs to the soil ammonium and nitrate pools are

$$I_{N,fert \rightarrow Namm} = f_{N,fert \rightarrow Namm} p_{ulse,fert}, \qquad I_{N,fert \rightarrow Nnit} = f_{N,fert \rightarrow Nnit} p_{ulse,fert},$$

$$f_{N,fert \rightarrow Nnit} = 1 - f_{N,fert \rightarrow Namm}, \tag{7.6c}$$

$$f_{N,fert \rightarrow Namm} = 0.9314.$$

The fraction $f_{N,fert \rightarrow Namm}$ of the applied N enters the soil ammonium pool; the remainder enters the nitrate pool. The default value of $f_{N,fert \rightarrow Namm}$ is as given above. For NH_4NO_3, $f_{N,fert \rightarrow Namm}$ would be set to 0.5. Clearly $f_{N,fert \rightarrow Namm}$ must lie between 0 and 1. The same fertilizer application function, $p_{ulse,fert}$, is applied to both ammonium and nitrate fertilizer.

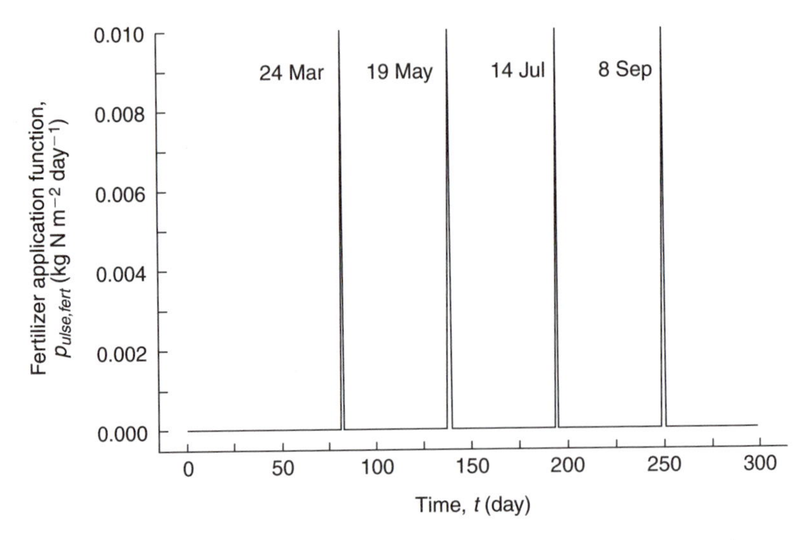

Fig. 7.5. Fertilizer application function, $p_{ulse,fert}$, of Eqn (7.6a), with Eqn (7.6b). Parameters are: $\sigma_{fert} = 1$, $n_{fert} = 4$, $t_{fert}[1..4] = 82, 138, 194, 250$, $f_{ertN}[1..4] = 0.01$ kg N m^{-2}. The quantities of N applied are the same at each application, namely 0.01 kg N m^{-2} (100 kg N ha^{-1}). Each application is at a constant rate for a period of one day.

The fertilizer application function is shown in Fig. 7.5 for four applications of fertilizer. Because the fertilizer applications take place at a uniform rate over one day, the application rate (kg N m^{-2} day^{-1}) is the same as the actual amount applied (kg N m^{-2}).

7.7.2 Harvesting

The default settings are for no harvesting. Harvesting is applied by using an ACSL PULSE function as in Eqn (7.6a) for fertilizer application:

$$p_{ulse,harv} = \sigma_{harv} \sum_{i=1}^{30} \left[\sigma_{harv,i} \text{ PULSE}\left(t_{harv,i}, 365, 1\right) \right],$$
$$\sigma_{harv} = 0, \quad \sigma_{harv,1...} = 0, \quad t_{harv,1,...} = 66, 73, 80, ..., 262, 269 \text{ day.} \tag{7.6d}$$

This function [see after Eqn (7.6b)] allows harvesting to be switched on in general ($\sigma_{harv} = 1$; program switch is s_harv); it also allows particular harvests out of the 30 which are set here to occur at weekly intervals, to be switched on (e.g. with $\sigma_{harv,1} = \sigma_{harv,3} = \sigma_{harv,5} = \sigma_{harv,7} = \sigma_{harv,9} = \sigma_{harv,11} = 1$, with the other $\sigma_{harv,i} = 0$ gives six harvests at fortnightly intervals). To avoid having double-valued functions (e.g. leaf area index is 6 and 2 at time $t = 80$ day), it is assumed that each harvest takes 1 day rather than occurs instantaneously, so that material is removed at a constant rate for 1 day. The harvesting function is shown in Fig. 7.6.

The harvest height h_{harv} (m) is also a management parameter, and is given by (standard value)

$$h_{harv} = 0.03 \text{ m.} \tag{7.6e}$$

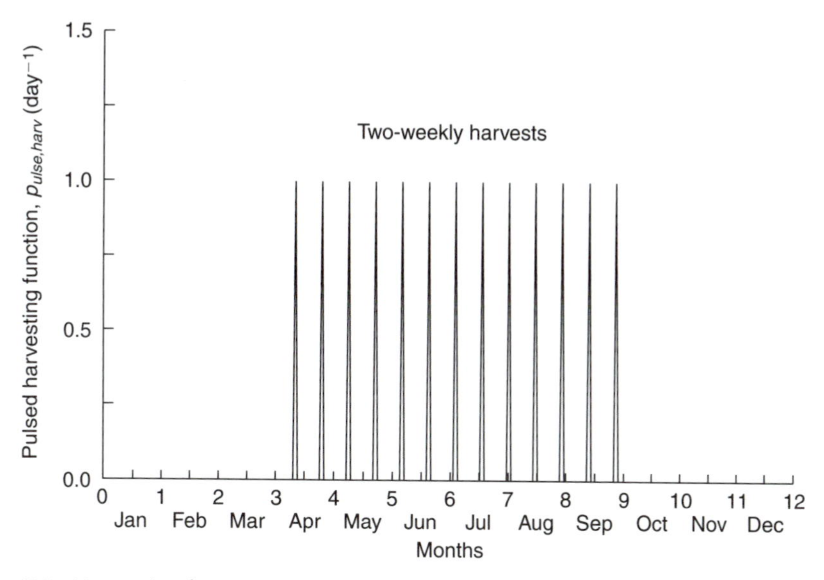

Fig. 7.6. Harvesting function, $p_{ulse,harv}$, of Eqn (7.6d), with parameters given by $\sigma_{harv} = 1$; $\sigma_{harv,6} = \sigma_{harv,8} = , \ldots , \sigma_{harv,28} = \sigma_{harv,30} = 1$, with the other $\sigma_{harv,i} = 0$ as given, to give 13 two-weekly harvests, starting $t = 101$ day (12 April).

7.7.3 Stocking

The default stocking pattern is for 12 sheep ha^{-1} (8 for upland northern Britain) for 7 months from 1 April to 31 October. The number of animals (assumed here to be sheep) grazing the pasture, $n_{animals}$ (animals m^{-2}), is calculated by

$$n_{animals} = \sigma_{stock}\Big\{(1-\sigma_{seasnl})n_{animals,c} + \sigma_{seasnl}\sum_{i=1}^{n_{stock,i}}\Big[n_{stock,i}\text{PULSE}\big(t_{stock,i}, 365, \tau_{stock,i}\big)\Big]\Big\},$$

$$\sigma_{stock} = 1, \quad \sigma_{seasnl} = 1, \quad n_{animals,c} = 0.0007 \text{ animals m}^{-2}, \quad (7.6\text{f})$$

$$n_{stock} = 1, \quad n_{stock,1} = 0.0012, \quad n_{stock,2\ldots} = 0 \text{ animals m}^{-2},$$

$$t_{stock,1} = 90, \quad t_{stock,2\ldots} = 10^{20} \text{ day}, \quad \tau_{stock,1} = 214, \quad \tau_{stock,2\ldots} = 0 \text{ day}.$$

Stocking can be switched off entirely with $\sigma_{stock} = 0$ (program switch s_stock). This formulation can be used to represent intermittent (cf. 'rotational') grazing (e.g. Fig. 8.9) and continuous grazing. The ACSL PULSE function is used to switch grazing on and off [see after Eqn (7.6b)] during the year. n_{stock} is the number of stocking periods during the year (maximum 26). $n_{stock,i}$ is the number of animals (sheep m^{-2}) during the ith stocking period which commences at time $t = t_{stock,i}$ (day) and is of duration $\tau_{stock,i}$ (day). The default values give seven months ($\tau_{stock,1} = 214$ day) grazing with $0.0012 \times 10000 = 12$ sheep ha^{-1} from 1 April ($t_{stock,1} = 90$) until and including 31 October; no animals graze from 1 November to 31 March. This summer grazing pattern is shown in Fig. 7.7.

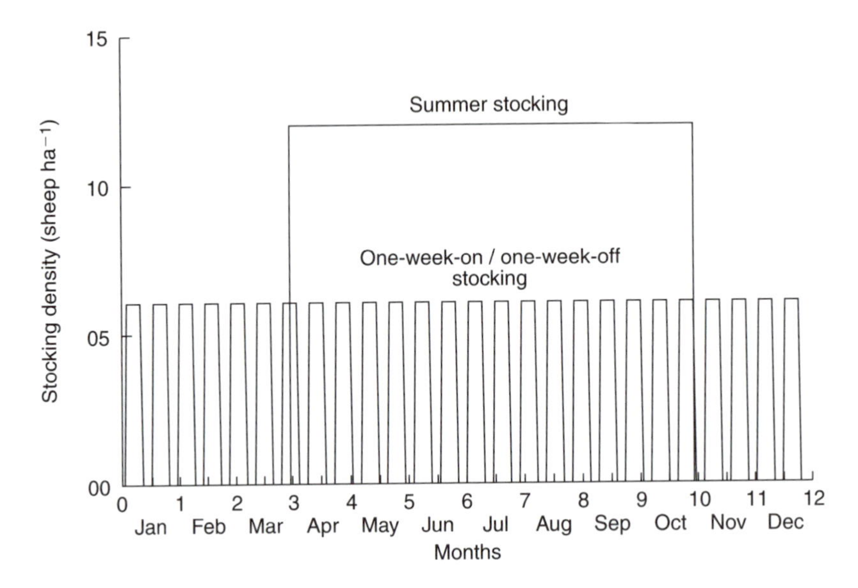

Fig. 7.7. Stocking density. Summer stocking, 1 April to 31 October (inclusive), as given by Eqn (7.6f) with 12 sheep ha^{-1}. Two-weekly rotation with 6 animals ha^{-1} present for 7 days followed by 7 days without grazing; this is obtained by Eqn (7.6f) with $\sigma_{stock} = \sigma_{seasnl} = 1$, $n_{stock}[1..26] = 0.0006$ animals m^{-2} and $\tau_{stock}[1..26] = 7$ day, $t_{stock}[1..26] = 2, 16, 30, \ldots, 352$ day.

If seasonality is switched off ($\sigma_{seasnl} = 0$), then a constant number of animals $n_{animals,c} = 0.0007$ sheep m^{-2} (7 sheep ha^{-1}) is present throughout the year.

The model allows for up to 26 grazing periods in each year, repeated every year during runs lasting more than one year. The dates on which the animals are introduced to the pasture, the time they spend on the pasture, and the numbers of animals can be set independently for each grazing period as in Eqn (7.6f). A scenario in which the animals are introduced every 2 weeks for equal periods of 7 days is shown in Fig. 7.7, alongside the default summer grazing pattern.

The default summer stocking density in Eqn (7.6f) is close to the recommended stocking densities for poor quality lowland grass (Holmes, 1989, p. 262), as is the 7-month grazing season (MAFF, 1976).

7.8 Climate Change

Climate change can cause changes in atmospheric CO_2 concentration, temperature and N deposition rates. There are two methods of simulating climate changes in the model.

7.8.1 Constant annual changes in CO_2 and temperature

A constant rate of change in CO_2 of ΔCO_2 vpm year^{-1} can be applied with the equation [using Eqn (7.1c)]

$$CO_2(y_{ear}) = CO_2(y_{ear,0}) + \sigma_{CO_2,cc}\, \Delta CO_2\,(y_{ear} - y_{ear,0}),$$
$$CO_2(y_{ear,0}) = 350 \text{ vpm}, \quad \sigma_{CO_2,cc} = 0, \qquad (7.7a)$$
$$\Delta CO_2 = 3 \text{ vpm year}^{-1}, \qquad y_{ear,0} = 0 \text{ year}.$$

When CO_2 increase is switched on (with $\sigma_{CO_2,cc} = 1$; program switch s_CO2cc), the above default values give a constant rate of increase of 3 vpm year^{-1} from a base value of 350 vpm which occurs in the year $y_{ear,0}$. If the model is run for, say, 500 years at 350 vpm CO_2, next a CO_2 increase is switched on before the simulation is continued beyond 500 years, then $y_{ear,0}$ takes the value 500 year.

```
'Climate data input file: climate.dat.'
23          nyrsofcd          Number of years for which there is climate data.
' year      delT (deg C)      CO2 (vpm)      Ndep (kg N ha-1 year-1)'
1850        -0.5              290            5
1890        -0.5              290            5
1900        -0.5              295            5
1910        -0.4              298            5
1920        -0.3              301            5
1930        -0.2              304            5
1940        -0.1              307            5
1950         0.0              310            5
1960         0.0              321            10
1970         0.0              333            15
1980         0.1              344            20
1990         0.3              355            20
2000         0.5              380            20
2010         0.7              400            20
2020         0.9              422            20
2030         1.1              450            20
2040         1.3              477            20
2050         1.5              510            20
2060         1.7              545            20
2070         1.9              580            20
2080         2.1              615            20
2090         2.3              650            20
2100         2.5              688            20
```

Fig. 7.8. Climate data file, climate.dat, for input of climate data to the ACSL program (set program switch s_cd = 1 at runtime, Section 10.2). These data are applied in Fig. 8.17. The program reads and dumps the first line of characters. In the second line the number of years for which climate data are provided is read into nyrsofcd. The third line of characters provides headings for the four columns and is read and dumped. The next 23 (nyrsofcd) lines contain, in successive columns, the year, the temperature relative to the chosen base climate (e.g. Table 7.3 for southern Britain in 1960), the CO_2 concentration, and the N deposition rate into the soil N ammonium pool.

A similar equation is used to give a temperature increase ΔT from climate change, with

$$\Delta T(y_{ear}) = \sigma_{T,cc} \Delta T_{cc} (y_{ear} - y_{ear,0}),$$
$$\sigma_{T,cc} = 0, \quad \Delta T_{cc} = 0.03° \text{ year}^{-1}, \quad y_{ear,0} = 0 \text{ year}. \tag{7.7b}$$

When temperature increase is switched on (with $\sigma_{T,cc}$ = 1; program switch s_Tcc), the default values give a constant rate of increase of 0.03°C year^{-1}. This temperature increment is added to both soil and air temperatures.

7.8.2 Variable annual changes in temperature, CO_2 and N deposition

A more flexible option for changes in temperature, atmospheric CO_2 and N deposition is provided in Fig. 7.8. Figure 7.8 is a typical climate data file, climate.dat, defining some climate variables between the years 1850 and 2100. It has four columns, successively the year, the temperature shift relative to the base climate chosen (e.g. see Table 7.3 for southern Britain in about 1960), CO_2 concentration, and the N deposition rate. This component of N deposition is assumed to enter the soil N ammonium pool (Fig. 5.1). Linear interpolation is used for the years for which data are not provided. This option can be employed by setting the program switch s_cd = 1, which causes the file climate.dat to be read in and interpolated in the INITIAL block (Fig. 10.1).

8 Dynamic Properties

8.1 Introduction

The grassland ecosystem simulator is described in the foregoing chapters. From here to the end of this book we evaluate the model and show what it can do. Our aim is to give results that illuminate the model and illustrate the different ways in which it can be used, rather than to tweak parameters to obtain better numerical agreement with particular observations. Using a slightly older version of the model (with minor differences), some of the results presented below have been reported by Thornley and Cannell (1997) and by Cannell and Thornley (1998).

The first step in model evaluation is to examine its dynamic characteristics, which is the subject of this chapter. We begin with diurnal changes over periods of a few days, then consider seasonal behaviour over periods from a week or two to a full year, and finally simulate responses spanning many years. Qualitative comparison with measured data is made where possible.

The most interesting predictions of a model are, usually, those that are unexpected, or those that lie outside our current experience. Such predictions can stimulate further work, experimental or theoretical, and lead to valuable progress. A major advantage of a deterministic simulator is that it enables the pathways between cause and effect to be unravelled, although this is not always easy. Therefore I do not apologize for presenting predictions for a variety of scenarios, knowing that where there are discrepancies between prediction and expectation, these will often point to over-simplifications or inadequacies in the model, but occasionally, they may point to something of greater significance.

Diurnal behaviour arises from the diurnally varying components of the environment (radiation, air temperature, relative humidity: Section 7.6; Fig. 7.4). These environmental factors influence the pools with the shortest turnover times: the plant water pools, the plant substrate C and N pools, and to a lesser extent the soil mineral pools.

Seasonal behaviour is determined by the average values of the fast pools and slower pools such as plant structural pools, metabolic and cellulose litter pools and soil biomass. Processes such as plant structural growth rates, shoot:root partitioning, senescence and litter decay are important over a time scale of a few days to weeks and months.

Long-term behaviour over years to centuries depends on the slowest pools in the system, which are in the soil and litter submodel (Chapter 5): the lignin and soil organic matter pools. Generally these pools are imperfectly represented in models, as they are here. Also, processes which are outside the scope of the present model, adaptation and changes in species composition, assume increasing significance as the time scale increases.

The scientist working in the laboratory or growth cabinet may be most interested in diurnal events. The crop physiologist, agronomist and farmer often focus on the season, and on matters such as the timing and efficacy of fertilizer application, harvesting, or grazing management in relation to animal production. Current concerns with climate change, carbon sequestration, sustainability and the environment, lengthen the focus to one of several years to many centuries.

Dynamic responses always depend on where we start the system from, that is, the initial values. For much of the world's grasslands, we do not know where the system is in relation to a hypothetical equilibrium, or whether the very slow soil organic matter pools are increasing or decreasing. Indeed an equilibrium may not exist, or be so distant in time as to be effectively non-existent (Section 5.14). We must therefore concern ourselves with the effects of initial values, especially of litter and soil organic pools, on seasonal behaviour. For this purpose we can either: (i) assign to the initial values the default values given in the differential equations or the symbols tables in each chapter; or, we can (ii) use for initial values the long-term 'steady-state' values calculated for the southern Britain environment. The default values in (i) are subjective and their application may give rise to transients which may, or may not, be of interest. The standard southern Britain environment (Chapter 7) with 7 months of grazing at a constant intensity and low nitrogen inputs (50 kg N ha^{-1} $year^{-1}$) is also rather arbitrary.

The long-term steady-state values are 'objective' properties of the model, and they are usually, but not always, unique for a particular environment/management scenario (but see Fig. 4.4, and the text). The steady-state solutions thus provide an invaluable means of examining some fundamental responses of the model (Chapter 9), but they only tell part of the story.

The dynamic properties of the model can be investigated in several ways. One of these is to apply periodic changes in one or more driving variables, and observe how these changes affect various pools and fluxes; these changes in driving variables may represent diurnal and or seasonal weather (Chapter 7). Another method is to depart from the steady state in a well-defined manner, e.g. by applying step changes to atmospheric CO_2, or temperature, and then to observe how the system approaches its new steady state. A third method is to apply a pulse of say additional nitrogen to the soil, and observe how its effects are propagated through the system and then die away. In this chapter, several of these techniques are used to investigate the dynamic properties of the system.

8.2 Diurnal Behaviour

Figure 8.1 illustrates diurnal changes in some of the variables of the system. These are driven by the diurnally varying environmental variables given in Fig. 7.4: radiation, air temperature and relative humidity. Shoot carbon substrate C_{sh} (Fig. 8.1a) reaches a maximum in late afternoon, the results agreeing both qualitatively and quantitatively with some of the results reported by Holt and Hilst (1969) in bluegrass, by Greenfield and Smith (1974) in switchgrass, by Grange (1985) in pepper leaves, and by Gordon *et al.* (1987) in white clover. The root carbon substrate level C_{rt} is much less modulated and lags behind the shoot C substrate level, as would be expected from the way the root is connected to the shoot (Fig. 3.1). The shoot N substrate concentration N_{sh} is slightly modulated through the interaction of shoot C substrate affecting the rate of utilization of N for growth [Eqns (3.3c) and (3.4e)]. The modulation of root N substrate concentration N_{rt} is almost imperceptible.

Shoot water potential and its components also fluctuate greatly over the day, as illustrated in Fig. 8.1b. Figure 8.1c gives shoot relative water content θ_{sh} [Eqn 6.5d)] and the stomatal conductance g_s [Eqn (3.2u)]. These simulated water relations data are in satisfactory agreement with Jones (1978, Fig. 4) who measured the leaf water potential of wheat plants over a period of 24 h, and Hughes (1974, p. 29) in cocksfoot. But agreement is less good with van Bavel (1974, Fig. 6) who measured hourly values of leaf water potential for sunflower. Kameli and Lösel (1993) reported diurnal responses of leaf water potential, relative water content, glucose, fructose, sucrose and proline concentrations in stressed wheat plants.

Growth in new lamina area (Fig. 8.1d) shows a large diurnal variation – this is in the absence of high water stress in early April; comparison with data on leaf elongation rate in fescue (Volenec and Nelson, 1982, Fig. 1) indicates that the peak in leaf area growth in Fig. 8.1d is about 3 h too early. This suggests that it could be important to involve turgor pressure (Fig. 8.1b) explicitly into leaf area growth rates [possibly into Eqn (3.6c) for incremental specific leaf area]. There has been much interest in leaf elongation in relation to water potential (Boyer, 1968; Bunce, 1977; Passioura, 1994), although the relation of leaf area growth to growth in leaf structural dry mass is unclear. It is generally considered that leaf growth has two components: cell division and cell expansion. With two potentially contributing mechanisms, complex responses may be expected, and many models of these and related processes have been proposed (e.g. Silk and Erickson, 1979, Thornley, 1991a; Gastal *et al.*, 1992; Feng and Boersma, 1995).

The shoot:root growth ratio is the ratio of the growth rate of new shoot to that of new root [$r_{sh:rt,G}$, Eqn (3.3d)]. Figure 8.1d also illustrates that this ratio has a substantial diurnal modulation, with a steeper and shorter increasing phase than the declining phase. Shoot:root growth ratio leads growth in leaf area; this is due to increasing substrate C (Fig. 8.1a) depressing incremental specific leaf area [Eqn (3.6c)].

Although not shown in Fig. 8.1, the rate of substrate C export from shoot to root [$O_{CS,sh \rightarrow rt}$; Eqn (3.3b)] has an appreciable diurnal variation, as expected because this flux is driven by shoot substrate C concentration (Fig. 8.1a). Grange (1985, Fig. 1a,b) gives curves for *cumulative* ^{14}C export from mature green pepper leaves; his curves are sigmoidal and asymptotic over a period of 24 h; differentiation of

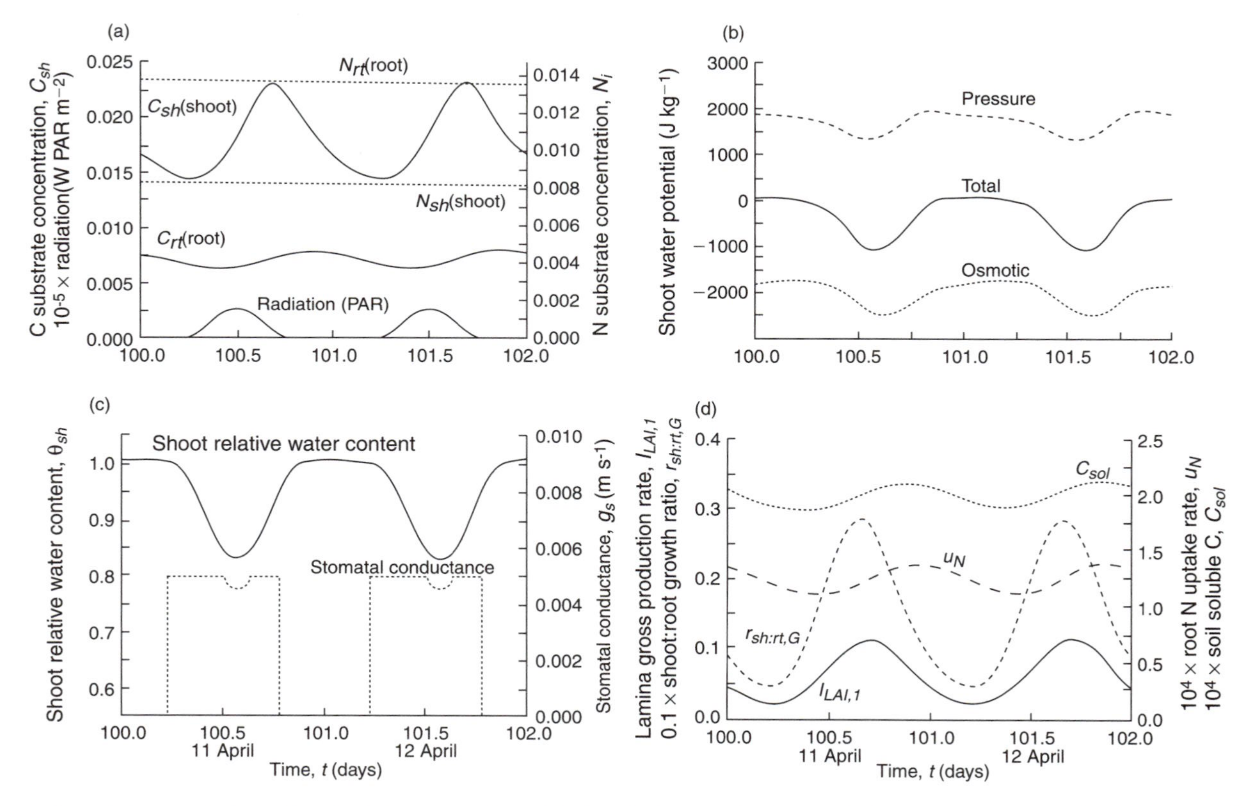

Fig. 8.1. Diurnally varying quantities. The model was run from 1 Jan with the standard south Britain environment (Chapter 3), standard parameters and steady-state initial values for the grazed sward. Results are shown for 11 and 12 April. Grazing animals were not introduced on 1 April. (a) Instantaneous photosynthetic radiation [$j_{PAR,sc}$, Eqn (7.5a)], shoot and root C and N substrate concentrations, C_{sh}, C_{rt}, N_{sh}, N_{rt}, [Eqns (3.1h)] [kg substrate C, N (kg structural dry mass)$^{-1}$]. (b) Shoot water potential (1 J kg^{-1} = 1 kPa): total, pressure, osmotic [ψ_{sh}, $\psi_{sh,pr}$, $\psi_{sh,os}$; Eqns (6.5c), (6.5b), (6.5a), $i = sh$. (c) Shoot relative water content [Eqn (6.5d) with $i = sh$] and stomatal conductance g_s [Eqn (3.2u)]. (d) New lamina area growth rate, $I_{LAI,1}$ [(m^2 lamina) (m^2 ground)$^{-1}$ day^{-1}, Eqn (3.6b)]; shoot:root growth ratio, $r_{sh:rt,G}$ [Eqn (3.3d)]; N uptake rate by plant root, u_N [kg N m^{-2} day^{-1}, Eqn (3.4h)]; soil soluble C pool C_{sol} [kg C m^{-2}, Eqn (5.3a)]. See also Fig. 7.4 for all diurnal environment variables.

his curves (to extract rates) yields patterns of export close to our simulations, although Grange's maximum export rate occurs around the beginning of the night period (but see Gordon *et al.*, 1987), which is later than we predict.

The rate of N uptake by the root u_N [Eqn (3.4h)] exhibits a 20% diurnal variation (Fig. 8.1d). The diurnal variation in N uptake rate leads to a barely perceptible modulation of soil mineral N pool [N_{min}, Eqn (5.5g)], which is not shown. There is appreciable diurnal modulation of the soil soluble C pool [C_{sol}, Eqn (5.3a)] (Fig. 8.1d).

8.3 General Seasonal Behaviour

Experimental work on grassland invariably starts from initial values which may be arbitrary, and some way removed from equilibrium values. First, in Figs 8.2 and 8.3, we illustrate the seasonal responses when the system is in a steady state (that is, the system is repeating the same time course every year). Then (Fig. 8.4), we ascertain, using the model, the extent to which the initial values on 1 January affect the seasonal dynamics over four subsequent years. Next, in Fig. 8.5, the summer stocking density is varied to show the range of predictions which can be generated. Finally, in Fig. 8.6, the seasonal consequences of grazing to a constant leaf area index, plus removing water stress, and supplying a constant soil mineral N concentration, are presented. This is to provide a comparison with experiments which follow this type of protocol, e.g. with regularly cutting to a constant height, and frequent applications of irrigation and fertilizer.

8.3.1 Steady-state seasonal dynamics

Figures 8.2 and 8.3 illustrate the seasonal dynamics for steady-state initial values with a southern Britain environment. Figure 8.2 focuses on the plant responses and Fig. 8.3 on the soil behaviour.

Leaf area index L_{AI}, stocking density (Fig. 7.7), net daily primary production (NDPP) [canopy photosynthesis less plant respiration; see after Eqn (8.1a)], and gross shoot growth rate G_{sh} are given in Fig. 8.2A. NDPP depends on radiation and temperature (Section 3.3.1), and on L_{AI} when light interception is incomplete. NDPP increases faster than L_{AI} in early summer because of increasing radiation and temperature (Fig. 7.1), and decreases more rapidly after midsummer. The shoot growth rate G_{sh} is partially decoupled from NDPP due to a varying shoot:root allocation (Fig. 8.2b). (See Section 8.3.4 for comparisons with experiment.) The whole-plant C and N substrate concentrations, $C_{S,pl}$, $N_{S,pl}$ [Eqns (3.1k)], and the shoot:root ratio are drawn in Fig. 8.2b. The substrate concentrations drive, through the resistance partitioning mechanism (Section 3.3.3; Section 3.4.1), the changes in shoot:root ratio. Garwood has described the seasonal pattern of root growth for a number of grass species (Garwood, 1967a, b). He found that, although many new roots were produced in autumn, the most active period for root growth (i.e. the time of the greatest increase in root mass) is the spring, which is consistent with Fig. 8.2b. Pollock and Jones (1979) observed that certain carbohydrates (e.g. fructans) accumulate in late spring and early summer, as in Fig. 8.2b. However, their data

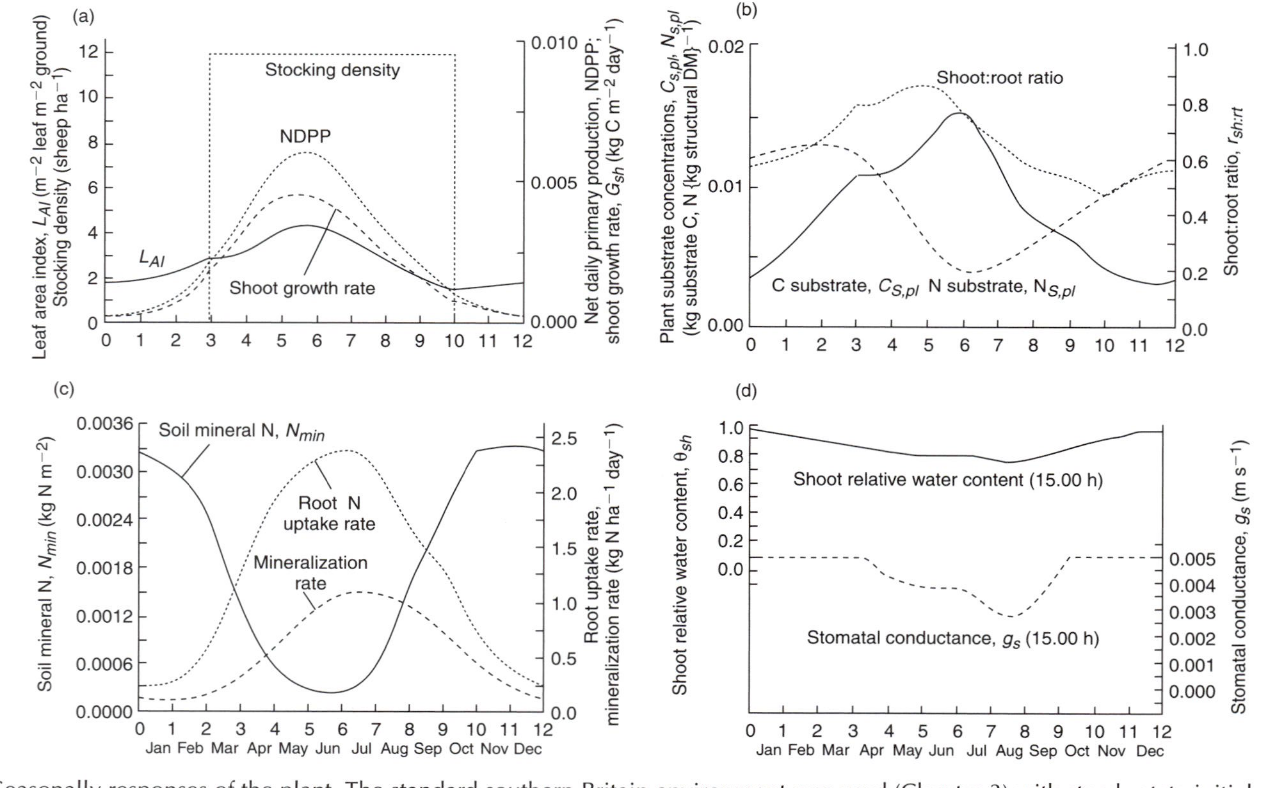

Fig. 8.2. Seasonally responses of the plant. The standard southern Britain environment was used (Chapter 3) with steady-state initial values. (a) Leaf area index, L_{AI} [Eqn (3.1b)]; net daily primary productivity (NDPP) [Eqn (8.1a)]; shoot growth rate, G_{sh} [Eqn (3.3c)]; stocking density [Eqn (7.6f) and Fig. 7.7]. (b) Shoot:root ratio (Eqn (3.1l)]; C and N substrate concentrations in the plant, $C_{S,pl}$, $N_{S,pl}$ [Eqns (3.1k)]. (c) Soil mineral N pool, N_{min} [Eqn 5.5g)]; root uptake rate, u_N [Eqn (3.4h)]; mineralization rate [Eqn (5.4d)]. (d) Shoot relative water content, θ_{sh} [Eqn (6.5d), i = sh]; stomatal conductance at 15 h, g_s [Eqn (3.2u)].

show that carbohydrate accumulation continues through the autumn and into the winter, indicating that more sophisticated mechanisms are operating, possibly via polysaccharide (fructans) synthesis and degradation, ensuring greater fitness and early success in spring. More detailed modelling of carbohydrate biochemistry would be required to represent these processes.

Figure 8.2c depicts a strong seasonal variation in soil mineral N, N_{min}, driven by the plant N uptake rate u_N [Eqn (3.4h)]. This is discussed in the next paragraph (Fig. 8.3a). The peak in N uptake occurs at around the time of maximum leaf area index (Fig. 8.2a). The seasonal changes in shoot relative water content θ_{sh} at 15 h and stomatal conductance at 15 h are shown in Fig. 8.2d (see Fig. 8.1 for diurnal changes). Partial stomatal closure in the afternoon occurs between April and September. The increased sugar content (Fig. 8.2b) and improved osmotic conditions (Section 6.6; Fig. 6.4) ameliorate water stress in midsummer.

Figure 8.3 focuses on the seasonal changes in the soil. In the summer (Fig. 8.3a) the soil mineral N pools are both depressed (Whitehead, 1995, p. 118), and there is a peak in the soil soluble C concentration, largely arising from litter decomposition and C exudation. Root N uptake [Eqn (3.4h)], driven by root mass and specific activity which is increased by high C substrate levels (Fig. 8.2b), as well as temperature, leads mineralization, driven by temperature alone [e.g. see Eqn (5.7d)]. This gives rise to low soil mineral N concentrations in early summer (Vaughn *et al.*, 1986). Substantial changes in ammonium:nitrate ratios have also been observed (e.g. Parsons *et al.*, 1991b). The surface litter metabolic and cellulose responses of Fig. 8.3b reveal, as expected, that the relative labile *met*abolic pool has the largest seasonal variation, followed by the *cel*lulose pool. The two soil litter pools are similar but the values are lower, as there is less soil litter and the decomposition rates are generally higher. The *lig*nin pools and the unprotected, protected, and stabilized SOM pools exhibit little seasonal variation and are not illustrated here. Soil microbial biomass (Fig. 8.3b) is seasonally very conservative although its growth and death rates [Eqns (5.6b), (5.6h)] vary greatly due to changing temperature. The magnitude of the soil microbial biomass in Fig. 8.3b agrees reasonably with reported observations. (e.g. Jenkinson and Powlson, 1980; Rowell, 1994, pp. 38–41). The seasonal data presented by Lundgren and Söderström (1983) are very variable and do not, arguably, show clear seasonal trends. N losses are illustrated in Fig. 8.3c. Leaching [Eqn (5.5e)] and denitrification [Eqn (5.5d)] have a similar seasonal dependence; they are linked by the common substrate, nitrate, and the processes, of drainage and nitrate reduction, both depend on the water status of the soil: when the soil water is at field capacity, runoff can occur and the soil is anaerobic. These processes can be greatly affected by particular weather events, but none the less our average weather scenario predicts the correct seasonality (e.g. Parsons *et al.*, 1991b; Whitehead, 1995).

8.3.2 Effect of initial values on seasonal dynamics

Figure 8.4 illustrates the dynamics over four seasons, comparing the standard (arbitrary) initial values with steady-state initial values. By the third year, leaf area index (L_{AI}, Fig. 8.4a) and shoot:root ratio (Fig. 8.4b) are following their normal seasonal time course. Both Figs 8.4a and 8.4b are showing evidence of damped oscillations

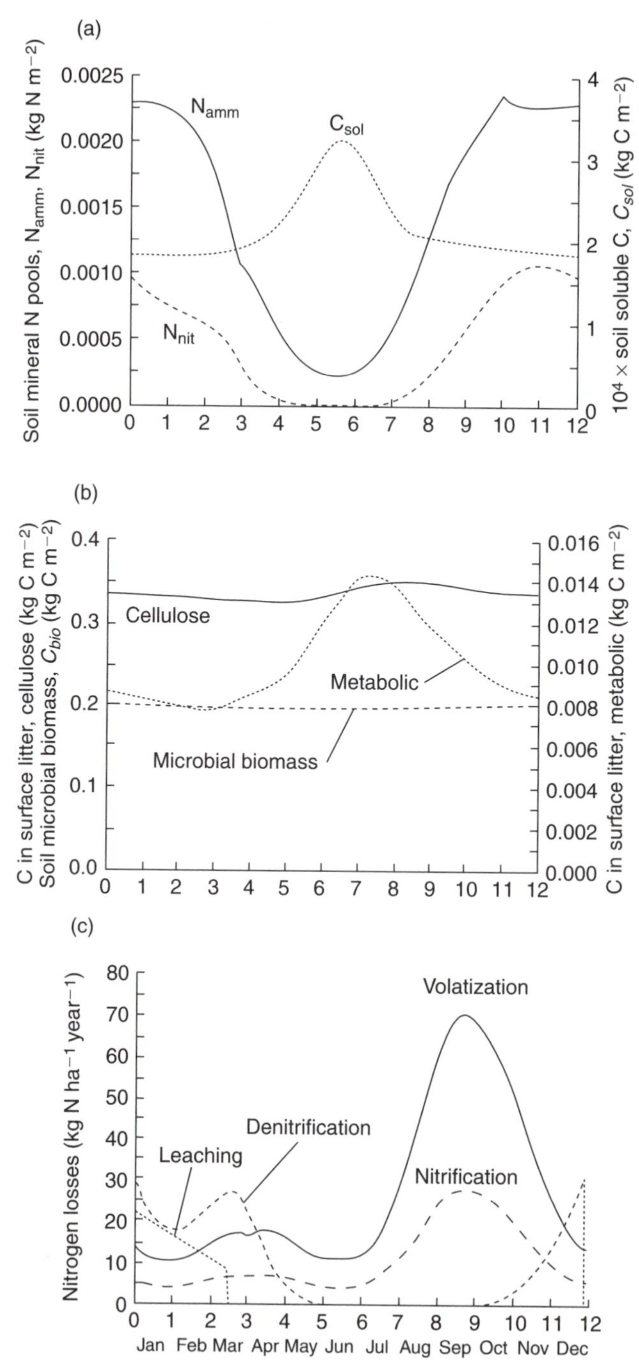

Fig. 8.3. Seasonal responses of the soil. The standard southern Britain environment was used (Chapter 3) with steady-state initial values. (a) Ammonium N, N_{amm} [Eqn (5.4a)], nitrate N, N_{nit} [Eqn (5.5a)], and soluble C, C_{sol} [Eqn (5.3a)]. (b) Metabolic and cellulose surface litter [Eqns (5.1a)]; microbial biomass [Eqn (5.6a)]. (c) N losses: volatilization [Eqn (5.4i)], nitrification [Eqn (5.4g)], leaching [Eqn (5.5e)], and denitrification [Eqn (5.5d)].

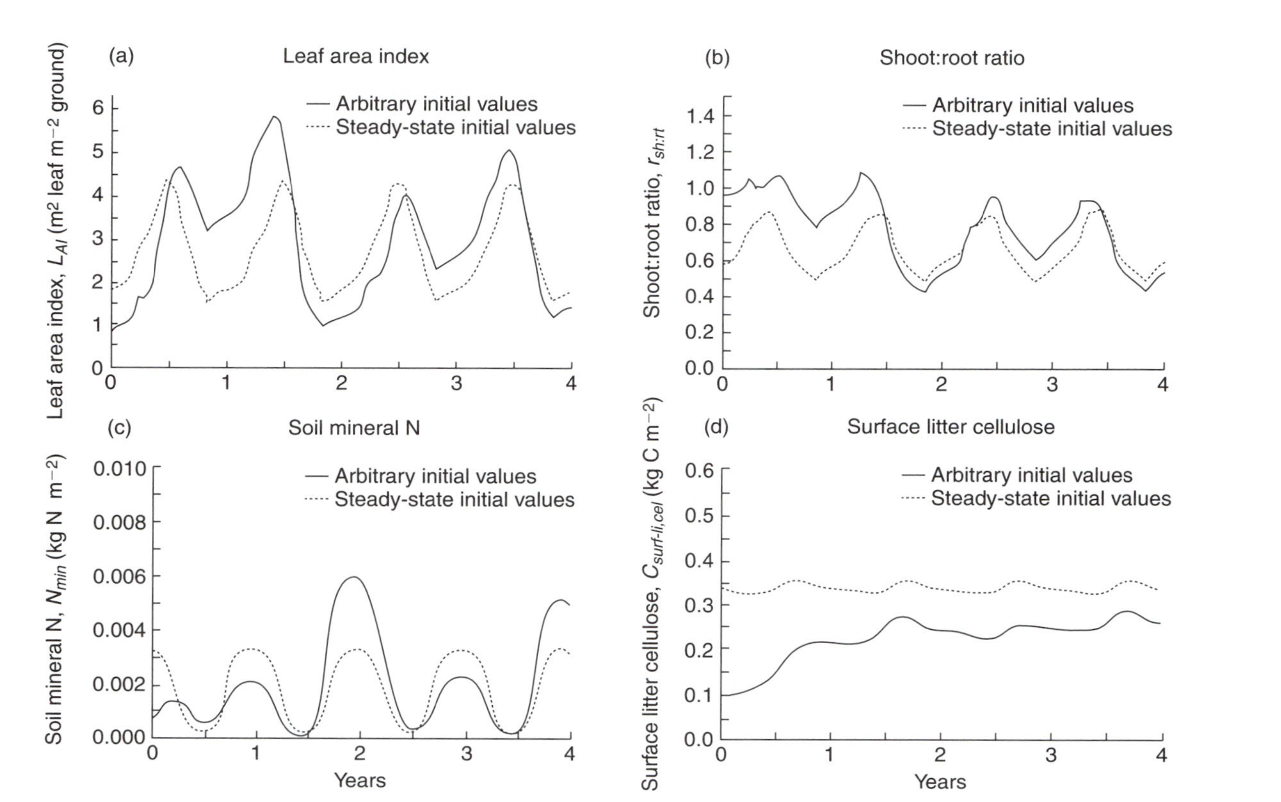

Fig. 8.4. Seasonally varying quantities over 4 years. The model was run from 1 January for 4 years with the standard southern Britain environment (Chapter 7) and parameters. Arbitrary initial values (as given in the equations and tables) are compared with steady-state initial values. (a) Leaf area index, L_{AI} [Eqn (3.1b)]. (b) Shoot:root ratio (Eqn (3.1l)]. (c) Soil mineral N pool, N_{min} [Eqn 5.5g)]. (d) Surface litter cellulose pool, $C_{surf\text{-}li,cel}$ [Eqn (5.1a)].

(Section 8.14): successive peaks and troughs lie above then below the steady-state curves. This damped oscillation is more apparent in Fig. 8.4c where soil mineral N, a labile variable, is shown. Slower variables, such as surface litter cellulose (Fig. 8.4d), move gradually towards their steady-state values. Although the first season has highly atypical seasonal dynamics, by the third year the seasonal dynamics is well-established in its normal pattern, although a slow baseline drift owing to changes in the litter and soil organic matter pools can continue for many years. The baseline drift is small in the present illustration because, fortuitously, the arbitrary initial values of the slower state variables are not too far from their long-term steady-state values. The baseline drift in L_{AI} seasonal dynamics can be considerable as soil organic matter levels shift upwards or downwards with consequent large changes in N nutrition. The lignin and soil organic matter pools (Fig. 5.1) have negligible seasonal variation.

Figure 8.4 and similar simulations underline some of the fundamental difficulties and likely futility of looking for quantitative agreement between experimental data over one or two seasons and the predictions of a simulator. These results suggest that experimental investigations of seasonal dynamics should extend over at least two seasons, that qualitative responses are more important than quantitative agreement, and that it is probably mistaken to imagine it will add to our understanding of ecosystem function to undertake excessive parameter adjustment to obtain a precise numerical agreement. While the sceptic might view this problem as indicating the inappropriateness of the initial conditions used in the simulation, the modeller's position is that the model is always a simplified representation of reality that can be understood and used, and in any case our knowledge of reality remains uncertain in many important areas, so it would be impossible to represent reality more exactly even if we wanted to.

8.3.3 Short-term seasonal responses to grazing intensity

Figure 8.5 demonstrates the general ability of the model to generate an appropriate rate of leaf area indices (and therefore dry matter intake or yield, Fig. 4.2), given a range of grazing regimes. The steady-state initial values corresponding to 12 sheep ha^{-1} (summer stocking) are used in every case, but the summer stocking density (1 April to 31 October) is given values of 5 and 20 sheep ha^{-1}. Thus, from 1 April, at 5 and 20 sheep ha^{-1}, the system is no longer in an annually repeating steady state. The simulations show that stocking density has a profound effect on the short-term properties of the plant–soil system (see Fig. 9.7 for long-term responses). Leaf area index (L_{AI}) is decreased by increased stocking (Fig. 8.5a), but while shoot:root ratio (Fig. 8.5b) is initially decreased, its subsequent time course is more complex. The immediate effect of increased stocking on soil mineral N, N_{min} (Fig. 8.5c) is to increase N_{min}. However, in the second year, the N_{min} values are decreased relative to the 12 sheep ha^{-1} simulation for some of the year. The litter pool (Fig. 8.5d) is showing the longer term dynamics which eventually result in a less productive system (in terms of net primary production) for a more severely grazed sward.

8.3.4 Seasonal response to constant leaf area index (LAI) grazing

In Fig. 8.5 the stocking density is constant during the summer, when the sward grows faster than animal intake can respond, leading to a variable L_{AI} (Fig. 8.5a). In

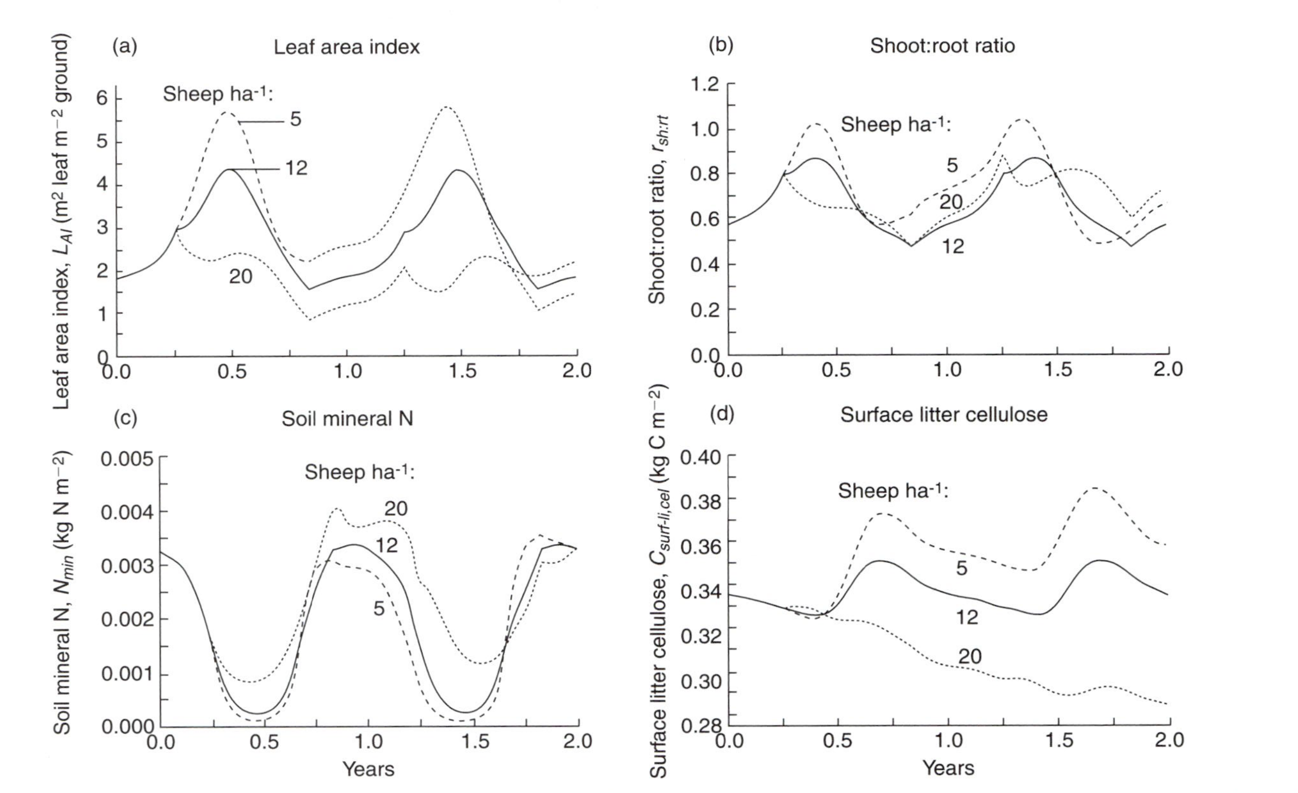

Fig. 8.5. Effect of stocking density over 2 years. The standard southern Britain environment (Chapter 3) with summer stocking of 12 sheep ha^{-1} was used to give steady-state initial values for all three simulations. The three values of summer stocking density are 5, 12 (standard steady-state) and 20 sheep ha^{-1}. (a) Leaf area index, L_{AI} [Eqn (3.1b)]. (b) Shoot:root ratio (Eqn (3.1l)]. (c) Soil mineral N pool, N_{min} [Eqn 5.5g)]. (d) Surface litter cellulose pool, $C_{surf\text{-}li,cel}$ [Eqn (5.1a)].

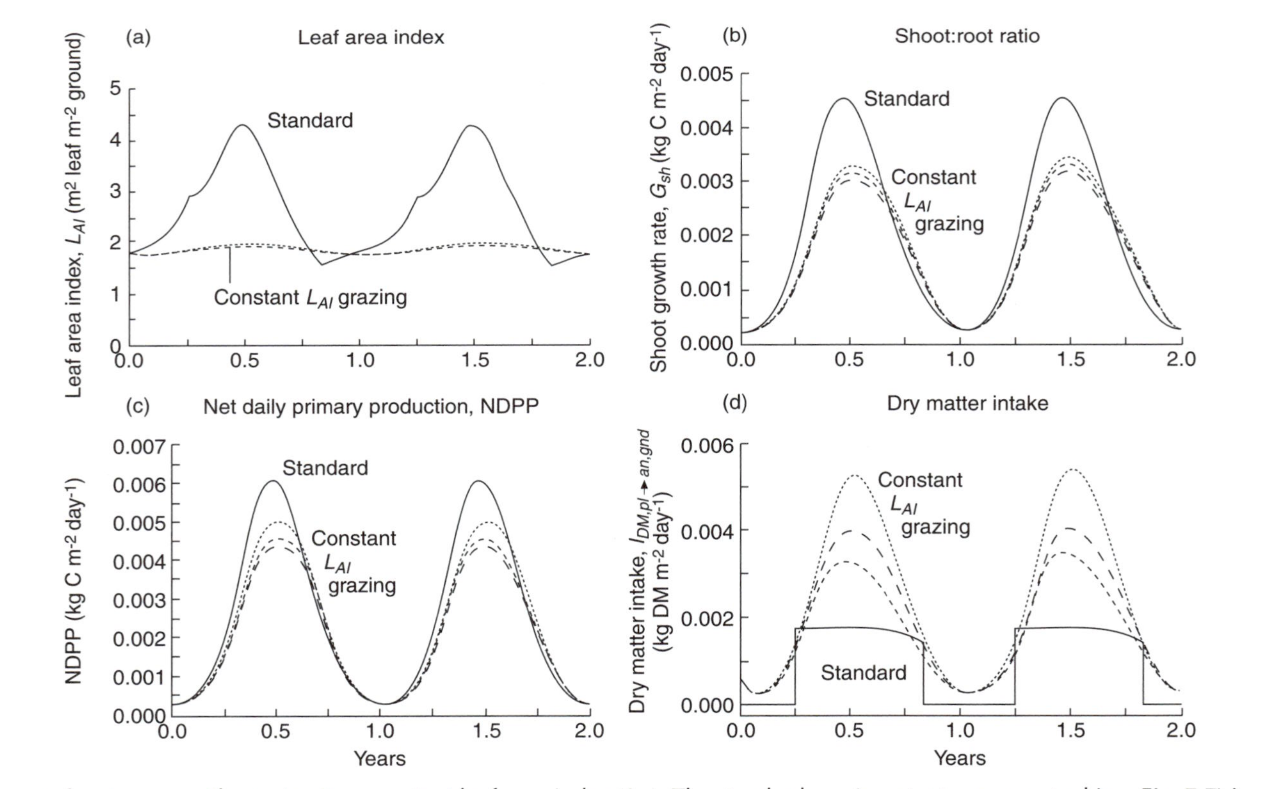

Fig. 8.6. Seasonal responses with grazing to a constant leaf area index (L_{AI}). The standard run (constant summer stocking. Fig. 7.7) is compared with three constant L_{AI} runs (see text): full model, (––); model with no water stress (---); model with no water stress and constant soil mineral N N_{min} (····). (a) Leaf area index, L_{AI} [Eqn (3.1b)]. (b) Shoot growth rate, G_{sh} [Eqn (3.3c)]. (c) Net daily primary production [see after Eqn (8.1a)]. (d) Dry matter intake by animals, $I_{DM,pl\rightarrow an,gnd}$ [Eqn (4.2b)].

practice, many farmers vary the stocking density so as to maintain a constant sward height, which, here, corresponds to a constant L_{AI} [Eqn (3.8a)], and many experimental studies are carried out with similar protocols (e.g. Orr *et al.*, 1988, 1995). In Fig. 8.6, the parameters of the animal intake function [Fig. 4.2, Eqns (4.2a)] have been altered so the sward L_{AI} stays close to 2, by taking $K_{LAI,an}$ = 2 m^2 leaf (m^2 ground)$^{-1}$, $q_{LAI,an}$ = 30, $I_{DM,pl \to an,max}$ = 10 kg DM animal^{-1} day^{-1}, and stocking throughout the year with [Eqn (7.6f)] $t_{stock,1} = 0$, $t_{stock,1} = 365$ at 10 sheep ha^{-1}. The intake function is now a step function with a high asymptote. To facilitate comparison with experiment, the simulations are also performed without water stress and additionally with a constant soil mineral N, N_{min}, to mimic the effects of frequent watering and fertilizer application.

Figure 8.6a confirms that L_{AI} is effectively held constant. In Fig. 8.6b, the shoot growth rate G_{sh} [Eqn (3.3c)) is effectively the intake to the first lamina compartment [Fig. 3.1; Eqn (3.6b)]. In Fig. 8.6c, net daily primary production, NDPP [see after Eqn (8.1a)] is drawn. The application of constant L_{AI} grazing decreases the peaks in G_{sh} and NDPP, and moves the peaks later towards midsummer. The three constant L_{AI} treatments (full model, no water stress, no water stress and constant soil mineral N) have little effect on G_{sh}, and a slightly larger effect on NDPP. However, in Fig. 8.6d, the three constant L_{AI} treatments have a substantial effect on dry matter intake by the animals, $I_{DM,pl \to an,gnd}$ [Eqn (4.2b)], indicating that the age structure of the canopy is changing substantially during this period (Parsons *et al.*, 1988b). In our units, Orr *et al.*'s (1988, Fig. 1b) intakes range from 0.002 to 0.009 kg DM m^{-2} day^{-1} during the April to October period.

The impacts of frequency and severity of defoliation on carbon fluxes have long been a controversial topic, and they have been much investigated with models (e.g. Johnson and Parsons, 1985b; Parsons *et al.*, 1988a), and by experimentation (e.g. Parsons *et al.*, 1983; Orr *et al.*, 1988, 1995). Some seasonal patterns of growth, measured under unusual or infrequent cutting regimes may generate phenomena dependent on the reproductive development of the sward (e.g. Corrall and Fenlon, 1978), which is not represented here (Section 3.1).

8.4 Fertilizer Responses to Nitrogen

Figure 8.7 gives the seasonal effects of N fertilizer application with harvesting on leaf area index and productivity. Fertilizer is applied three times and the sward is cut three times, as in Fig. 8.7a. The leaf area index curve (Fig. 8.7b) reveals how, in the absence of fertilizer, there is very little recovery from the first cut, which occurs at a time when soil mineral concentrations are low (Fig. 8.7c). Note how cutting without fertilizer increases midsummer mineral N concentrations due to the decreased plant uptake, whereas autumn mineral N concentrations are decreased due to the decrease in litter inputs. The accumulating yield is given in Fig. 8.7d. The timely addition of fertilizer makes a big difference to the harvested dry matter yield. The yield, at about 10 t DM ha^{-1}, is typical of southern Britain fertilized grassland (e.g. Whitehead, 1995, p. 225).

A fertilizer response over a growing season is given in Fig. 8.8. The same protocol as in Fig. 8.7 has been followed, but with different total amounts of N applied. The standard response agrees well with that given in Fig. 2.15 of Robson *et al.*

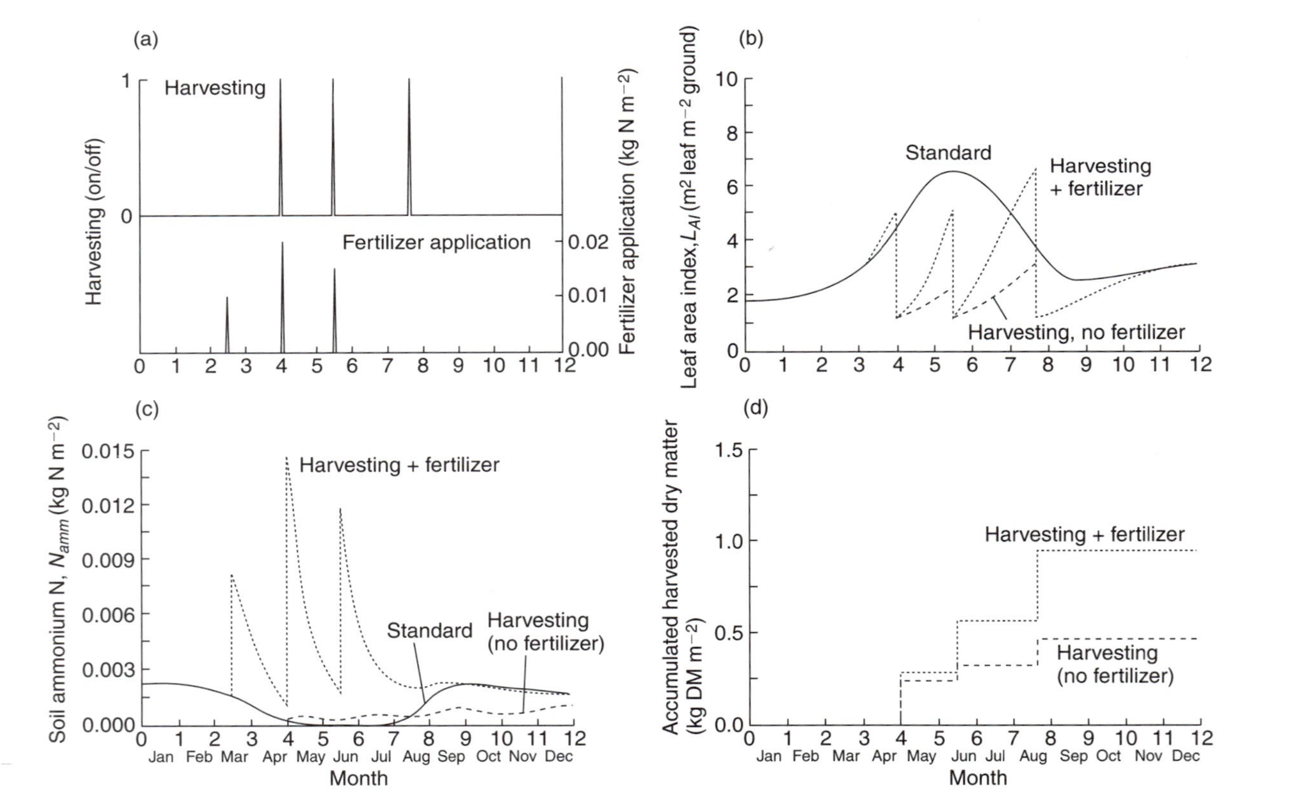

Fig. 8.7. Effects of fertilizer application. 100, 200, 150 kg N ha^{-1} is applied on 17 March, 3 May, 18 June; 70% of the N is ammonium N, the rest is nitrate N (Section 7.7.1). The sward is cut to 3 cm on 2 May, 17 June, 22 August (Section 7.7.2). The standard south Britain grazed environment (Chapter 7) and parameter values were used, with steady-state initial values on 1 January. (a) Fertilizer application and times of harvesting. (b) Leaf area index. (c) Soil ammonium pool. (d) Cumulative yield.

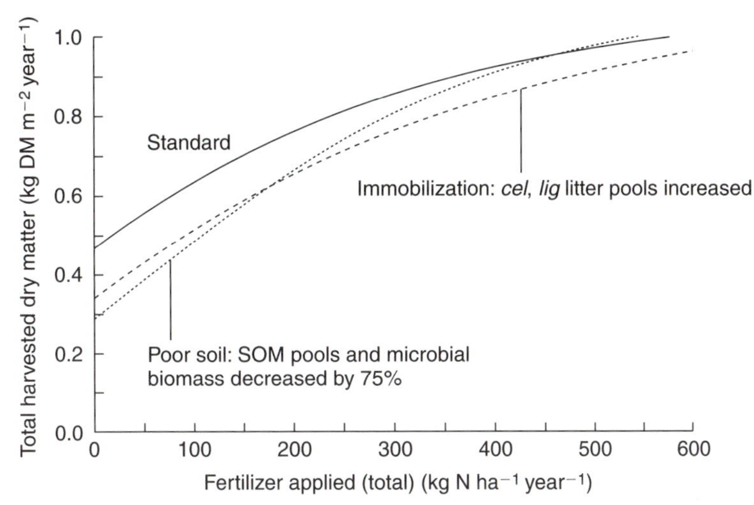

Fig. 8.8. Fertilizer responses. The environment, sward cutting and fertilizer regimes are as in Fig. 8.7 but using different total quantities of fertilizer as given. The standard curve is for the southern British environment steady-state initial conditions. For the second curve the initial values on 1 January of the litter cellulose and lignin pools (Fig. 5.1) were doubled to 0.67, 0.78 (surface), 0.8, 0.91 (soil) kg C m^{-2} to mimic the effects of adding straw. For the third curve the six initial values of the three SOM pools, and the microbial biomass pool (Fig. 5.1) on 1 January were decreased by 75%, to mimic the effects of a poor soil.

(1989, p. 56) for a grass–clover sward. The initial values are steady-state values for the southern British environment, which is a grazed system. Note that, the default model assumes a constant N input from the environment (wet and dry deposition) to the N_{amm} pool of 50 kg N ha^{-1} year^{-1} [$I_{N,env \to Namm}$, Eqn (5.4a), Table 7.2]. Also in the southern Britain environment steady state there is an assumed input from non-symbiotic N fixation by soil microorganisms of about 2 kg N ha^{-1} year^{-1} [$I_{N,fix \to Namm}$, Eqn (5.4c)].

The response to N addition depends on the initial values, both those of the litter pools and those of the SOM pools. This is illustrated in Fig. 8.8. In one simulated response the initial values of the three SOM pools and the microbial biomass pool (Fig. 5.1) are decreased by 75% from the standard values, to mimic pool soil conditions. In the other simulated response the initial values of the cellulose and lignin litter pools (Fig. 5.1) are increased, to simulate the effects of adding straw to the soil. These responses describe two types of 'immobilization'. Arable land has equilibrium SOM values well below those for grassland (Adger *et al*., 1992, Table 2). For many years after cultivated land is planted with grass, mineralization rates will be low, and there will be a net flow of C and N into the SOM pools which will slowly move towards the steady-state levels. This can decrease the response to added N especially at low N. In the other simulation of Fig. 8.8, extra cellulose and lignin ('straw') is added. This pulse of high C:N material can be seen to predict a temporary reduction in the availability of mineral N for plant growth. This has major relevance for some agricultural situations (e.g. in the New Zealand dairy industry) where previously unfertilized pastures begin to receive external mineral N inputs as the intensity of agriculture is stepped up, and there can be uncertainty and confusion about the expected response to N fertilizer.

Whitehead (1995, p. 223) reviews fertilizer responses in different grasses. His tables of recommended fertilizer N levels for different soils in the UK (pp. 212–213) range from 250 to 420 kg N ha^{-1}.

8.5 Responses to Stocking Regimes

Many grazing regimes for grassland have been used, each providing various degrees of control of pasture and animal performance, and each requiring various inputs of management and capital (Holmes, 1989, Fig. 4.6, p. 152). Often grazing is combined with cutting and conservation of forage. Using Eqn (7.6f) to define stocking, any grazing pattern can be simulated, and this can be combined with any pattern of cutting (Section 7.7.2), fertilizer application (Section 7.7.1) and rainfall/irrigation (Fig. 7.2).

Parsons *et al.* (1988a) have given an in-depth modelling study of defoliation patterns and productivity. Using a precursor of the present model which assumes a constant environment and no limitations from water or nutrients, they concluded that maximum yield is achieved when the leaf area index (L_{AI}) is maintained close to the value which gives maximum yield under continuous grazing. They suggested that a target L_{AI} could provide a basis for rational pasture management. Using our model, decoupled from water and the soil and in a constant temperate summer environment, Fig. 3.9c shows that net shoot growth rate is maximum when L_{AI} is *c.* 2.6. Arguably, this L_{AI} value for maximum shoot growth rate is relatively independent of environment, being affected largely by the balance between light interception, photosynthesis and respiration, and therefore may not vary greatly during the season.

Additionally, Fig. 3.9c indicates that as long as the L_{AI} is between about 2 and 4, shoot growth rate is quite close to its maximum. However, note that in Fig. 8.6d, where the pasture is maintained at a constant L_{AI} through grazing, dry matter intake is affected considerably by water and nutrient limitations.

8.5.1 Comparison of two stocking regimes

Here, as an example of many possibilities, we compare a continuous grazing regime where the sward is grazed at 12 sheep ha^{-1} from 0 h on 1 April for 204 days until 24 h on 31 October (this is the standard default regime), with a 1-week-on/1-week-off regime from 1 April for 30 weeks (210 days) where the week-on stocking density is 24 sheep ha^{-1}. This gives the same average stocking density. The two stocking patterns are illustrated in Fig. 8.9a. Note that there are no fertilizer applications, although the steady-state initial conditions simulate a fertile soil, with annual mineralization of about 200 kg N ha^{-1} $year^{-1}$. None the less, during the productive early summer months, growth is decreased by low soil mineral N concentrations during this time (Fig. 8.2c). The effect on leaf area index is given in Fig. 8.9b, and it may be noted that, although there is no adjustment of grazing intensity in order to keep L_{AI} constant, the L_{AI} falls mostly within the 2 to 4 band of maximum shoot growth

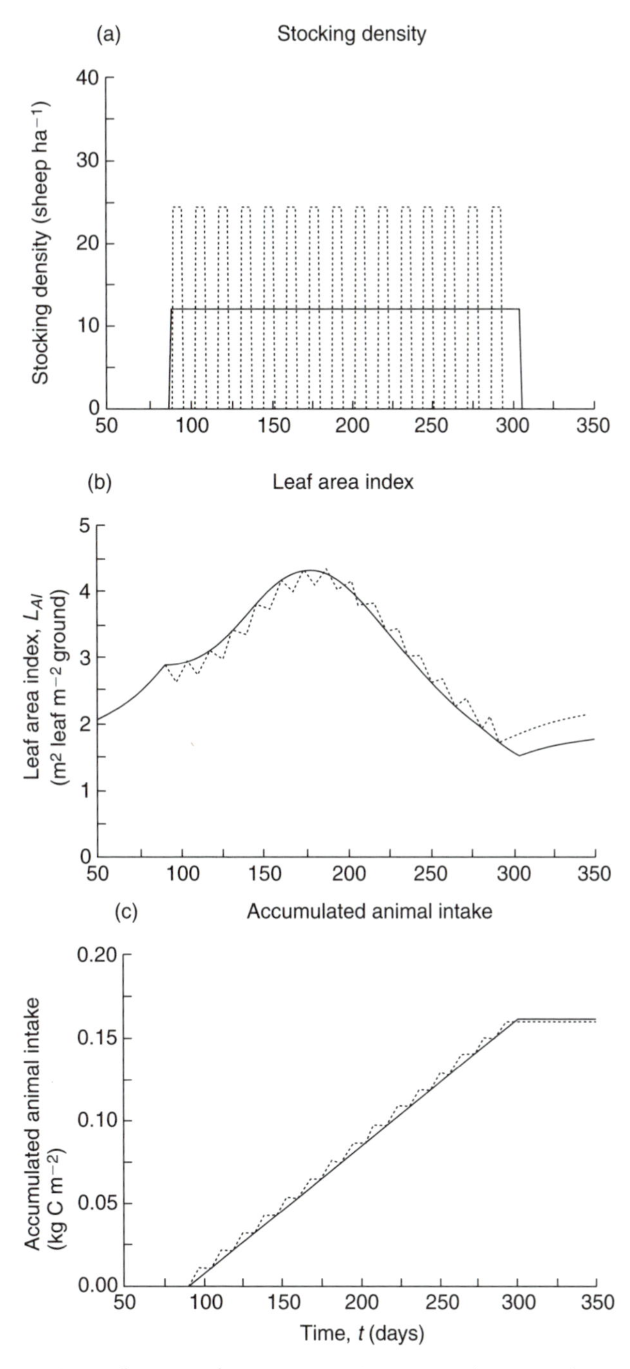

Fig. 8.9. Comparison of two stocking regimes. Seven months (1 April to 31 October) of constant stocking density of 12 sheep ha^{-1} is compared with week-on/week-off rotational grazing at 24 sheep ha^{-1}. Otherwise, default values are used, with steady-state initial values and no fertilizer application. (a) Stocking density. (b) Leaf area index. (c) Accumulated animal intake.

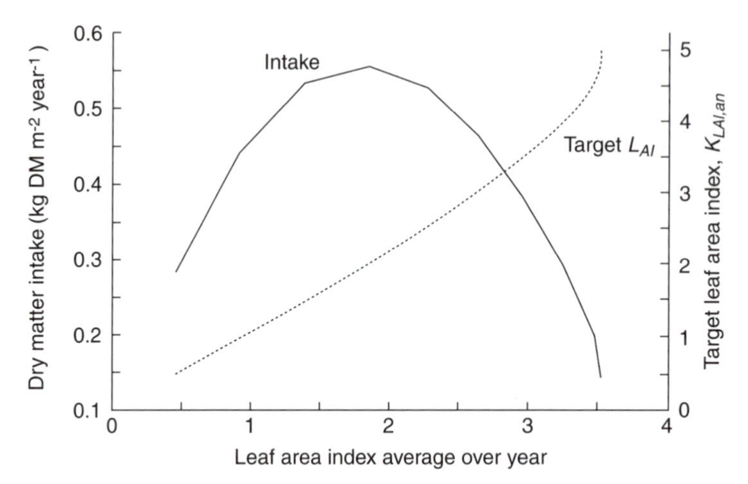

Fig. 8.10. Yearly intake at different maximum leaf area indices (L_{AI}). The sward is grazed with a step-function intake equation (see text). The target leaf area index ($K_{LAI,an}$) is a maximum value which is not always achieved owing to environmental factors. Dry matter intake is shown as a function of average L_{AI} over the year. Except for the intake and grazing functions, default values are used, with steady-state initial values and no fertilizer application.

rate (see above and Fig. 3.9c). The accumulating C input to the animal (kg C m^{-2}) in Fig. 8.9c reveals that there is very little difference between the two grazing systems. With a 4-week-on/4-week-off rotational pattern, the rotational system is about 5% ahead in accumulated animal intake after the third 8-week period.

8.5.2 Intake response to leaf area index with variable grazing

Similarly to the simulations of Fig. 8.6, the parameters of the animal intake function [Fig. 4.2, Eqns (4.2a)] are changed so the sward L_{AI} stays close to the value of $K_{LAI,an}$ which is ascribed a range of values. We assign $q_{LAI,an} = 30$, $I_{DM,pl \rightarrow an,max} = 10$ kg DM animal^{-1} day^{-1}, and apply stocking throughout the year with [Eqn (7.6f)] $t_{stock,1} = 0$, $\tau_{stock,1} = 365$ at 10 sheep ha^{-1}. The intake function is now a step function with a high asymptote. However, seasonal variation in leaf area index (L_{AI}) does not always allow L_{AI} to stay close to its target value of $K_{LAI,an}$, as shown in Fig. 8.10. The graph of L_{AI} averaged over the year against target L_{AI} (the dashed line in Fig. 8.10) deviates from linearity above an averaged L_{AI} of about 1.8. The target L_{AI} functions as a maximum L_{AI}. Figure 8.10 shows that maximum intake is achieved when the average L_{AI} over the year is about 1.8, which occurs when the target L_{AI} is 2. A target L_{AI} of 5 only secures an annual average L_{AI} of 3.5 with low intake and little grazing.

8.6 Simulated Multi-harvest Experiment with Temperature Increase and CO_2 Enrichment

Some investigations of grassland involve repeatedly cutting the sward to a constant height and using the dry mass of the cuttings as a measure of growth, and sometimes

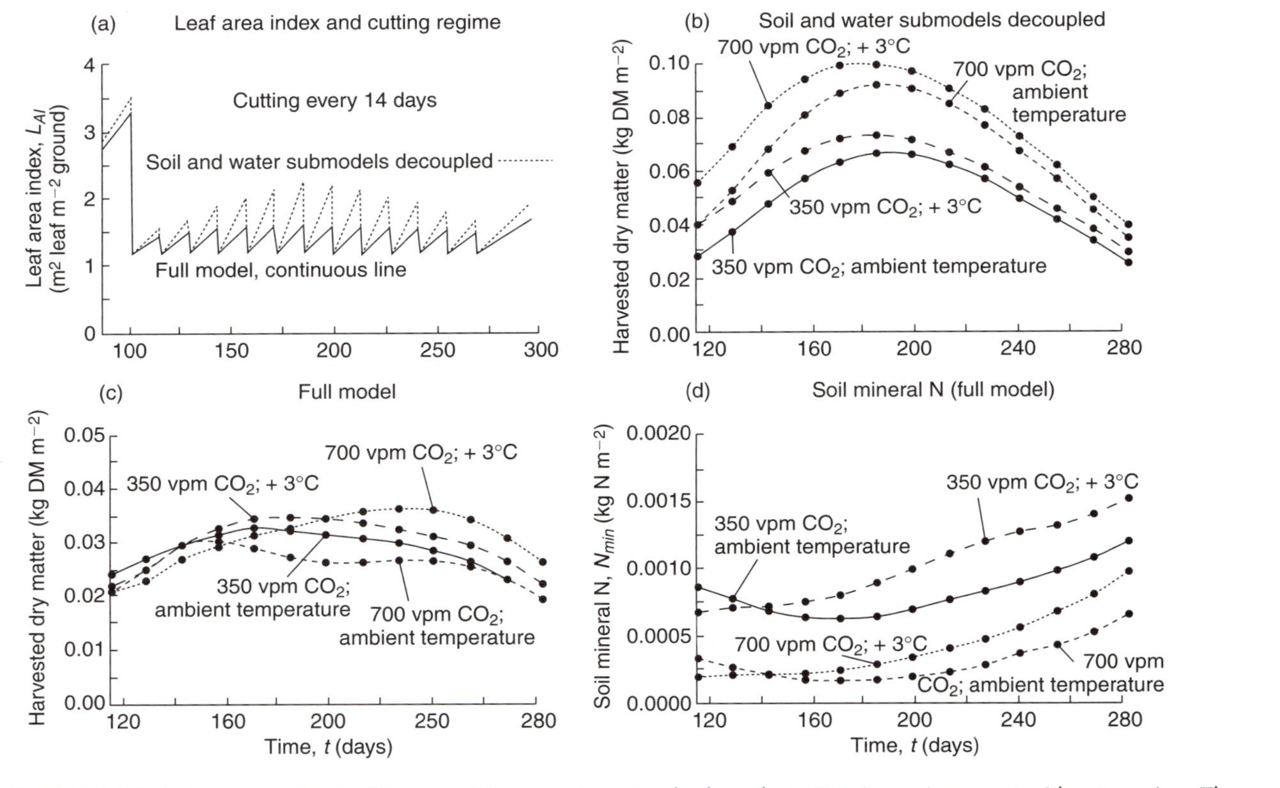

Fig. 8.11. Simulated multiple-harvest experiment. The sward is grown in a standard southern Britain environment without grazing. The sward is cut to a height [Eqn (3.8a)] of 3 cm on 12 April (t = 101 day) and at 2-week intervals until 27 September. Four treatments applied from 1 January: (i) standard environment (350 vpm CO_2, ambient temperatures); (ii) 350 vpm CO_2, mean air and soil temperatures increased by 3°C; (iii) 700 vpm CO_2, ambient temperatures; (iv) 700 vpm CO_2, mean air and soil temperatures increased by 3°C. The radiation values have been reduced by 20% (see text). Initial values on 1 January are for the steady state under standard conditions with grazing. (a) Leaf area index under the harvesting regime used, standard conditions of temperature and CO_2, using the full model and with the soil and water submodels decoupled. (b) Dry mass collected at each harvest for the four treatments, with the soil and water submodels decoupled. (c) As for (b) but with the full model. (d) Soil mineral N concentration for the four treatments running the full model.

of net primary production. The methodology of such experiments has been widely described (e.g. Jones *et al.*, 1994, 1996). Figure 8.11a illustrates the simulation of such an experiment, and the effect on the time course of leaf area index (L_{AI}) of cutting every 2 weeks. This is shown for the full model, where the soil mineral nitrogen concentration varies and water stress occurs (Figs 8.2c, d), and also where the soil and water submodels are disabled so that soil mineral N remains constant at its high 1 January value (Fig. 8.2c) and there is no water stress.

Four treatments are applied: (i) no change (350 vpm CO_2, ambient temperature); (ii) 350 vpm CO_2, mean air and soil temperatures increased by 3°C; (iii) 700 vpm CO_2, ambient temperatures; (iv) 700 vpm CO_2, mean air and soil temperatures increased by 3°C. The four treatments are applied to the standard southern British climate (Chapter 7). The radiation values have been reduced by 20% to allow for the shading effect of the chambers. The first cut, which has a very high yield, is not shown.

Not only does removing nutrient and water stress greatly enhance plant growth (Fig. 8.11a; cf. Fig. 8.11b to Fig. 8.11c), but it changes the qualitative nature of the response substantially (cf. Fig. 8.11b to Fig. 8.11c). It is probably realistic to assume that the effects of frequent watering and fertilizer application are to partially decouple the soil and water submodels, giving a response which is somewhere between Fig. 8.11b and Fig. 8.11c. The extent of the decoupling will depend on other factors of management and the environment. While the experimental results obtained in such investigations are often quite variable, our predictions agree with many observations in experiments of this type (e.g. Baxter *et al.*, 1994; Newton *et al.*, 1995, 1996; Wolfenden and Diggle, 1995; Casella *et al.*, 1996; Jones and Jongen, 1996; Jones *et al.*, 1996).

The consequences of allowing soil nutrient status and water status to vary are marked, leading to increased complexity of the predictions. The dry matter yields decrease by some 50%, but, compared with Fig. 8.11b, there is no longer such a clear pattern to the response. There are several cross-overs during the season. Some of this complexity derives from the seasonal behaviour of the soil submodel, interacting with the plant submodel. In Fig. 8.11d, the soil mineral N concentration N_{min} is shown. Increased CO_2 decreases N_{min} greatly, but increased temperature increases mineralization rates and N_{min} (e.g. Rogers *et al.*, 1994; Soussana *et al.*, 1996; Cannell and Thornley, 1998). The processes giving rise to Fig. 8.11c can be understood in terms of the mechanisms represented in the model by examining plant internal substrate concentrations which affect growth rates and shoot:root allocation.

Table 8.1 gives the total harvested dry mass over such an experiment, and also total plant dry mass at the end of the year. For the full model (columns 3 and 4), increasing temperature decreases end-of-year plant dry mass (column 4) at both CO_2 concentrations, but increases total harvested dry matter (column 3) by 18% (350 vpm CO_2) and 27% (700 vpm CO_2). Increasing CO_2 at either temperature increases total harvested dry mass (column 3) by 5% (ambient temperature) and 13% (ambient + 3°C), but increases total end-of-year plant dry mass (column 4) by 34% and 35% respectively. When the soil and water submodels are decoupled (columns 5 and 6), increasing temperature now increases harvested dry mass (column 5) by 22% (350 vpm CO_2) and 23% (700 vpm CO_2), and continues to decrease plant dry mass (column 6) at both CO_2 concentrations. Increasing CO_2 increases

Table 8.1. A simulated cutting experiment.*

		With soil and water submodels†		Without soil and water submodels‡	
Temperature	CO_2¶	Total harvested dry matter‖	Total plant dry matter**	Total harvested dry matter‖	Total plant dry matter**
Ambient	350 vpm	0.497	0.333	0.778	0.366
+3°C§	350 vpm	0.585	0.265	0.948	0.293
Ambient	700 vpm	0.521	0.446	1.066	0.518
+3°C§	700 vpm	0.662	0.359	1.309	0.404

* See Fig. 8.11 and text.
† Full model including soil and water submodels.
‡ Soil and water submodels are disabled, to simulate the effects of frequent fertilizer application and watering.
§ Mean annual values are increased by 3°C with the seasonal variation unchanged.
¶ CO_2 concentration in the atmosphere.
‖ Total over the cutting season: 12 April to 27 September (kg dry mass m^{-2}), excluding the first harvest.
** Total plant dry mass (standing shoot + root) at the end of the year (kg dry mass m^{-2}).

harvested dry mass (column 5) by 37% (ambient temperature) and 38% (ambient + 3°C), and increases end-of-year plant dry mass (column 6) by 42% and 38% at the respective temperatures. The results in Table 8.1 are consistent with the yield ratios (usually defined as the ratio of yield for twice-ambient CO_2 to that for ambient CO_2) reported by Newton (1991, table 1), which range from 1.10 to 1.76, from observations made over periods of 21 to 129 days. Note that increased CO_2 always has a large positive effect on total plant dry matter, defined as root plus standing shoot (Table 8.1, columns 4, 6), but when availability of nutrients and water may be influencing response, the effect of increased CO_2 on harvested shoot dry matter may be small (column 3). Indeed, in some simulations the model has predicted a negative effect of increased CO_2 on harvested dry matter for the fully coupled model. Although it is not shown here, the higher temperature leads to higher mineralization rates and a higher soil mineral N concentration (Fig. 8.11d); also higher CO_2 leads to larger roots which are more active per unit of root dry mass in N uptake and decrease soil mineral N. A negative response of harvested shoot dry matter to +CO_2 can arise from the large changes in shoot:root allocation and root activity that occur in N limited conditions (Lutze and Gifford, 1995). It is to be emphasized that the predictions of Fig. 8.11 and Table 8.1 are of short-term responses, and the long-term responses can be quite different (Figs 8.15, 8.16; Chapter 9; Cannell and Thornley, 1998).

The only direct effects of increased CO_2 are to increase leaf photosynthesis [Eqns (3.2s), (3.2t)], and to decrease stomatal conductance [Eqn (3.2u)]. Temperature on the other hand affects most rate parameters in the system (Section 3.11; Section 5.12). It is far more difficult to disentangle cause and effect when temperature is changed than when CO_2 is changed. For instance, increased temperature may increase or decrease leaf photosynthesis depending on the irradiance level: low light emphasizes initial slope α which decreases with temperature above 15°C [Eqn

(3.2s)]; high irradiance emphasizes the asymptote P_{max} [Eqn (3.2t)] which increases with temperature over the range applied here [Fig. 3.7]. A 3-cm cutting height with a short interval between cuts maintains low leaf area indices where most leaves are fully illuminated. The between-harvest time interval may also affect the results substantially: a short time interval measures the initial growth rate after cutting, whereas growth during a longer time interval may be more influenced by the position of the asymptote.

The simulated experiments in Fig. 8.11 and Table 8.1 underline the difficulties of drawing valid conclusions from these apparently straightforward measurements. Experimentally it may be presumptious to assume that frequent watering and fertilizer application removes seasonal effects arising from water status and soil processes.

8.7 Long-term Time Course from Poor Initial Soil Conditions

The dynamics over many seasons determine the long-term consequences of changes in the environment and management. Some variables of particular interest are: the net annual primary production (NAPP), the total carbon sequestered in the soil-plant system (C_{sys}), leaf area index [L_{AI}, Eqn (3.1b)], and the soil mineral concentration [N_{min}, Eqn (5.5g)].

Instantaneous net primary production (N_{PP}) is defined as gross canopy photosynthesis less plant respiration

$$N_{PP} = P_{can} - R_{pl}, \tag{8.1a}$$

where the variables all have units of kg C m^{-2} day^{-1} and we have used Eqns (3.2f) or (3.2n) for canopy gross photosynthetic rate P_{can}, and Eqn (3.3m) for total plant respiration R_{pl} . Integration of N_{PP} over a day gives the net daily primary production, NDPP, and integration over a calendar year gives the net annual primary production, NAPP.

Although widely used, there are doubts concerning the value of this definition of net primary production, especially for grassland (e.g. Parsons, 1994). For a cut or a grazed sward, NAPP does not equate to potential agricultural yield, or to the inputs to the litter and soil submodels. This is because cut material may not be returned to the soil, and under grazing, a large fraction of the carbon ingested by the animal is respired (Fig. 4.3) and a further fraction is retained in animal growth and may be removed from the system. Parsons (personal communication) has suggested that it may be preferable to define (i) canopy photosynthesis, P_{can}; (ii) gross tissue production (GTP) as canopy photosynthesis less plant respiration (R_{pl}), GTP = $P_{can} - R_{pl}$; (iii) and 'yield', $Y = P_{can} - (R_{pl} + S)$, where S comprises the plant inputs to the litter pools from tissue turnover, degradation (includes grazing by other fauna) and exudation (Fig. 3.1). Therefore yield equals gross tissue production less litter production, with Y = GTP $- S$. Yield is the agricultural yield, obtained from cutting or grazing the grassland and removing material from the system. We can also write $S = P_{can} - R_{pl} - Y$: the higher the yield, the smaller is litter production S. A grassland may have zero agricultural yield, but be heavily grazed; in this case the litter flows to the soil can be appreciably decreased, as illustrated in Fig. 4.5a.

There is a continuum between green plant tissue and senesced or degraded tissue. There is also an arbitrariness about just which tissues still belong to the plant, and may contribute to plant respiration, and exactly when the senescing or degrading tissues are ascribed to the soil and litter submodel, and may contribute to soil respiration. Given these difficulties, I shall use the 'ecological' definition of net primary production of Eqn (8.1a), because it is easy to calculate, and it has been widely used although it may be misleading especially for grassland. It is important that the limitations of Eqn (8.1a) for N_{PP} as a measure of the inputs to the soil system are appreciated.

C_{sys} is the total C in all the C-containing pools in the plant and soil submodels. C_{sys}, the corresponding quantity for N, N_{sys}, and the C:N ratio in the whole system $r_{C:N,sys}$, are defined by

$$\begin{aligned} C_{sys} &= M_{CS,pl} + f_{C,plX} M_{X,pl} + C_{sol} + C_{bio} + C_{SOM} + C_{surf-li} + C_{soil-li}, \\ N_{sys} &= M_{NS,pl} + f_{N,plX} M_{X,pl} + N_{\min} + \frac{C_{bio}}{r_{C:N,bugs}} \\ &\quad + N_{SOM} + N_{surf-li} + N_{soil-li}, \\ r_{C:N,sys} &= \frac{C_{sys}}{N_{sys}}. \end{aligned} \tag{8.1b}$$

The state variables and other variables in these equations are referenced in Tables 3.1 and 5.1.

The yearly averages of leaf area index L_{AI}, soil mineral N N_{min}, net annual primary production NAPP, total system C C_{sys}, and total system C:N ratio $r_{C:N,sys}$ are plotted in Fig. 8.12, over 10 years, 100 years and 1000 years. The initial values for the soil and litter variables (Chapter 5) correspond to a poor soil. The graphs begin at the end of the first year because yearly averages cannot be calculated until then. Over the first 10 years (Fig. 8.12a) there is a transient in leaf area index (L_{AI}), soil mineral N (N_{min}) and NAPP, with these three quantities increasing then decreasing. These early transients (Fig. 8.12a) can be identified with the lignin and unprotected soil organic matter pools (Fig. 5.5a). The build-up in surface and soil lignin with high C:N ratio (Fig. 5.1) greatly increases the system C:N ratio because the larger SOM pools with low C:N have not yet had time to increase. Over 10 to 100 years (Fig. 8.12b), leaf area index L_{AI} increases by some 20% although soil mineral N N_{min} only increases slightly; these changes have little effect on net primary production. Total system C C_{sys} increases substantially and the total system C:N ratio $r_{C:N,sys}$ decreases as the long-lived SOM pools further dominate the system. Over the 10 to 100 year time scale, the protected SOM pools (*pSOM*, Figs 5.1 and 5.5b) have the major dynamic effect on the system. As the SOM pools fill, the associated components of net N immobilization decreases towards zero, allowing mineral N levels to increase. The 1000-year run in Fig. 8.12c shows that total system C C_{sys} approaches an asymptote after about 1000 years, with a further small increase in soil mineral N N_{min}. However, leaf area index has reached a plateau after some 300 years, and NAPP does not change appreciably after 70 years, when light interception by the canopy has saturated. The slow build-up of soil organic matter is is one of the most important characteristics of grassland, and has been observed in many long-term studies (e.g. Whitehead, 1995, p. 84).

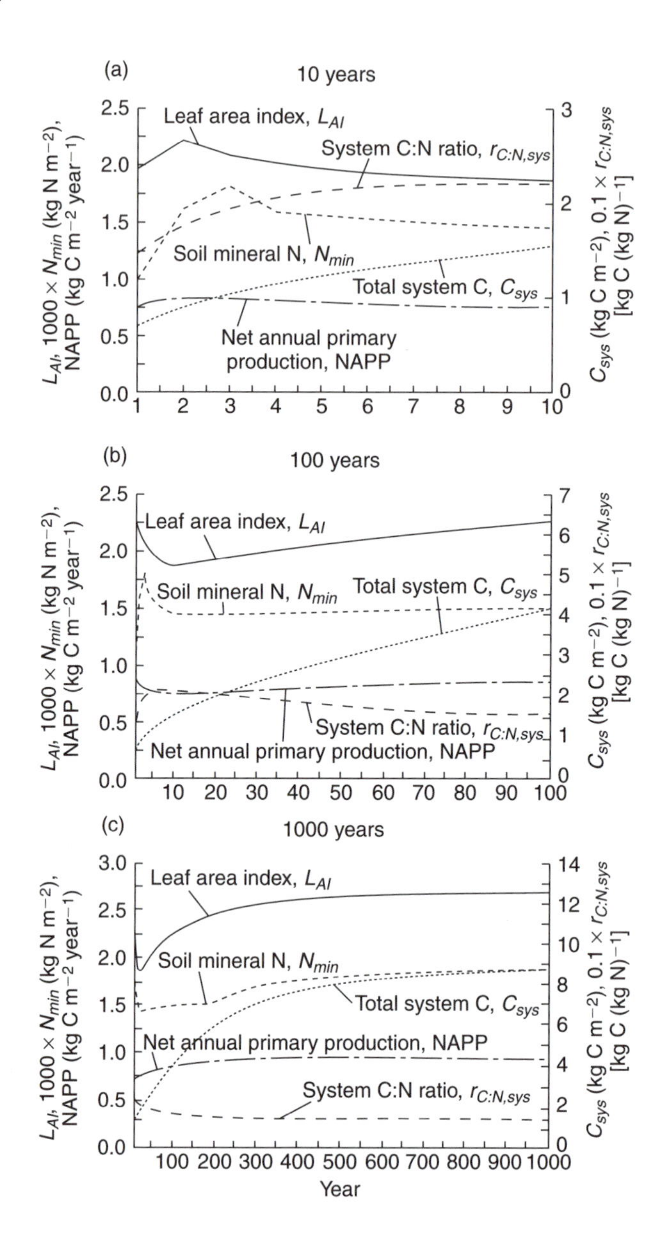

Fig. 8.12. Grassland dynamics over many years starting with a poor soil. Standard southern Britain environment. Initial values correspond to a poor soil, with (see Table 5.1) $C_{bio} = 0.02$, $C_{surf\text{-}li,cel} = C_{surf\text{-}li,lig} = C_{soil\text{-}li,cel} = C_{soil\text{-}i,lig} = 0.01$, $C_{uSOM} = C_{pSOM} = C_{sSOM} = 0.1$ kg C m^{-2}, $N_{uSOM} = N_{pSOM} = N_{sSOM} = 0.01$ kg N m^{-2}. The four variables: leaf area index [L_{AI}, Eqn (3.1b)], soil mineral N [N_{min}, Eqn (5.5g)], total system C [C_{sys}, Eqn (8.1b)], total system C:N ratio [$r_{C:N,sys}$, Eqn (8.1b)], are annual averages. Net annual primary production NAPP is instantaneous primary production [N_{pp}, Eqn (8.1a)] accumulated over a year. (a) Over 10 years. (b) Over 100 years. (c) Over 1000 years.

8.8 Responses to Gradual Climate Change

Figure 8.13 illustrates the predicted consequences of a gradual change in climate on five ecosystem variables: total system carbon C_{sys} [Eqn (8.1b)], net annual primary production NAPP [derived from N_{PP}, Eqn (8.1a)], leaf area index L_{AI} [Eqn (3.1b)], soil mineral N_{min} [Eqn (5.5g)], and the system C:N ratio $r_{C:N,sys}$ [Eqn (8.1b)]. The annual averages of these quantities are plotted. Three scenarios are considered: air

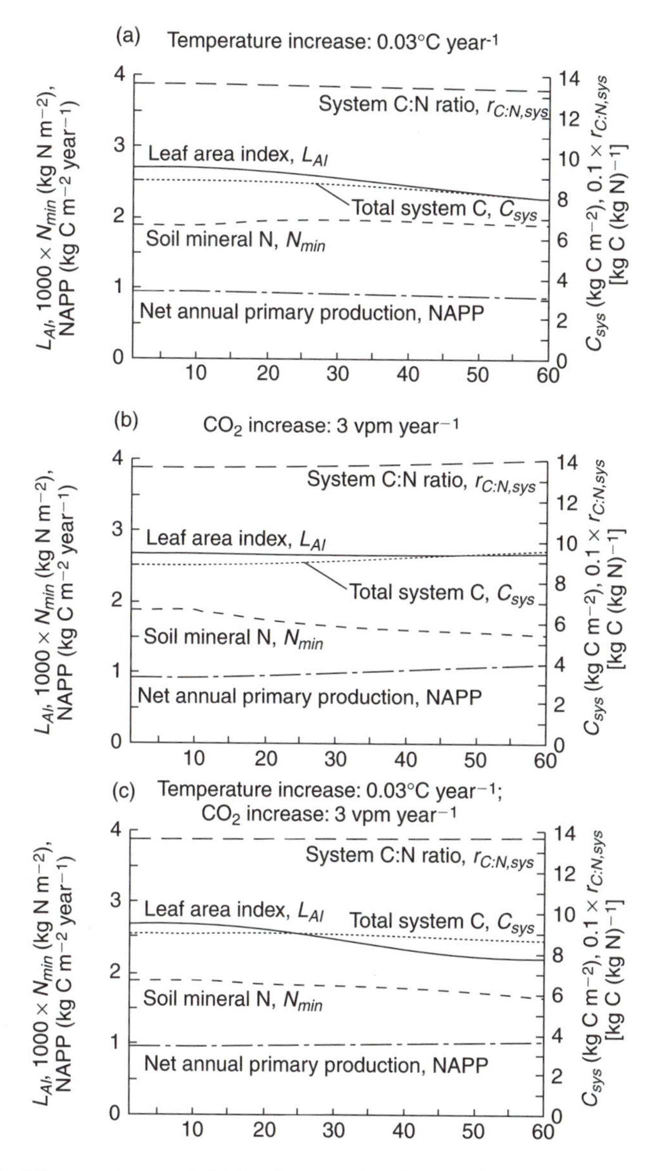

Fig. 8.13. Climate change. Initial values are for the steady-state southern Britain environment. The changing climate is imposed from year 10 onwards. Variables as in Fig. 8.12. (a) Temperature increase of 0.03°C year^{-1}. (b) CO_2 increase of 3 vpm year^{-1}. (c) Temperature and CO_2 increases of 0.03°C and 3 vpm year^{-1}.

and soil temperatures are increased by 0.03°C year^{-1} (Fig. 8.13a); atmospheric CO_2 is increased by 3 vpm year^{-1} (Fig. 8.13b); both temperature and CO_2 increases are applied together (Fig. 8.13c). These changes are imposed on the system, which is in a steady state for the first 10 years of the simulation. It can be seen that, with temperature increase alone (Fig. 8.13a), net primary production increases very slightly and then decreases, although there is a substantial decrease in leaf area index. Soil mineral N first increases, as mineralization increases relative to immobilization, and then decreases as the soil organic matter C and N pools decrease. In the long term, soil mineral N pools always decrease with increased temperature because the rate constants for the loss processes increase with temperature, and the N losses from the system, which are proportional to rate constant × pool size, remain substantially unchanged (see Figs 9.1 and 9.2). In Fig. 8.13b with CO_2 increase alone, the patterns are simpler than with temperature. Leaf area index decreases before increasing. Net primary production and the soil organic matter pools all increase. Soil mineral N decreases owing to increased root activity and an increase in immobilization relative to mineralization as the SOM pool increases. In Fig. 8.13c, temperature and CO_2 increase. Increasing primary production is insufficient in this example to prevent sequestered C from decreasing. Finally, we note that the system C:N ratio decreases slightly under temperature increase alone (Fig. 8.13a), increases under CO_2 increase alone, and is virtually unchanged when both are increased together.

8.9 Responses to Step Changes in the Environment

Figure 8.14 depicts how the system responds to a step change in temperature, CO_2, and atmospheric N input. These are applied at 10 years to a system in a steady state. The same system variables as in Fig. 8.13 are plotted.

An increase in temperature (Fig. 8.14a) causes a large but temporary increase in the soil mineral N concentration N_{min} due to higher mineralization rates of the SOM pools, but N_{min} then decreases towards a lower steady-state value (cf. Fig. 9.1a). In spite of this increase in soil mineral N, leaf area index L_{AI} falls substantially, reflecting the sensitivity of plant dry matter values to tissue turnover rates which increase with temperature. Net annual primary production decreases slightly with increasing temperature (cf. Fig. 9.1b; see discussion in penultimate paragraph of Section 8.6). The total C in the system C_{sys} decreases markedly as does the total system N (not shown). The system C:N ratio decreases initially as mineral N increases and then increases as mineral N decreases [e.g. Eqn (5.7b)].

The response to a step increase in CO_2 is easier to interpret (Fig. 8.14b). Again the soil mineral pool N_{min} exhibits a large transient, in this case a decrease followed by partial recovery. Leaf area index L_{AI} decreases initially but eventually recovers to a higher value than at 350 vpm CO_2. The positive response of net annual primary production NAPP is due to the leaf response to CO_2 [Eqns (3.2s), (3.2t)] which more than makes up for the initial decrease in L_{AI} which is already high enough to capture much of the incident light. The positive response of system C C_{sys} is substantial, and is accompanied by an increase in C:N ratio.

The response to a doubling in atmospheric N inputs (Fig. 8.14c) from 50 to 100 kg N ha^{-1} year^{-1} demonstrates the lack of response of a lightly grazed system that is already N rich, to further increases in N inputs. Mineral N levels increase, as they

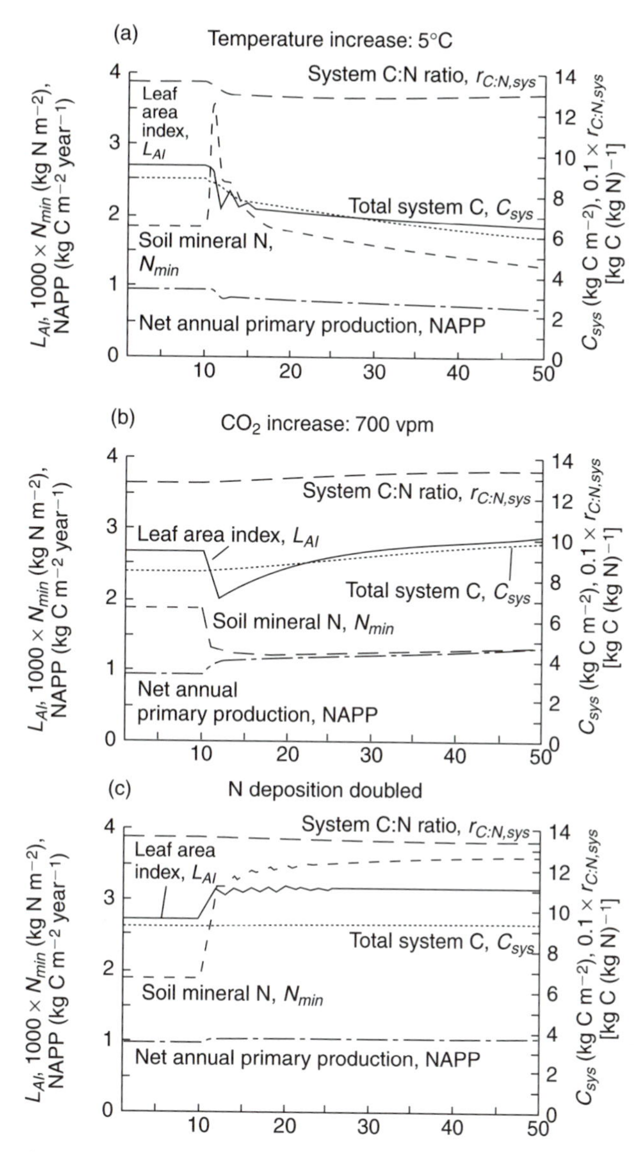

Fig. 8.14. Step changes in environmental factors. Initial values are for the steady-state southern Britain environment. Variables as in Fig. 8.12. The step changes are applied after 10 years of simulation. (a) Air and soil temperatures increased by 5°C. (b) CO_2 concentration doubled to 700 vpm. (c) Environmental N deposition [$I_{N,env \rightarrow Namm}$, Table 7.2, Eqn (5.4a)] doubled to 100 kg N ha^{-1} year^{-1}.

must, and the system C:N ratio decreases. Leaf area index responds via increased shoot allocation and lower plant sugar concentrations, although net primary production (NAPP) and sequestered carbon (C_{sys}) are not changed appreciably. Doubling the N inputs increases the instability of the system, as can be seen in the damped biennial oscillations of L_{AI} in Fig. 8.14c (see also Section 8.14). A high N grassland is more susceptible to water stress, and the water submodel is primarily responsible for these instabilities.

8.10 CO_2 Enrichment in N-poor and N-rich Grassland

In Fig. 8.15 the model is used to examine the short-term and long-term responses of a grazed southern Britain grassland with low and high N inputs to a step increase in ambient CO_2 from 350 vpm to 700 vpm. It is interesting to compare the short-term and the long-term responses of the N-poor grassland with those of the N-rich grassland. This account is based on a recent investigation of a northern Britain grassland (Cannell and Thornley, 1998), where further discussion can be found.

First note that the N-poor grassland undergoes a prolonged transient, and takes more than a thousand years to approach the new steady state after CO_2 doubling. This arises from the fact that the N-poor system can only accumulate N at a rate of a few kg N ha^{-1} $year^{-1}$ (Fig. 8.15h), and it therefore takes many years for the total system N (not shown) to increase by, say, 2000 kg N ha^{-1} to its new equilibrium value. In the N-rich system, the rate of accumulation of N is much faster as a fraction of total system N. Note also that immobilization of N (shown by a decrease of N_{min}) results from CO_2 doubling in both N-poor and N-rich grassland (Fig. 8.15d, l), but it is proportionately greater in the N-poor grassland and persists for much longer.

Second, it can be seen that although the short-term response of N-poor grassland can be very modest (e.g. system C, Fig. 8.15c; root uptake rate, Fig. 8.15f), in the long term, the response can be greater than that of the N-rich grassland to CO_2 enrichment. The large long-term response of N-poor grassland is partly due to the increased N nutrition of the grass, resulting from decreased N losses and increased non-symbiotic N fixation, which amplifies the CO_2 response much more in N-poor grassland than in N-rich grassland. The N-poor grassland is, as it were, on the steep portion of an N fertilizer response curve, and can thus respond to improved N nutrition, whereas the N-rich grassland is on the asymptote, so that improvements in N nutrition make no difference.

Third, note that, particularly in the N-poor grassland, the short-term response can be totally misleading if used to infer long-term behaviour (e.g. leaf area index, Fig. 8.15a).

Several important conclusions can be deduced from these simulations.

1. Plant ecosystems may employ the extra C acquired at higher CO_2 concentrations to acquire and retain nutrients. This may also apply to nutrients other than N.
2. In the long term, the response of nutrient-poor ecosystems may be proportionately greater than those of nutrient-rich ecosystems.
3. Short-term experiments, e.g. less than 5–10 years, on elevated CO_2 responses may be misleading, especially on nutrient-poor systems.
4. The temporal response of an ecosystem may not be determined by its internal rate constants, but by external factors. For example, it is often assumed that ecosystem response to environmental change is determined by the rate constants of soil organic matter pools. But here the limiting factor is the rate at which N is added to the system each year, compared with the amount of extra soil organic matter N required to reach equilibrium.

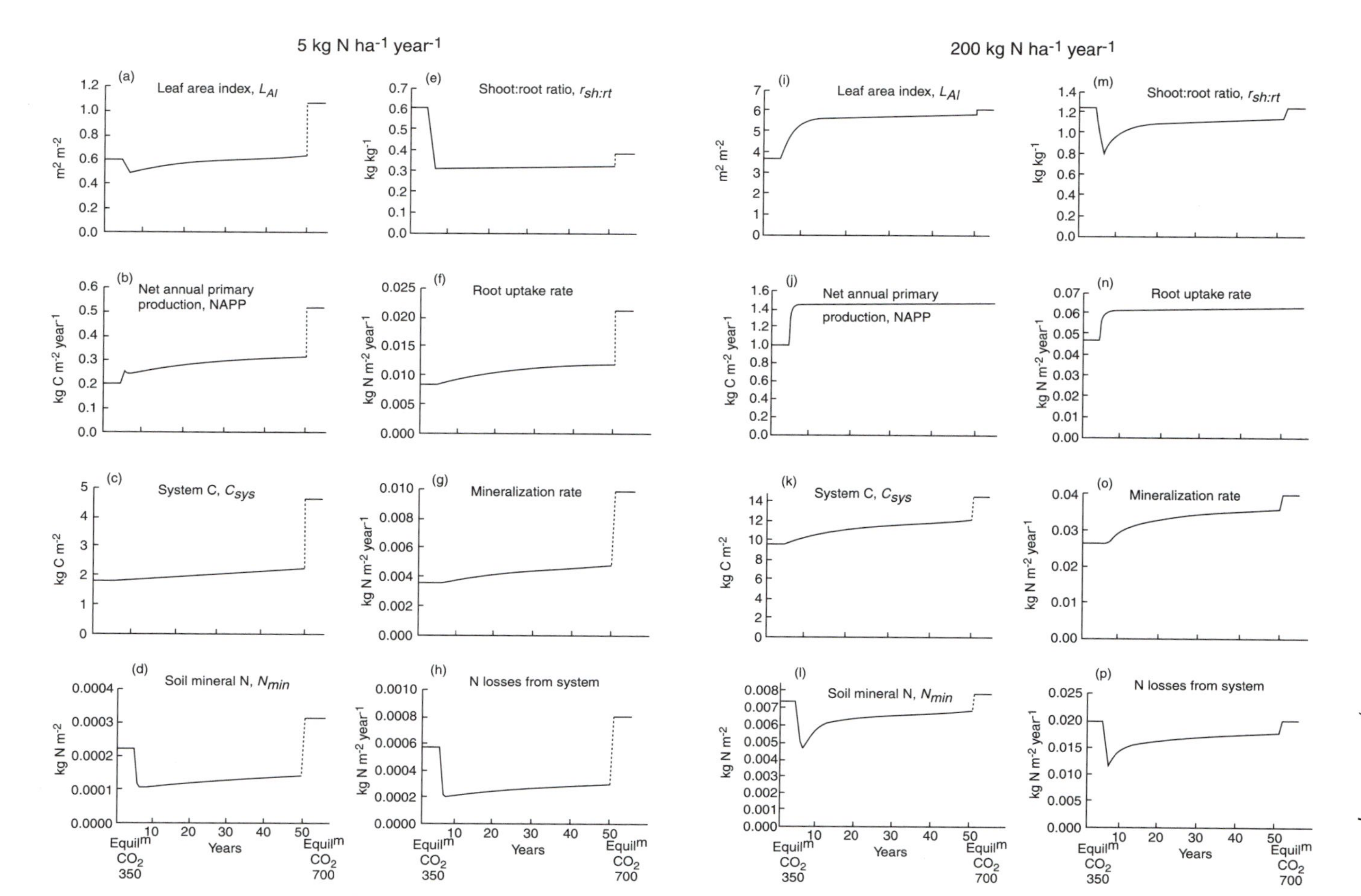

Fig. 8.15. CO_2 enrichment in N-poor and N-rich grassland. The standard southern Britain environment is used (Chapter 7), except that, for N-poor grassland, the N deposition rate ($I_{N,env \to Namm}$, Table 7.2) is set to 5 kg N ha^{-1} year^{-1}, and to 200 kg N ha^{-1} year^{-1} for N-rich grassland. After 5 years in the appropriate steady state at 350 vpm CO_2, the CO_2 concentration is doubled, and the simulation is continued until 50 years. The equilibrium values at 700 vpm CO_2 are indicated. All quantities are annual averages or annual sums.

5. Simple ecosystems models may not be able to simulate CO_2 responses adequately. While simple models can be elegant and transparent, they will be misleading if critical processes are omitted.

8.11 CO_2 Enrichment at Ambient Temperatures and at Ambient Temperatures + 5°C

Analogously to Fig. 8.15 and the last section, we now compare in Fig. 8.16 the short- and long-term responses to CO_2 doubling of southern Britain grassland at current temperatures to the responses at current temperatures +5°C. In each case the response begins from the steady state at 350 vpm CO_2.

First, compare Fig. 8.16 with Fig. 8.15: it is seen that the lower (ambient) temperature is more like the N-poor grassland, and the higher temperature is more like the N-rich grass in its responses. Although a sudden increase in temperature (Fig. 8.14a) temporarily increases mineralization and soil mineral N, long term, both mineralization rate and soil mineral N are decreased (compare the initial values of Fig. 8.16g with Fig. 8.16o, and of Fig. 8.16d with Fig. 8.16l). At ambient +5°C, soil mineral N is half its ambient value (Fig. 8.16l, d), and some pools such as system C (Fig. 8.16c, k) decrease to at least one-third, as does system N (from 0.74 at ambient to 0.21 kg N m^{-2} at ambient +5°C). The temperature factor on the biochemical rate constants (Fig. 3.6, cubic) is 0.7 at ambient +5°C compared with 0.4 at ambient temperature (annual averages), so that the turnover rate (proportional to flux/pool size) of most pools is much increased.

At both ambient and ambient +5°C the N deposition rate is 50 kg N ha^{-1} $year^{-1}$ (Fig. 8.16h, p), and whereas this constrains the temporal response of the ambient temperature grassland with 7400 kg N ha^{-1}, it does so much less in the ambient +5°C grassland with only 2100 kg N ha^{-1}.

Finally, we note that, although the temporal responses of the ambient +5°C grassland (Fig. 8.16i–p) resemble those of the N-rich grassland at ambient temperatures (Fig. 8.15i–p), the magnitude of the response of the ambient +5°C grassland to CO_2 enrichment is much closer to that of the N-poor grassland (Fig. 8.15a–h), with many important ecosystem variables such as system C, NAPP and mineralization rate being more than doubled.

In this case (cf. the discussion of Section 8.10), we can conclude that: (i) the proportional responses to CO_2 enrichment can be greater at higher temperatures; (ii) short-term experiments at modest temperatures can give misleading results.

8.12 Simulated Climate Change in Southern Britain from 1850

The effects of climate change from pre-industrial times (1850) on a southern Britain grassland are simulated in Fig. 8.17. The climate data applied are drawn in Fig. 8.17a, and are tabulated in Fig. 7.8. At 1850 the system is in the steady state for the 1850 climate. The three components of climate considered, CO_2, temperature and N deposition, are applied additively, showing the consequences of CO_2 alone, CO_2 and temperature, and finally, CO_2 plus temperature plus N deposition.

Because the pre-industrial environment has N deposition of only 5 kg N ha^{-1} $year^{-1}$, by far the greatest effect results from the quadrupling of N deposition which

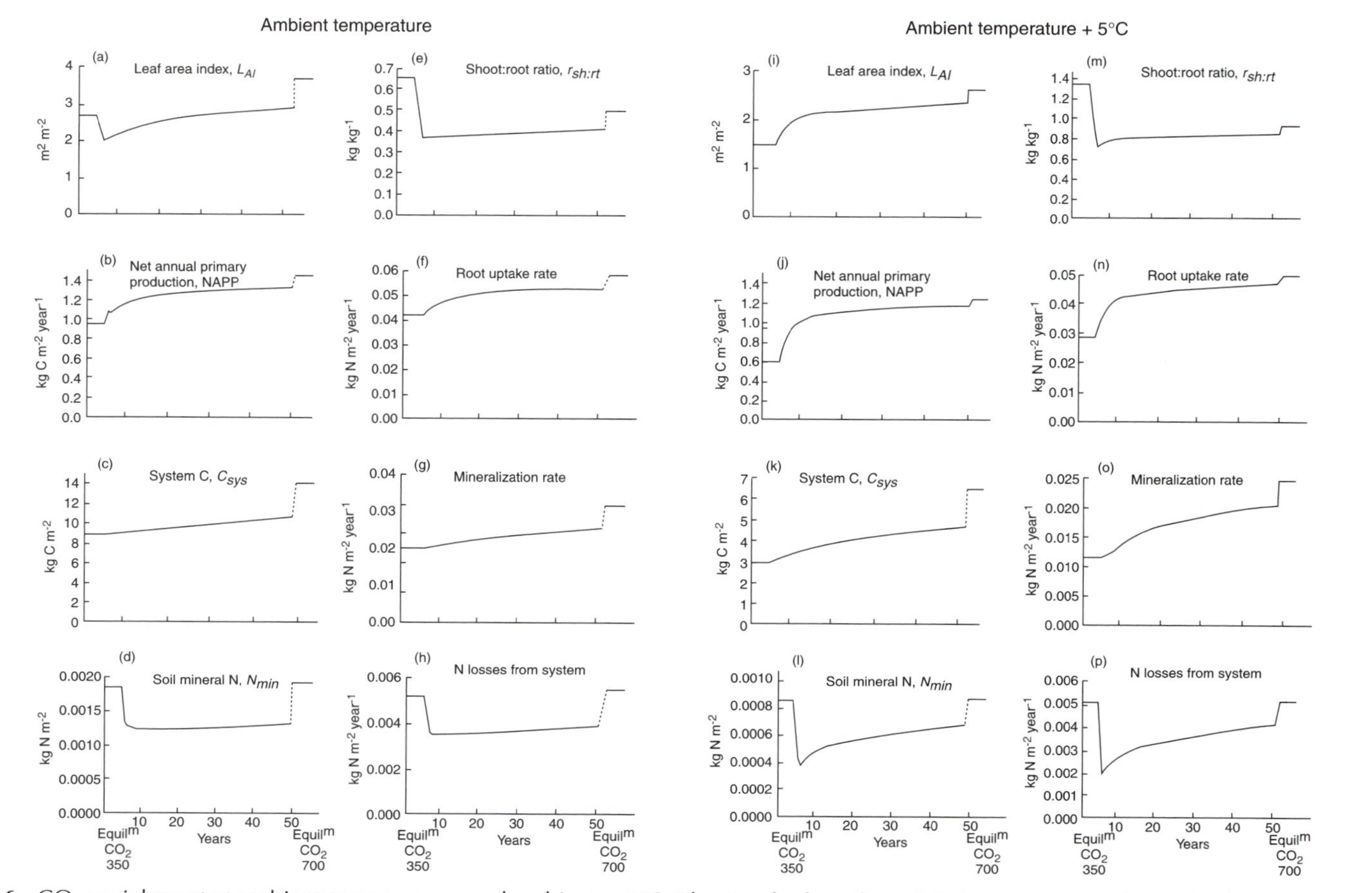

Fig. 8.16. CO_2 enrichment at ambient temperatures and ambient + 5°C. The standard southern Britain environment is used (Chapter 7). For ambient + 5°C, all temperatures are increased by 5°C. After 5 years in the appropriate steady state at 350 vpm CO_2, the CO_2 concentration is doubled, and the simulation is continued until 50 years. The equilibrium values at 700 vpm CO_2 are indicated. All quantities are annual averages or annual sums.

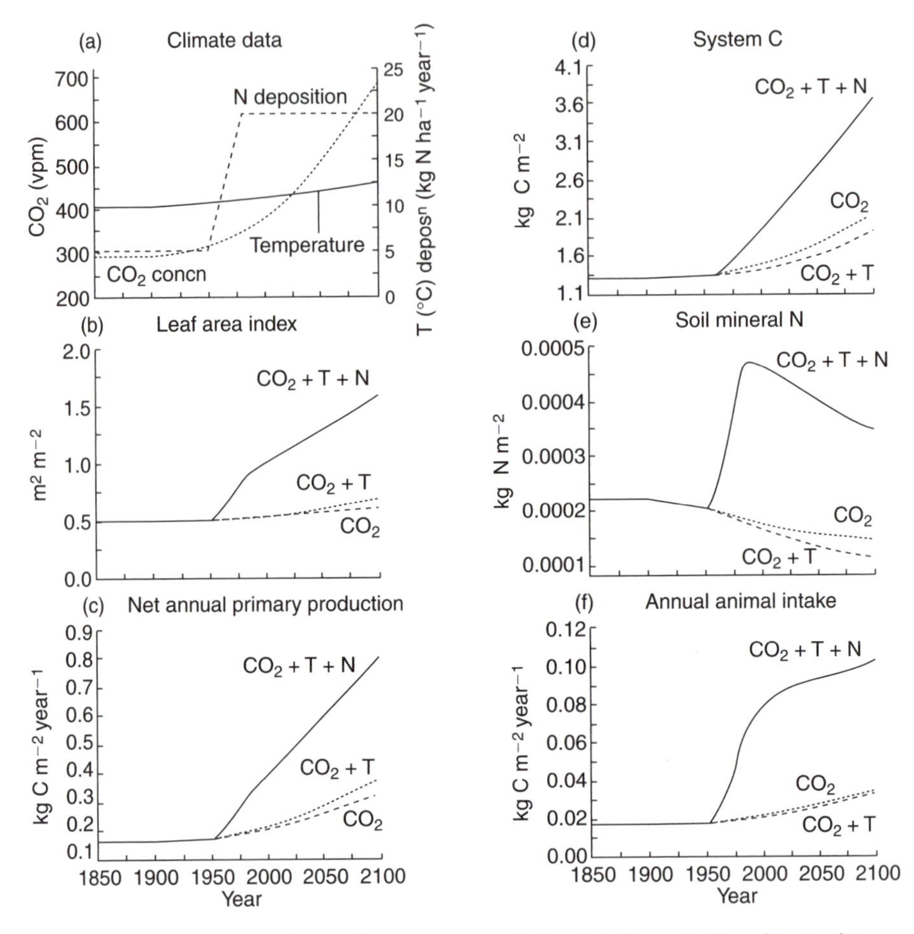

Fig. 8.17. Simulated climate change from 1850 (pre-industrial climate). The climate data shown in (a) (also Fig. 7.8) are used to modify the standard southern Britain environment. The effects of the components of the climate are shown: with CO_2 alone, CO_2 + temperature, and CO_2 + temperature + N deposition. N deposition is placed in the ammonium soil pool, and replaces the standard rate of 50 kg N ha^{-1} year^{-1} (Table 7.2). Quantities shown are annual averages or sums.

is assumed to occur between 1950 and 1980. In such an N-poor environment, the pasture is over-grazed with 12 sheep ha^{-1} during the grazing period (Section 7.7.3), giving very low leaf area indices (L_{AI}). In a hard-grazed low leaf area index pasture, net annual primary production (NAPP) responds strongly to changes in L_{AI}, and growth rates (which respond positively to temperature) are more important than the balance between the rates of growth and senescence (which may respond quite differently to temperature).

In Fig. 8.17b,c it can be seen that L_{AI} and and NAPP respond positively to CO_2, temperature and N deposition, with NAPP (a rate, rather than a state, such as L_{AI} or system C) responding particularly strongly. Although NAPP is increased by increasing temperatures, system C (Fig. 8.17d) is decreased. Soil mineral N is pulled in two directions: downwards by increasing CO_2 and temperature, and upwards by increasing N deposition. Animal intake responds most positively to N deposition, especially in the region where LAI has a value where the intake:L_{AI} curve has steepest slope (Fig. 4.2).

It is of interest to compare the responses of Fig. 8.17 with those of Fig. 8.14 for step changes in temperature, CO_2 and N deposition. Figure 8.14 is based on the standard environment with 350 vpm CO_2 and N deposition of 50 kg N ha^{-1} year^{-1}, where 12 sheep ha^{-1} is a modest grazing intensity leading to a L_{AI} of about 2.7. In this case, there is comparatively little response to doubling N deposition (Fig. 8.14c).

8.13 Decay of a Single Pulse of Mineral Nitrogen

The two favoured methods for looking at the transient characteristics of dynamic systems are the application of step changes in conditions (as in Fig. 8.14), or to apply a short pulse injection of substance to one of the pools of the system. In the first case the time course by which the system approaches its new equilibrium is followed; in the second case, it is how the system re-assumes its initial equilibrium that is of interest. Here we apply a single application of nitrogen to the ammonium N pool of the soil organic matter submodel (Fig. 5.1), equivalent to 1000 kg N ha^{-1} (or 0.1 kg N m^{-2}). Annual averages are plotted to avoid the extra complications of seasonality.

Figure 8.18 illustrates how plant growth is first affected with a sharp increase in leaf area index. In the same year there is an increase in litter nitrogen, which is followed by a maximum in the soil organic matter N a year later. The soil organic matter N takes many years to return to its steady-state value. The tendency of the system as presently parameterized towards biennial oscillations (see next section) is obvious in the leaf area index and the litter N pool.

8.14 Biennial Oscillations in Grassland

Any cyclical system is potentially capable of oscillatory behaviour. Indeed, such behaviour is not only of academic interest, but it could also be of great practical importance, bearing on questions of ecosystem stability and the management required for sustainable productive grassland systems. It has been established that oscillatory systems with non-linear terms can easily give rise to chaotic or unpredictable dynamics (e.g. Baker and Gollub, 1996; Olsen and Degn, 1985).

There has been much recent interest in grass–legume systems, focusing on their dynamics and stability. Using a non-spatial model, consisting of highly simplified plant and soil submodels, Thornley *et al.* (1995) were able to show that the dynamics can be highly complex, but their model did not give sustained oscillations. Using an extension of Thornley *et al.*'s model which includes spatial aspects and selective grazing, Schwinning and Parsons (1996a, b) demonstrated the existence of sustained oscillations, but did not find solutions giving autonomous chaotic behaviour. Relative to the present model of a spatially homogeneous monoculture, a grass–legume model with spatial variation offers many more possibilities for cycles, delays and non-linear processes.

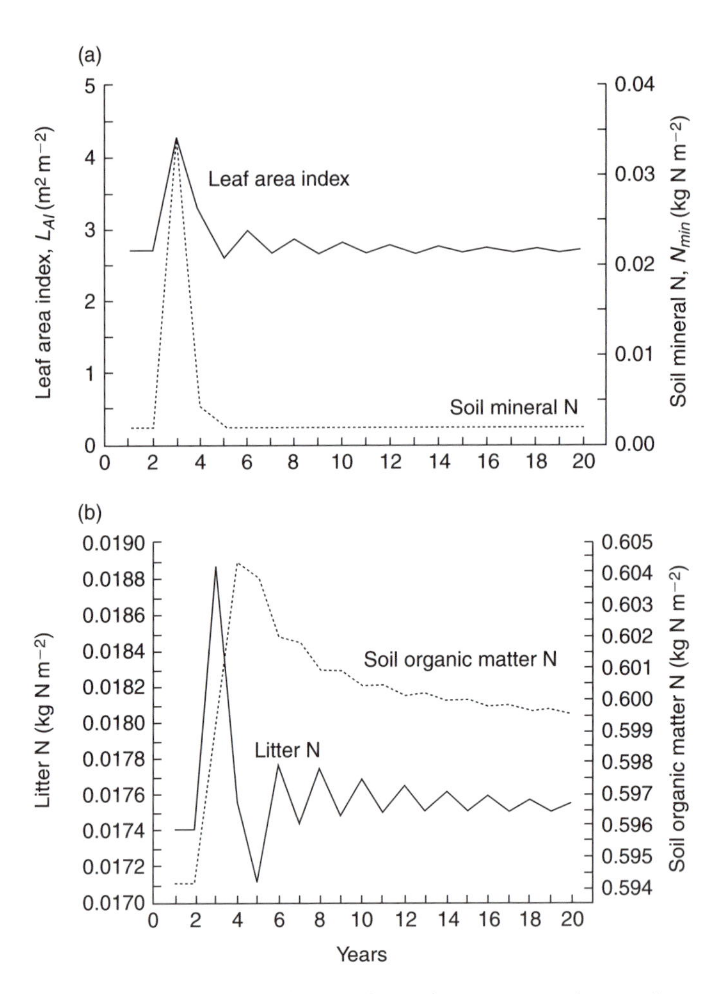

Fig. 8.18. Decay of a single pulse of mineral N. After 2 years in the steady state under standard southern Britain conditions (Chapter 7), 1000 kg N ha^{-1} are placed in the soil ammonium pool (Fig. 5.1). Annual averages of leaf area index, soil mineral N (ammonium + nitrate N), litter N (above- and below-ground litter), and soil organic matter N (the unprotected, protected and stabilized SOM N pools, Fig. 5.1) are plotted in (a) and (b).

In Fig. 8.19, it is demonstrated that the present model, with just a single parameter change from its default parameterization, exhibits sustained biennial oscillations. These oscillations persist, at increased amplitude, if the soil submodel is decoupled. However, if the water submodel is decoupled, the system undergoes damped biennial oscillations, with the oscillations dying out after some 15 years (cf. Thornley *et al.*, 1995, Figs 7 and 8, where similar behaviour in a grass–legume model without a water submodel is exhibited). Biennial oscillations are also caused by decreasing the rainfall (southern Britain) by 30% (Fig. 9.11), although, in the higher-rainfall lower-radiation northern Britain environment (Table 7.3), sustained oscillations do not occur, even when the rainfall is decreased by 30%. The reason

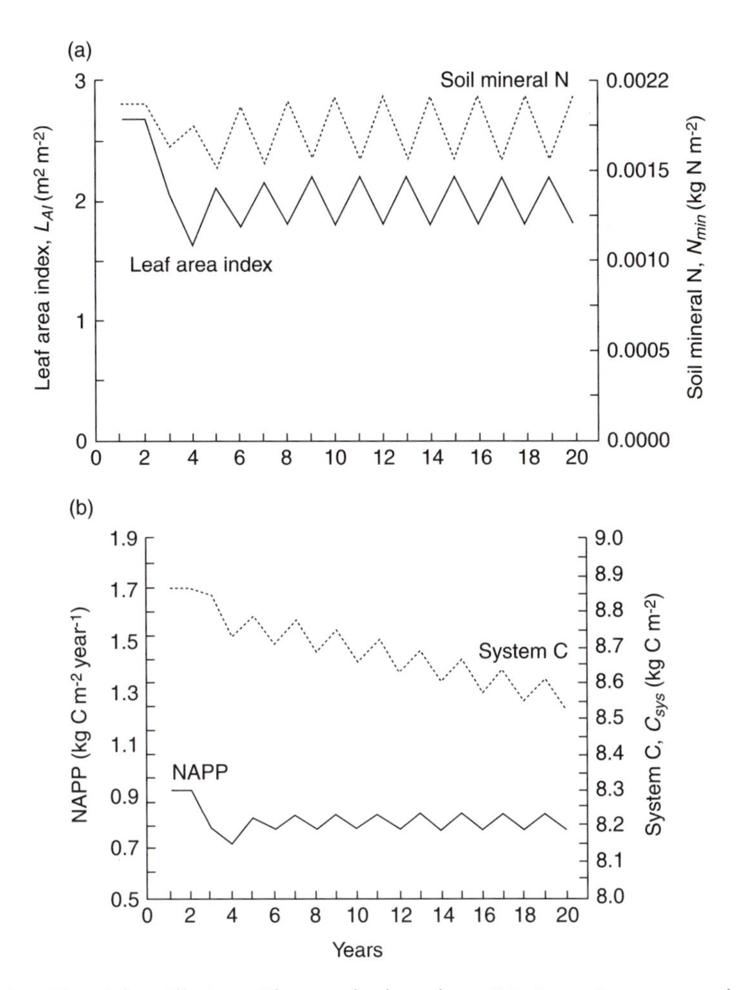

Fig. 8.19. Biennial oscillations. The standard southern Britain environment and management are used. A single parameter, the root turnover rate, $k_{turn,rt,20}$ [Eqn (3.5c)], is halved from its default value of 0.08 to 0.04 day^{-1}, doubling the half-life of the senescing roots. The simulation begins in the steady state, and the parameter change is effected at the beginning of the third year. Quantities plotted are annual averages or sums.

for the oscillations appears primarily to be an interaction between the plant and water submodels, which, interestingly, is able to occur even though the soil reaches field capacity in the winter. As currently formulated, the water submodel loses all memory of the previous season when it attains field capacity, but during this period, memory of the previous season is carried by the plant submodel.

While the oscillatory characteristics illustrated in Fig. 8.19 merit further investigation, we close this section by remarking that such behaviour could be a significant factor in the variability present in many grassland systems, and the consequent difficulties in constructing rational management regimes.

9 Steady-state Simulations

9.1 Introduction

Dynamic simulations for a range of time scales and conditions are described in Chapter 8. The 'real' world, whether of grassland farming or of scientific investigations of grassland, is always dynamic. Nevertheless, it is useful to examine steady states, because this tells us where our grassland ecosystems are headed. Indeed, in some situations short-term responses are in the opposite direction to long-term responses (Fig. 8.15). Steady-state solutions do not, usually, depend on initial conditions, which may be uncertain or arbitrary. Therefore they can be easier to understand and interpret. In this chapter the responses of the steady state to changes in environmental and management variables are examined (some of these have been reported by Thornley and Cannell, 1997, using a slightly different version of the model). Using a similar model of a forest ecosystem, Thornley and Cannell (1996) presented analogous responses to temperature, nitrogen and CO_2 in a forest plantation.

There are several important aspects of grassland response. The fluxes of carbon between the atmosphere and the vegetation–soil system, and the quantity of carbon sequestered within vegetation and soil are of interest to the climatologist and ecologist. Leaching and other nitrogen losses in relation to climate and grassland management concern the environmental scientist. Forage and animal production are a central issue to the farmer. The ecophysiologist wants to make the connections between mechanisms and the responses of the system.

We focus on important output variables such as leaf area index [L_{AI}, Eqn (3.1b)], net primary production [N_{PP}, Eqn (8.1a)], total system carbon [C_{sys}, Eqn (8.1b)] and soil mineral N concentration [N_{min}, Eqn (5.5g)]. The carbon input to the grazing animals [$I_{C,pl \to an}$, Eqn (4.2c)] is important for animal production considerations. N leaching [$O_{Nnit \to lch}$, Eqn (5.5e)] is also of environmental interest. Because all these quantities vary seasonally, annual averages are calculated for 'property' variables (e.g. L_{AI}, C_{sys}, N_{min}) and 'process' or flux variables are accumulated over one year (e.g. N_{PP}, $I_{C,pl \to an}$, $O_{Nnit \to lch}$). The principal fluxes in the carbon, nitrogen and water budgets are also examined.

The default environmental regime, which includes seasonal and diurnal variation, is explained in Chapter 7. The environmental parameters for southern Britain are listed in Tables 7.2 and 7.3. The components of environment and management considered here are: temperature, annual atmospheric N input, CO_2 concentration, grazing and rainfall. Arguably, temperature changes are the most important component of climate change; their effects are complex, they affect all rate constants in the ecosystem, and the consequences are difficult to disentangle. Atmospheric N inputs can vary by a factor of about ten in different environments, from 10 to 100 kg N ha^{-1} $year^{-1}$, and may have a marked effect in some situations. Atmospheric CO_2 is assumed in the model to affect directly only photosynthesis and stomatal conductance; this leads to substantial changes in the plant, soil and water subsystems, which, in the absence of a temperature change accompanying the CO_2 change, are usually easy to interpret. Grazing is important because grazing intensity is easily managed and altered, and in unmanaged grassland grazing is likely to change as conditions change. Grazing regimes have a major impact on temperature responses (Fig. 9.2). Radiation is perhaps the least likely component of the environment to change, and responses to radiation are not simulated. Predictions of rainfall from meteorological climate models are highly uncertain. The effects of changing rainfall depend very much on the environment to which those changes are applied (Section 9.6), and whether the environment is more water-limited or more nitrogen-limited.

Responses to changes in a lowland environment in southern Britain (Hurley) are examined in detail. We also contrast the effects of changes made to the lowland southern Britain environment, with the same changes made to an upland environment in northern Britain (Eskdalemuir). Sometimes, the same environmental changes may have opposite effects when applied to these different localities (Fig. 9.11).

9.2 Responses to Temperature

In Fig. 9.1, air and soil temperatures have been decreased and increased by up to 5°C relative to their usual values for the southern Britain environment for grazed grassland. Figure 9.1a illustrates some key variables: leaf area index, shoot:root ratio, soil mineral N, and total C in the system. Increasing temperature decreases leaf area index (L_{AI}) substantially due to the decreasing effect of increasing temperature on leaf longevity. Total system C (C_{sys}) and N (not shown) decrease markedly as increasing temperature increases mineralization rates. Although increasing temperature leads to a short-term increase in mineral N (Fig. 8.14a), steady-state mineral N concentration decreases greatly with increasing temperature. The shoot:root ratio responds positively to increasing temperature. Fig. 9.1b depicts the C balance sheet. Net annual primary production (NAPP) increases up to about 8°C but decreases thereafter, mostly because leaf area index decreases. The N balance sheet (Fig. 9.1c) has some surprises. Total system N input changes slightly because non-symbiotic N fixation changes (Section 5.5.2). Volatilization losses increase: this is a purely physical process where the increased rate constant at high temperatures (Section 5.5.3) more than compensates for the decreased value of the N ammonium pool. Nitrification and denitrification both depend on soil microbial biomass as well as temperature-dependent rate constants and the ammonium and nitrate substrate pools

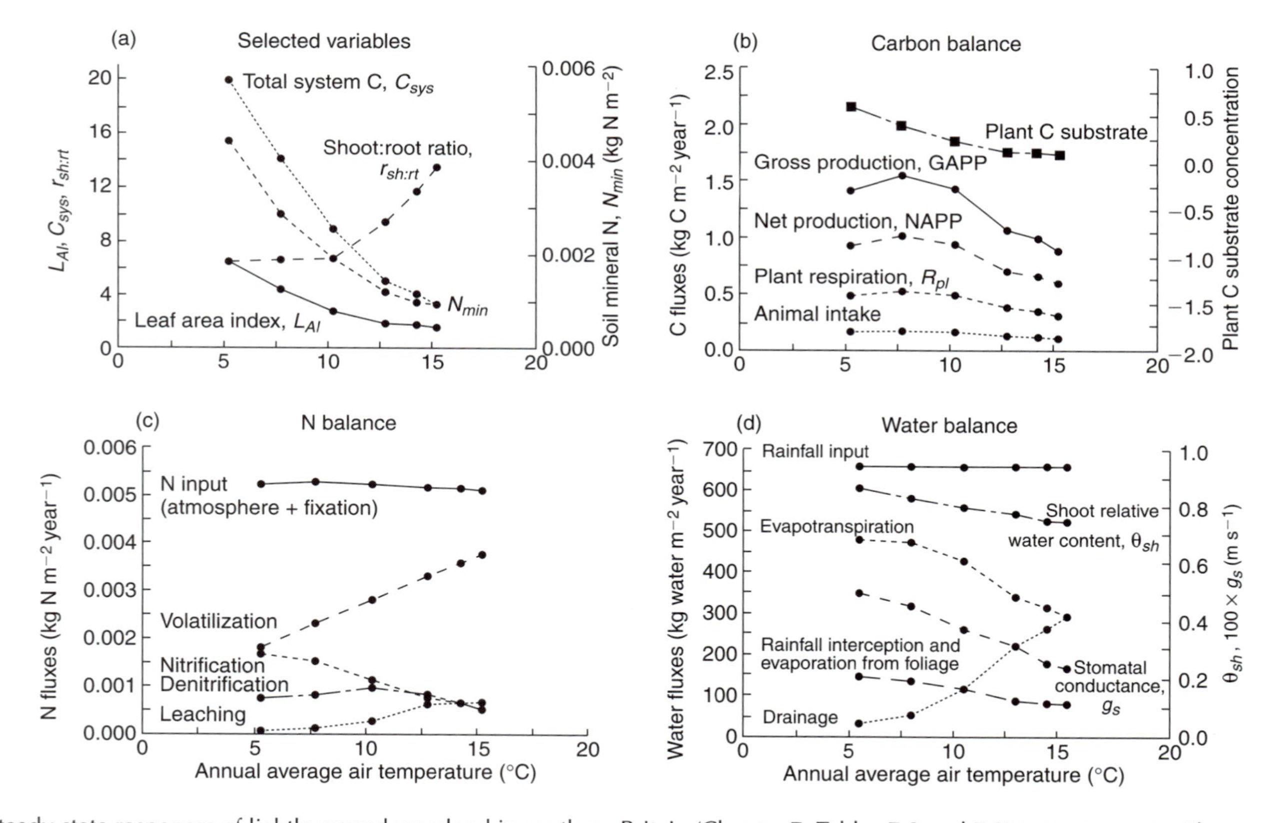

Fig. 9.1. Steady-state responses of lightly grazed grassland in southern Britain (Chapter 7, Tables 7.2 and 7.3) to temperature. The annual average air and soil temperatures (standard values, 10.25°C and 10°C) are varied by ±5°C maintaining the same seasonal temperature variations. Quantities shown are annual averages (e.g. L_{AI}), or for fluxes, annual sums. (a) Leaf area index [Eqn (3.1b)], total system C [kg C m^{-2}, Eqn (8.1b)], shoot:root ratio [Eqn (3.1l)], soil mineral N [Eqn (5.5g)]. (b) Carbon balance fluxes, and the plant C substrate concentration [kg C substrate (kg structural DM)$^{-1}$, $C_{S,pl}$ of Eqn (3.1k)]. (c) Nitrogen balance fluxes. (d) Water balance fluxes, shoot relative water content, θ_{sh} [Eqn (6.5d), $i = sh$] and stomatal conductance, g_s [Eqn (3.2u)]; the last two quantities are for 1 July, 15.00 h when the water stress is relatively high.

(Sections 5.5.3 and 5.6.3). Additionally, denitrification depends on the anaerobicity and water status of the soil. Their behaviour can therefore be complex. Denitrification actually decreases at the lowest temperatures (contrast this with nitrification); this is due to more aerobic conditions: high leaf area index, high rainfall interception and evaporation giving a drier soil. Leaching losses increase with increasing temperature. Leaching depends on the product of nitrate concentration and drainage (Section 5.6.3). In this case, drainage increases strongly from a low value (Fig. 9.1d). This increase occurs because decreasing leaf area index (Fig. 9.1a) leads to lower values of evapotranspiration and rainfall interception given that the radiation receipt is assumed constant. Drainage at a temperature of 10.25°C (southern Britain) compares well with observed values (Smith, 1984). The *average* soil relative water content over the year increases with increasing temperature (not shown) as drainage increases, although the higher temperatures are giving rise to increased summer water stress and stomatal closure (Fig. 9.1d; the values shown for the shoot relative water content and the stomatal conductance are for 1 July 15.00 h).

Many of the responses in Fig. 9.1 stem from the strong response of leaf area index (L_{AI}) to temperature (Fig. 9.1a). This L_{AI} response can be suppressed by grazing continuously to maintain the L_{AI} below a selected value, in this case an L_{AI} of unity. This is achieved by adjusting the parameters of the intake function [Eqn (4.2a); Fig. 4.2; $K_{LAI,an}$ = 1 m^2 leaf (m^2 ground)$^{-1}$, $q_{LAI,an}$ = 30, $I_{DM,pl \to an,max}$ = 10 kg DM animal^{-1} day^{-1}], and stocking throughout the year at 12 sheep ha^{-1} [with $t_{stock,1}$ = 0, $\tau_{stock,1}$ = 365 in Eqn (7.6f)]. The responses are illustrated in Fig. 9.2 for the standard southern Britain environment as in Fig. 9.1, with and without grazing to a constant L_{AI}, and with and without CO_2 enrichment to 700 vpm. Also, for net annual primary production (Fig. 9.2b) the consequences of doubling the photosynthetically active component of radiation are illustrated.

In Fig. 9.2a it is seen that grazing to a maximum L_{AI} of unity does indeed result in a constant L_{AI}: the three constant L_{AI} treatments superimpose in the bottom curve but appear separately in Fig. 9.2b. The temperature responses of net annual primary production (NAPP) (Fig. 9.2b) and of system C (Fig. 9.2c) are greatly altered by constant L_{AI} grazing. The addition of severe grazing shifts the temperature optimum of NAPP to higher temperatures, as does increasing photosynthesis, either by a high CO_2 treatment of high light.

These temperature responses can be understood in terms of how canopy photosynthesis (P_{can}, Fig. 3.4), which drives NAPP, is related to the leaf photosynthetic response curve (Fig. 3.3), and its two most important parameters: the initial slope [α, photosynthetic efficiency, Eqn (3.2s)] and the light-saturated rate of photosynthesis [P_{max}, Eqn (3.2t)]. These have different responses to temperature: α decreases slightly with increasing temperature above 15°C, whereas P_{max} responds positively to temperature (Fig. 3.7) up to 30°C. Most important for the temperature response of canopy photosynthesis and NAPP is that the lower L_{AI} in the hard-grazed sward means the leaves on average are better illuminated, emphasizing the asymptote P_{max} with its positive temperature coefficient over the initial slope α with its zero of negative temperature coefficient. Also, both the irradiance above the canopy and the CO_2 concentration affect the relative influence of P_{max} and α on canopy photosynthesis.

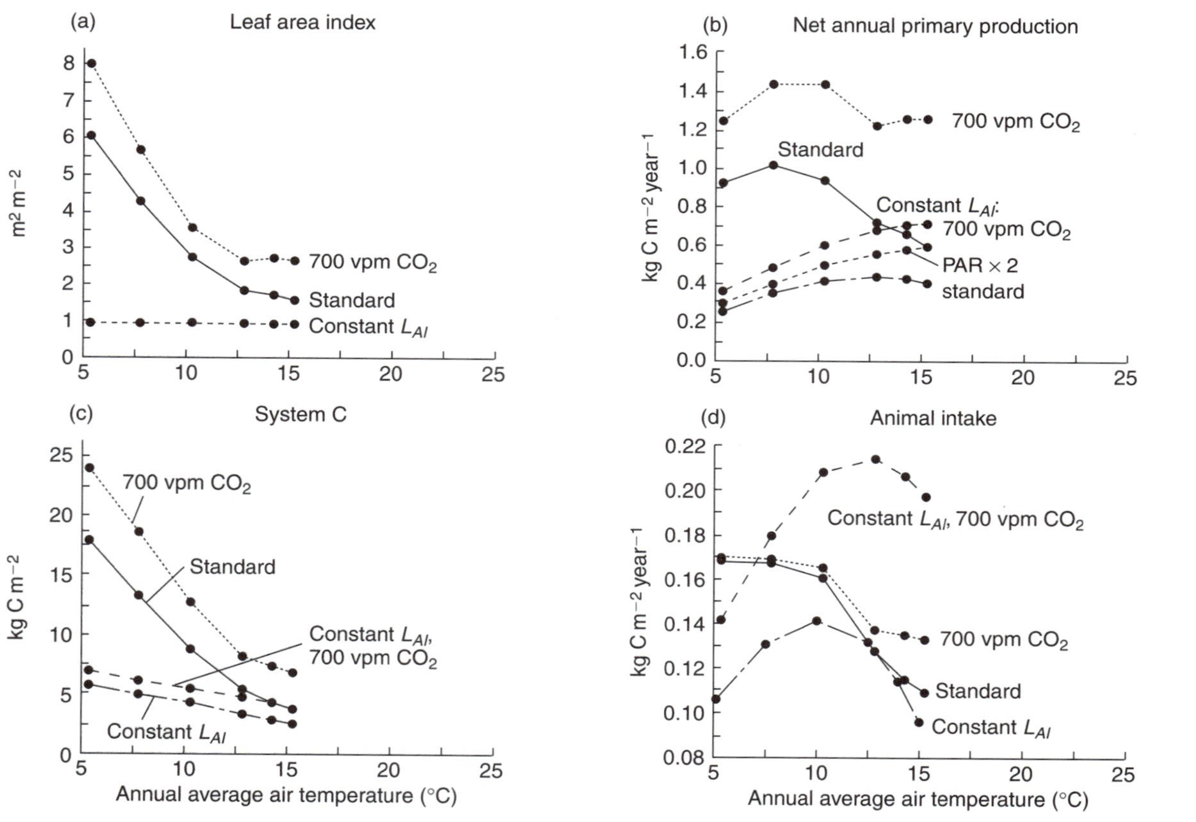

Fig. 9.2. Steady-state responses to temperature of grassland with four principal treatments: (i) standard light grazing as in Fig. 9.1; (ii) grazed to a maximum leaf area index of unity; (iii) as in (i) but with 700 vpm CO_2; (iv) as in (ii) but with 700 vpm CO_2. The effect of doubling the photosynthetically active radiation on net annual primary production is given in (b). The standard southern Britain climate is used (Chapter 7, Tables 7.2 and 7.3). The annual average air and soil temperatures (standard values, 10.25°C and 10°C) are varied by ±5°C, maintaining the same seasonal temperature variations. Quantities shown are annual averages (e.g. L_{AI}), or for fluxes, annual sums. (a) Leaf area index [Eqn (3.1b)]. (b) Net annual primary production [from Eqn (8.1a)]. (c) Total system C [kg C m^{-2}, Eqn (8.1b)]. (d) Animal intake [$I_{C,pl \rightarrow an}$, Eqn (4.2c)].

9.3 Responses to Nitrogen

Figure 9.3 illustrates the responses to varying the atmospheric N input into the ammonium N pool ($I_{N,env \rightarrow Namm}$, Table 7.2). Under the standard southern Britain conditions, the N deposition rate of 50 kg ha^{-1} year^{-1} (Table 7.2) on lightly grazed grassland produces a steady state with moderate N nutrition. Thus a doubling of the N deposition rate from 50 to 100 kg ha^{-1} year^{-1} produces only modest rises in leaf area index (L_{AI}), system C (C_{sys}, Fig. 9.3a), and net primary production (NAPP, Fig. 9.3b). This result may be compared to the simulated fertilizer response of Fig. 8.8, where fertilizer is applied during the growing season at a time when soil mineral N concentrations are low (Fig. 8.7). However, at lower N deposition rates than 50 kg N ha^{-1} year^{-1}, L_{AI}, C_{sys}, and NAPP decrease sharply. Shoot:root ratio also responds strongly to the low N deposition rates (Fig. 9.3a). The soil mineral N concentration responds linearly (Fig. 9.3a) to N deposition rate. The N loss fluxes are unexceptionable (Fig. 9.3c): in the steady state N losses must equal N input. In the water loss fluxes (Fig. 9.3d), drainage losses decrease as L_{AI} increases and consequently evapotranspiration and rainfall interception and evaporation increase. Under other conditions, the model has surprisingly predicted a decrease in gross production and evapotranspiration with increasing N deposition, arising from an increase in N supply to the plant, increasing shoot:root ratio and decreasing plant C substrate concentrations, which lead to decreased shoot relative water content, increased stomatal closure, and finally decreased photosynthesis and evapotranspiration.

Figure 9.4 describes the response to N deposition rate when the sward is subject to four treatments: the standard regime with light grazing as in Fig. 9.3; an ambient CO_2 concentration of 700 vpm; a temperature increase of 5°C; and last, grazing to a constant maximum leaf area index of unity (explained in Section 9.2). In Fig. 9.4a, it can be seen that high CO_2 increases leaf area index (L_{AI}) at all N deposition rates, and is especially effective in maintaining L_{AI} at the lowest level of N deposition of 10 kg N ha^{-1} year^{-1}. At the lowest N deposition rate under constant maximum L_{AI} grazing, L_{AI} (annual average) falls below its maximum value of unity due to insufficient growth. Net annual primary production (NAPP) exhibits a much flatter response to N deposition than L_{AI} because at an L_{AI} above about four, light interception is saturating (Fig. 9.4b). Increased temperature has a massive negative effect on system C (Fig. 9.4c) which is especially large at low N deposition rates. The animal intake responses (Fig. 9.4d) can be understood in terms of the assumed intake equation [Eqn (4.2a); Fig. 4.2]. Above an L_{AI} of two, intake is saturated, so any further increases in L_{AI} which occur as a result of high CO_2 or increased N deposition are ineffective in promoting increased intake. At the higher temperature, the lower L_{AI} values give rise to a positive intake response to N deposition. Similarly when the sward is grazed to a constant L_{AI}, there is a positive intake response to N deposition. Fig. 9.4 illustrates the point which has been recently made in greater detail by Cannell and Thornley (1998): that it is in N-poor systems where the greatest proportional and sometimes absolute responses to CO_2 enrichment may be expected to occur.

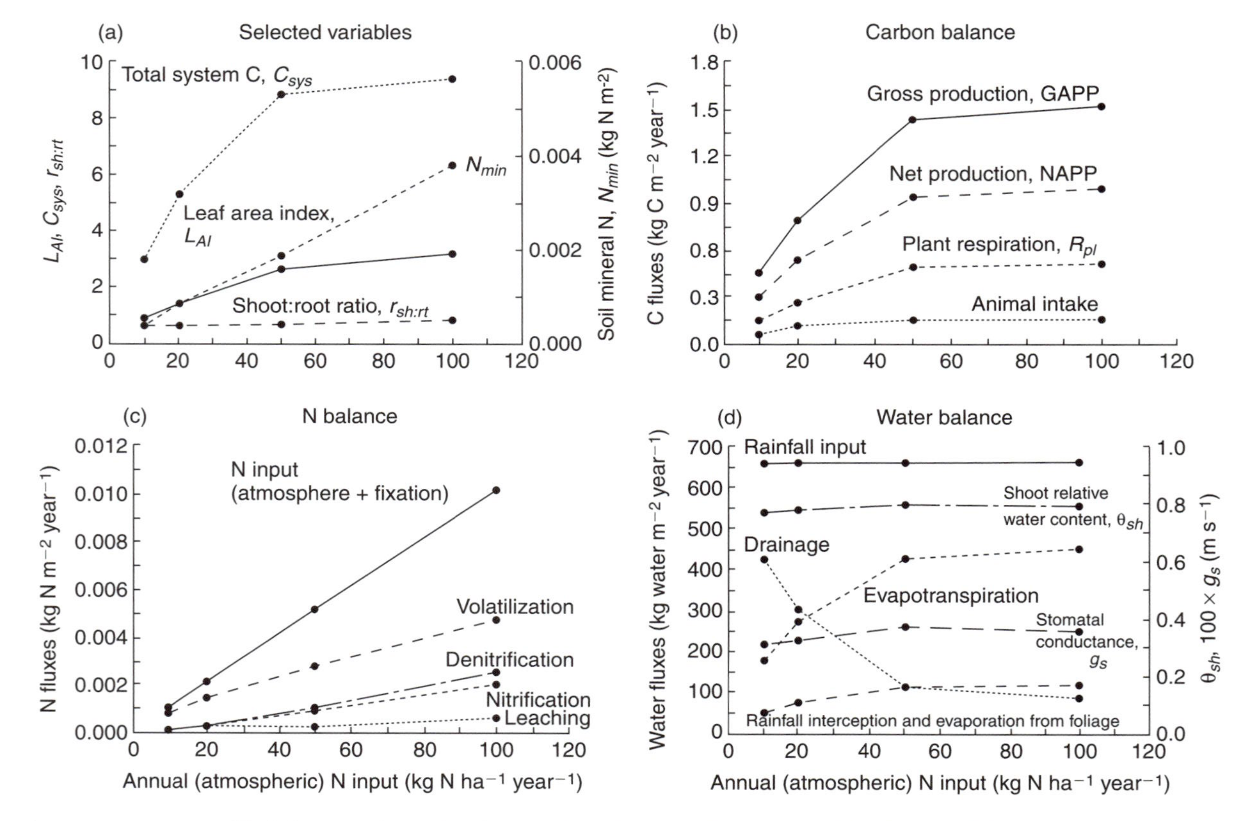

Fig. 9.3. Steady-state responses of lightly grazed grassland in southern Britain (Chapter 7, Tables 7.2 and 7.3) to atmospheric N deposition, which is varied about its default value (equivalent to 0.005 kg N m^{-2} $year^{-1}$, Table 7.2). Quantities shown are annual averages (e.g. L_{AI}), or for fluxes, annual sums. (a) Leaf area index [Eqn (3.1b)], total system C [kg C m^{-2}, Eqn (8.1b)], shoot:root ratio [Eqn (3.1l)], soil mineral N [Eqn (5.5g)]. (b) Carbon balance fluxes. (c) Nitrogen balance fluxes. (d) Water balance fluxes, shoot relative water content, θ_{sh} [Eqn (6.5d), $i = sh$] and stomatal conductance, g_s [Eqn (3.2u)]; the last two quantities are for 1 July, 15.00 h when the water stress is relatively high.

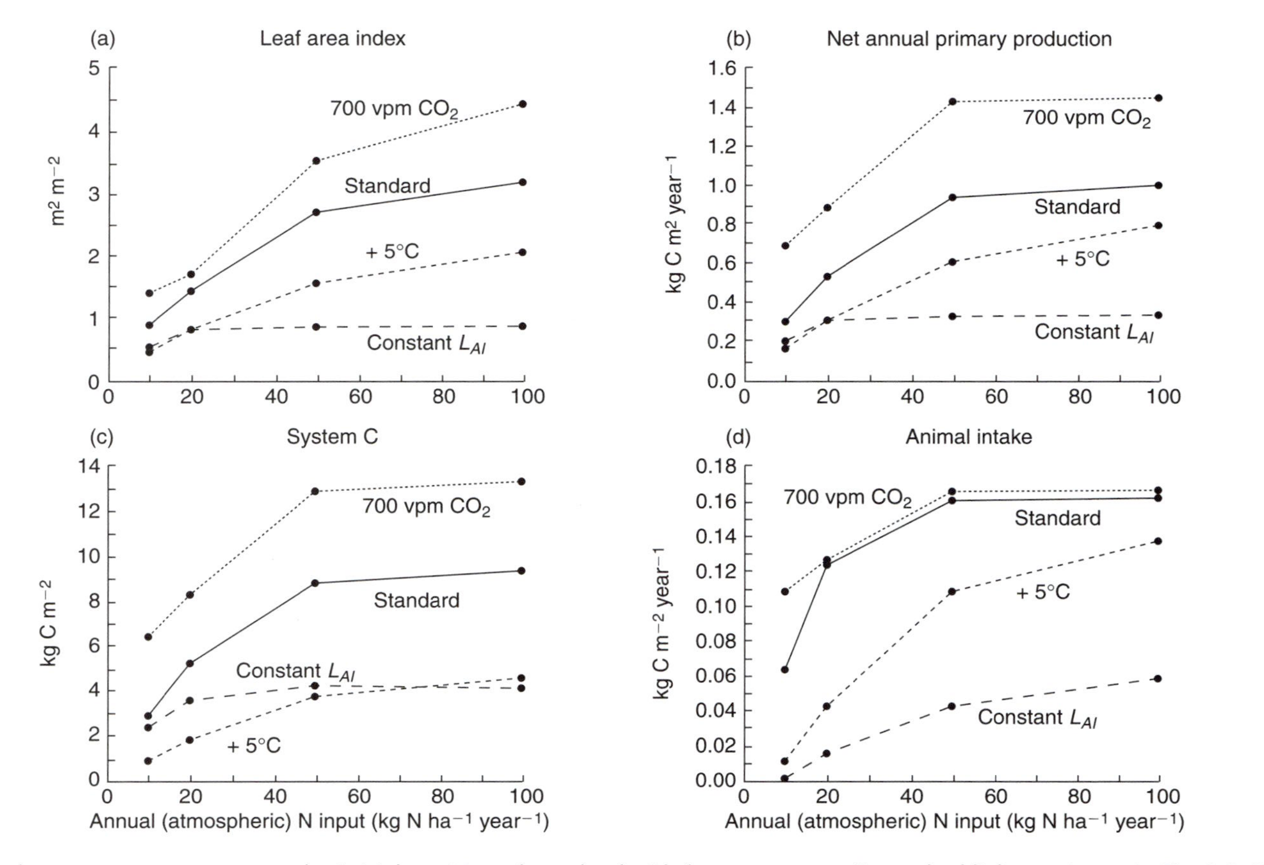

Fig. 9.4. Steady-state responses to atmospheric N deposition of grassland with four treatments: (i) standard light grazing as in Fig. 9.3; (ii) grazed to a maximum leaf area index of unity; (iii) 700 vpm CO_2; (iv) with temperature increased by 5°C. The standard southern Britain climate is used (Chapter 7, Tables 7.2 and 7.3). Quantities shown are annual averages (e.g. L_{AI}), or for fluxes, annual sums. (a) Leaf area index [Eqn (3.1b)]. (b) Net annual primary production [from Eqn (8.1a)]. (c) Total system C [kg C m^{-2}, Eqn (8.1b)]. (d) Animal intake [$I_{C,pl \to an}$, Eqn (4.2c)].

9.4 Responses to CO_2

The CO_2 responses are the easiest to understand. In the model CO_2 only affects photosynthesis [Eqns (3.2s), (3.2t)] and stomatal conductance [Eqn (3.2u)] directly. Although Fig. 8.14b shows a large transient decrease in soil mineral N which then partially recovers over many years, Fig. 9.5a shows that in the steady state the average soil mineral N concentration decreases by about 15% over the range 350–1000 vpm. However, this average decrease does not reveal the seasonal changes that are significant for processes such as leaching and denitrification which occur at critical times of the year (Fig. 8.3c). Shoot:root ratio decreases although there is a positive response in leaf area index (L_{AI}). The increase in L_{AI} does not affect animal intake (Fig. 9.5b) which is already saturated (Fig. 4.2). There is a large increase in sequestered C, C_{sys}. The principal C fluxes all increase (Fig. 9.5b). As a fraction of gross production, plant respiration at three CO_2 concentrations of 350, 600 and 1000 vpm is 0.34, 0.36 and 0.37 respectively. Leaching losses are sharply down (Fig. 9.5c), owing to decreased drainage (Fig. 9.5d) and to lower soil mineral N concentrations (Fig. 9.5a). Nitrification increases as a result of increased biomass (not shown). Volatilization, which does not depend on soil microbial activity, decreases with increasing CO_2. However denitrification first increases as microbial activity increases and then decreases as drainage decreases (Fig. 9.5d), and the soil becomes more aerobic. Evapotranspiration increases slightly (Fig. 9.5d) as a result of the higher L_{AI} (Fig. 9.5a), although stomatal conductance is much decreased (Fig. 9.5d). The large decrease in drainage (Fig. 9.5d) arises from the increased interception losses and the higher evapotranspiration. Plant water relations, as indicated by the shoot relative water content (Fig. 9.5d), improve with increasing CO_2 concentrations.

Figure 9.6 describes the response to ambient CO_2 concentration when the sward is subject to four treatments: the standard regime with light grazing (as in Fig. 9.5); an atmospheric N deposition rate of 10 kg N ha^{-1} $year^{-1}$ (instead of 50); a temperature increase of 5°C; and last, grazing to a constant maximum leaf area index of unity (explained in Section 9.2). In Fig. 9.6a, it can be seen that under lightly grazed cool conditions, high CO_2 is especially effective in increasing L_{AI}. In Fig. 9.6b, the constant L_{AI} grazing treatment gives the residual effects of non-L_{AI} factors on primary production. The non-L_{AI} factors are mostly operating through the leaf photosynthetic response to light. At 10 kg N ha^{-1} $year^{-1}$ there is an excellent response of net primary production (Fig. 9.6b), system C (Fig. 9.6c) and animal intake (Fig. 9.6d) to increased CO_2 (Cannell and Thornley, 1998). It can be seen in Fig. 9.6c that the CO_2 response of system C is strongly depressed by high temperatures and severe grazing. Animal intake is increased by high CO_2 under severe grazing (Fig. 9.6d, constant L_{AI} curve); it is seen in Fig. 9.2d (the constant L_{AI} grazing curves) that the temperature optimum for intake moves to higher temperatures with raised CO_2 concentrations (Section 9.2).

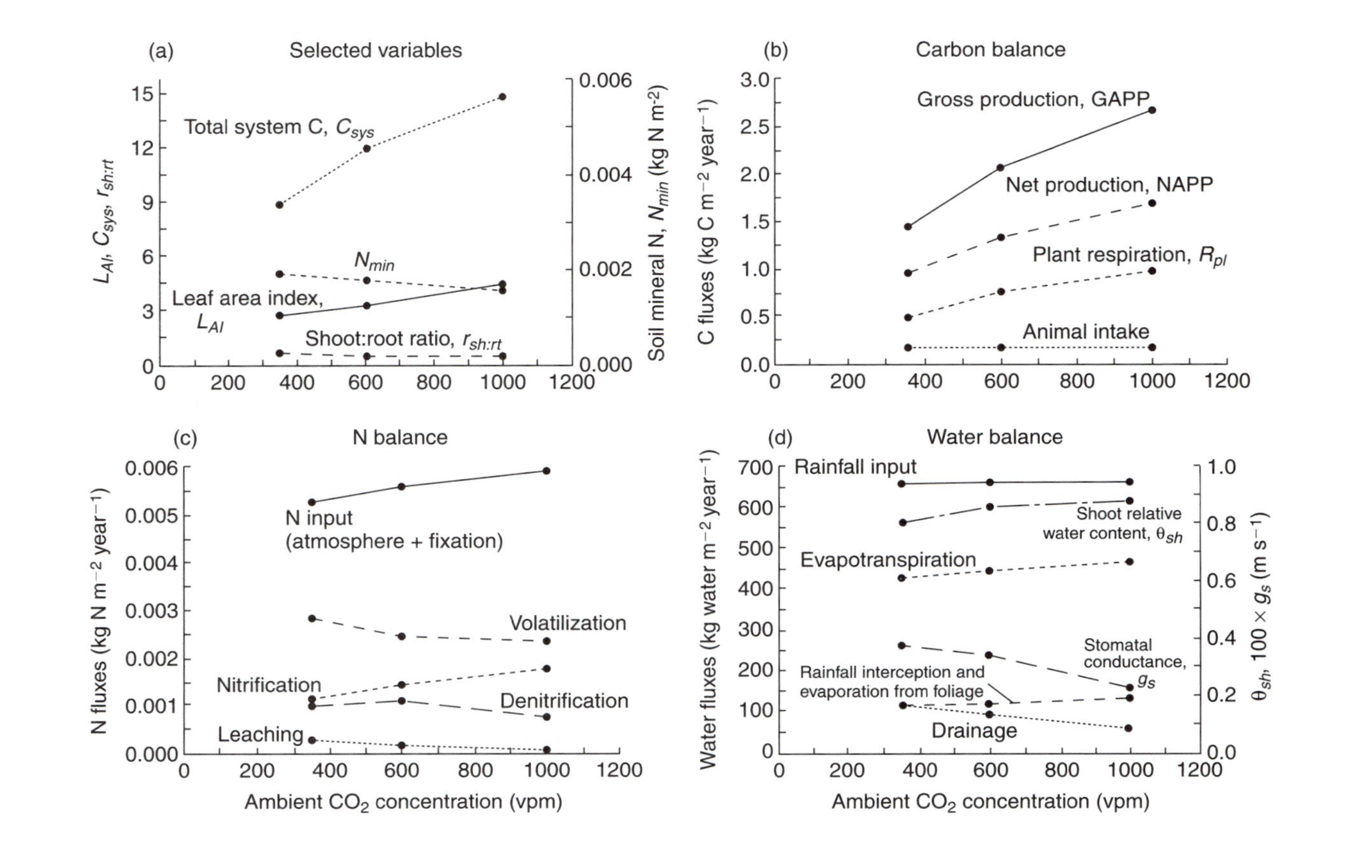

Fig. 9.5. Steady-state responses of lightly grazed grassland in southern Britain (Chapter 7, Tables 7.2 and 7.3) to ambient CO_2 concentration. Quantities shown are annual averages (e.g. L_{AI}), or for fluxes, annual sums. (a) Leaf area index [Eqn (3.1b)], total system C [kg C m^{-2}, Eqn (8.1b)], shoot:root ratio [Eqn (3.1l)], soil mineral N [Eqn (5.5g)]. (b) Carbon balance fluxes. (c) Nitrogen balance fluxes. (d) Water balance fluxes, shoot relative water content, θ_{sh} [Eqn (6.5d), $i = sh$] and stomatal conductance, g_s [Eqn (3.2u)]; the last two quantities are for 1 July, 15.00 h when the water stress is relatively high.

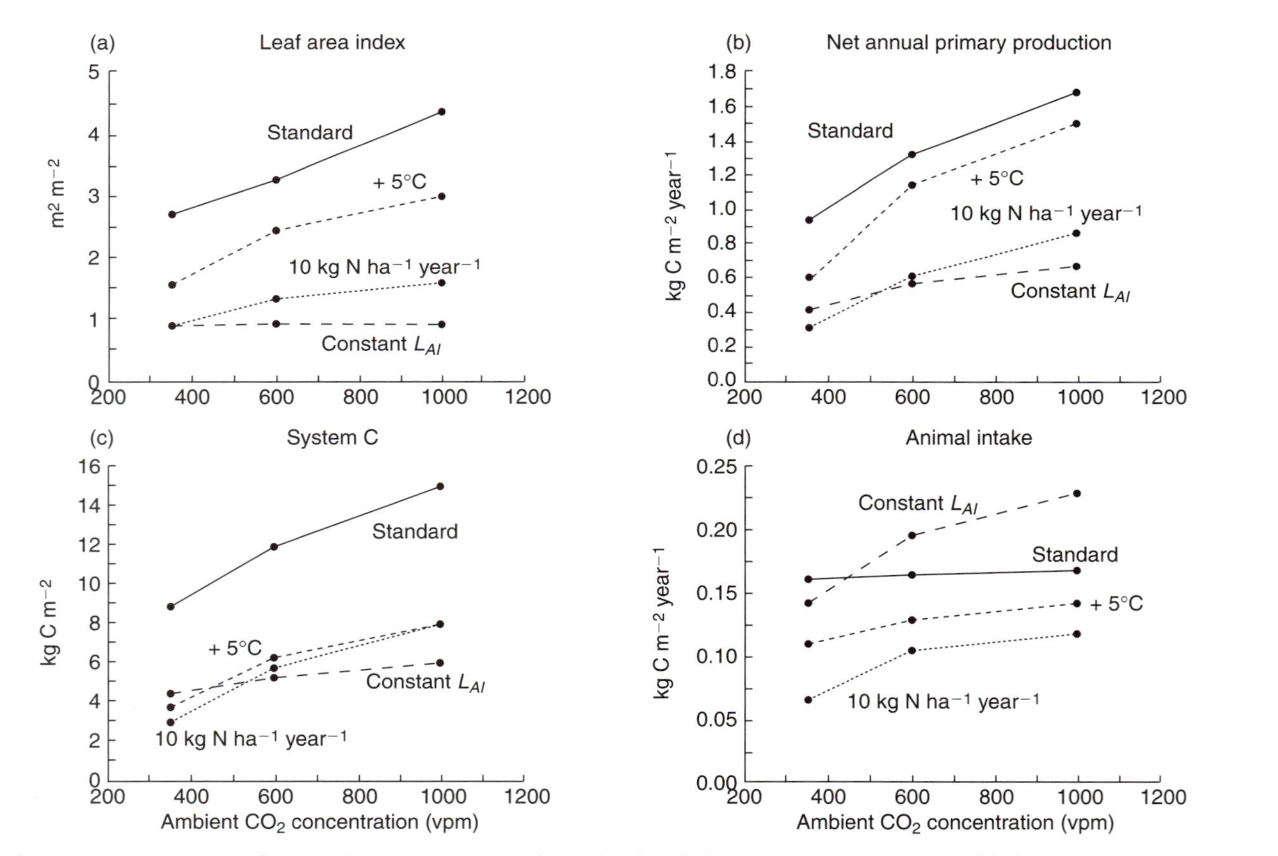

Fig. 9.6. Steady-state responses to ambient CO_2 concentration of grassland with four treatments: (i) standard light grazing as in Fig. 9.5; (ii) grazed to a maximum leaf area index of unity; (iii) 10 kg N ha^{-1} $year^{-1}$ (instead of 50); (iv) temperature increased by 5°C. Otherwise the standard southern Britain climate is used (Chapter 7, Tables 7.2 and 7.3). Quantities shown are annual averages (e.g. L_{AI}), or for fluxes, annual sums. (a) Leaf area index [Eqn (3.1b)]. (b) Net annual primary production [from Eqn (8.1a)]. (c) Total system C [kg C m^{-2}, Eqn (8.1b)]. (d) Animal intake [$I_{C,pl\rightarrow an}$, Eqn (4.2c)].

9.5 Responses to Stocking

The stocking density over the 7-month period during which the pasture is stocked (Section 7.7.3) is varied in Fig. 9.7. Note that our treatment of grazing assumes that the pasture remains uniform: in practice grazed grassland can become very patchy, as a result of animal selection, excreta deposition, and changes in species composition (e.g. Schwinning and Parsons, 1996a, b).

Predictably increased grazing decreases leaf area index (Fig. 9.7a). Shoot:root ratio (Fig. 9.7a) decreases as stocking is increased from 0 to 10 sheep ha^{-1}, and then increases as stocking is further increased from 10 to 20 sheep ha^{-1}. Increased stocking has a depressing effect on total system C and soil mineral N (Fig. 9.7a), with severe grazing leading to a marked decline in system C. Although the introduction of a moderate number of sheep (10 ha^{-1}) depresses gross production appreciably (Fig. 9.7b), the decrease in plant respiration arising from the decreased maintenance component of a smaller plant, leads to almost no change in net production. Note that our 'ecological' definition of net production, which is equal neither to the inputs to the litter and soil submodel, nor to agricultural production, can be misleading (Section 8.7). The limitations of this definition are not always realized. Leaching increases (Fig. 9.7c) as drainage losses increase (Fig. 9.7d). Nitrification losses decrease owing to decreased biomass (not shown) and decreased soil mineral N (Fig. 9.7a). Denitrification first increases due to an increased period of anaerobicity which is linked to drainage (Fig. 9.7d), and then decreases as a result of the opposing effects of decreased biomass and soil mineral N. Evapotranspiration and rainfall interception decrease (Fig. 9.7d) due to the decreasing leaf area index (Fig. 9.7a). These quantites are both per unit ground area; the response per unit leaf area can be different. Shoot relative water content and stomatal conductance on 1 July at 15.00 h both decrease, indicating greater water stress of grazed grassland relative to ungrazed grassland; this occurs in spite of an increase in average soil water content with increased grazing.

Figure 9.8 describes the stocking density response with the sward subject to four treatments: the standard regime (as in Fig. 9.7); an atmospheric N deposition rate of 10 kg N ha^{-1} year^{-1} (instead of 50); a temperature increase of 5°C; and fourth, high ambient CO_2 of 700 vpm (instead of 350). Increasing stocking decreases leaf area index (L_{AI}) in all treatments (Fig. 9.8a). The impact of grazing on the low N treatment is severe. In Fig. 9.8b at 700 vpm CO_2, light grazing increases net annual primary production (NAPP). This occurs because, although gross production decreases, plant respiration decreases even further. Increased temperature has little effect on NAPP when the sward is not grazed, or is hard grazed (Fig. 9.8b, comparing 'standard' with '+ 5°C', 0, 20 sheep ha^{-1}), although total system C (Fig. 9.8c) is much decreased by increased temperature. At 10 kg N ha^{-1} year^{-1}, animal intake (Fig. 9.8d) is maximum with light grazing (10 sheep ha^{-1}), although for the other three treatments grazing at 20 sheep ha^{-1} leads to higher animal intake.

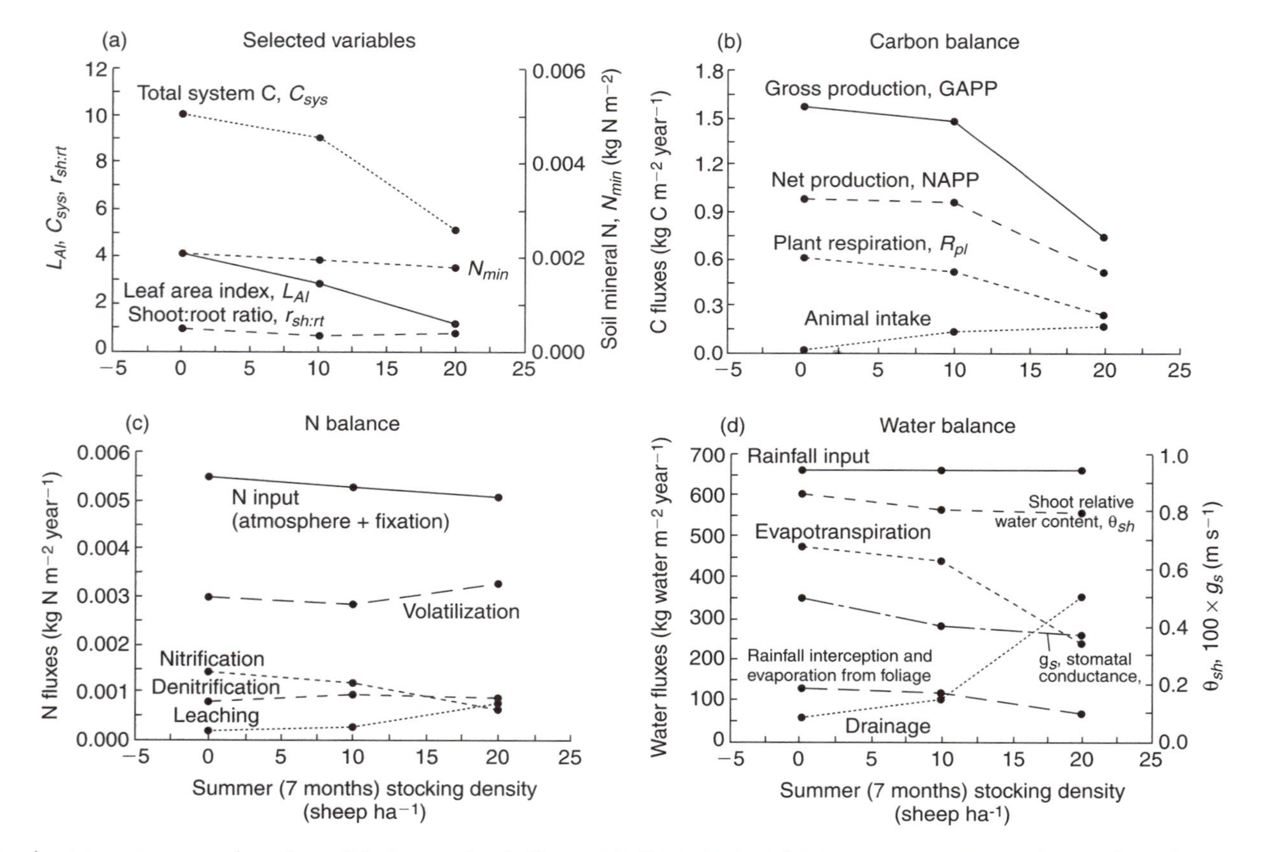

Fig. 9.7. Steady-state responses of southern Britain grassland (Chapter 7, Tables 7.2 and 7.3) to summer (7 months) stocking density. Quantities shown are annual averages (e.g. L_{Al}), or for fluxes, annual sums. (a) Leaf area index [Eqn (3.1b)], total system C [kg C m^{-2}, Eqn (8.1b)], shoot:root ratio [Eqn (3.1l)], soil mineral N [Eqn (5.5g)]. (b) Carbon balance fluxes. (c) Nitrogen balance fluxes. (d) Water balance fluxes, shoot relative water content, θ_{sh} [Eqn (6.5d), $i = sh$] and stomatal conductance, g_s [Eqn (3.2u)]; the last two quantities are for 1 July, 15.00 h.

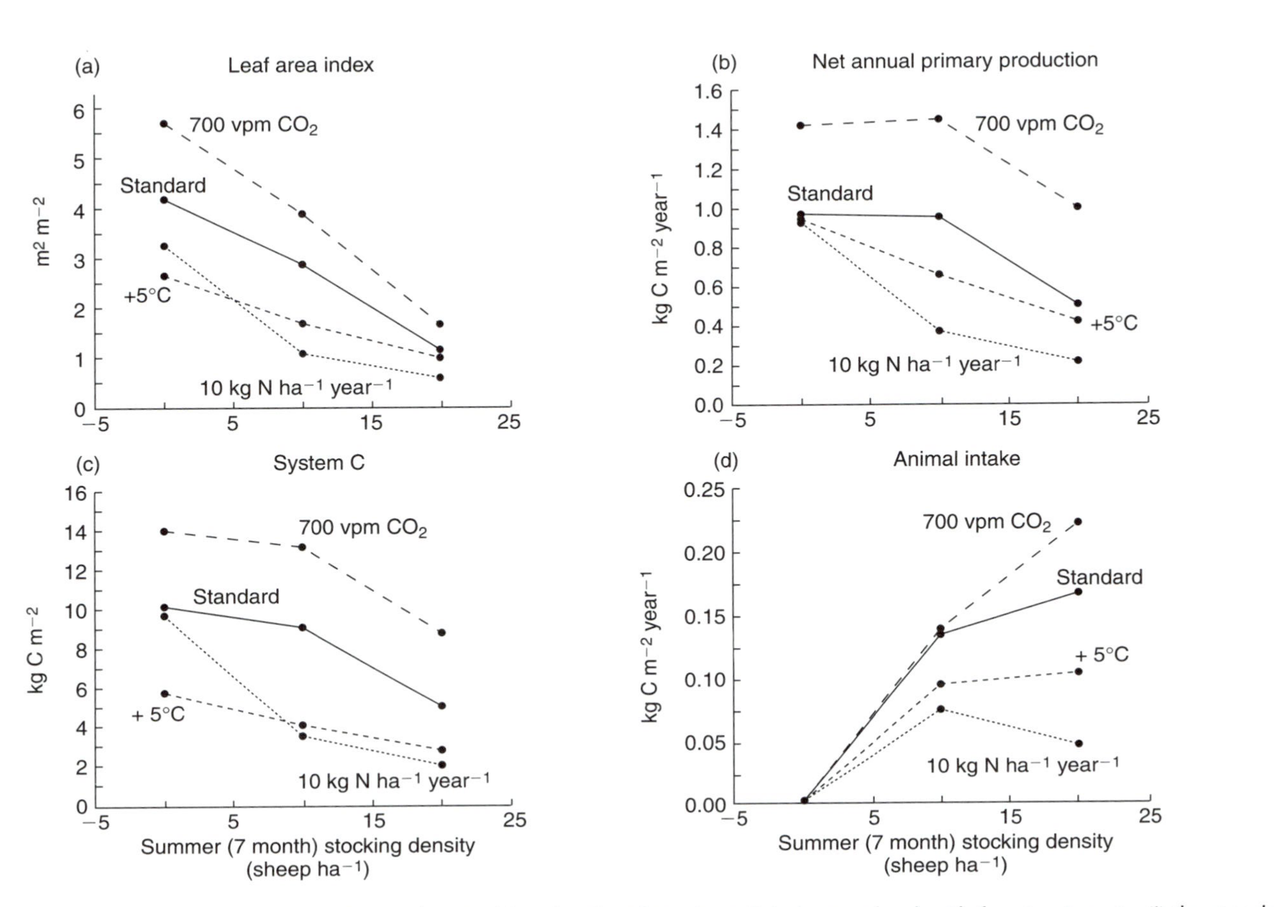

Fig. 9.8. Steady-state responses to summer (7 months) stocking density of southern Britain grassland with four treatments: (i) the standard environment; (ii) 700 vpm CO_2 (instead of 350); (iii) 10 kg N ha^{-1} $year^{-1}$ (instead of 50); (iv) temperature increased by 5°C. Otherwise the standard southern Britain climate is used (Chapter 7, Tables 7.2 and 7.3). Quantities shown are annual averages (e.g. L_{Al}), or for fluxes, annual sums. (a) Leaf area index [Eqn (3.1b)]. (b) Net annual primary production [from Eqn (8.1a)]. (c) Total system C [kg C m^{-2}, Eqn (8.1b)]. (d) Animal intake [$I_{C,pl\rightarrow an}$, Eqn (4.2c)].

9.6 Responses to Rainfall

In Fig. 9.9 annual rainfall has been varied by ±30%, without changing its seasonal pattern (Fig. 7.1c). Because summer water stress is significant in southern Britain, there is a substantial positive response of leaf area index (L_{AI}), sequestered C (C_{sys}) (Fig. 9.9a) and net annual primary production (NAPP) (Fig. 9.9b). Soil mineral N (N_{min}) decreases. Shoot:root ratio increases slightly with increasing rainfall (Fig. 9.9a): this is the net result of being driven down by decreasing N_{min}, (Fig. 9.9a), and upwards by increasing shoot relative water content (Fig. 9.9d). The increased C fluxes (Fig. 9.9b) result from decreased water stress and stomatal closure (Fig. 9.9d), and the virtuous circle of + L_{AI}, + light interception, + gross production, + growth, + L_{AI}. Although drainage increases greatly (Fig. 9.9d), leaching increases proportionately initially (Figs. 9.9c), and then decreases slightly. Note that leaching depends strongly on seasonal factors (Fig. 8.3c). Whitehead (1995, p. 142) comments that rainfall or irrigation applied during the growing season generally decreases N leaching during the following winter, by increasing grass growth and N uptake. When rainfall is decreased by 30% the steady state is a biennial oscillation in L_{AI} (Section 8.14), with yearly average L_{AI}'s of 2.07 and 1.97. There are small errors mostly of a few percent in the low rainfall data in Fig. 9.9 because the final-year averages and accumulations are given rather than averages and accumulations over the final 2 years. However, in the case of drainage, leaching and denitrification, the biennial oscillations are (yearly averages) from 22 to 25 kg water m^{-2} $year^{-1}$, 0.6 to 1.4 kg N ha^{-1} $year^{-1}$ and from 7 to 13 kg N ha^{-1} $year^{-1}$.

Figure 9.10 describes the rainfall response with the sward given four treatments: the standard regime (as in Fig. 9.9); an atmospheric N deposition rate of 10 kg N ha^{-1} $year^{-1}$ (instead of 50); a temperature increase of 5°C; and fourth, high ambient CO_2 of 700 vpm (instead of 350). The response of leaf area index (L_{AI}) to rainfall (Fig. 9.10a) is large for the 700 vpm CO_2 and standard treatments, quite small for +5°C, and negligible for the low N treatment. The impact of increased rainfall on NAPP (Fig. 9.10b) and system C (Fig. 9.10c) is similar to Fig. 9.10a but flattened, as would be expected because gross production does not respond linearly to increases in L_{AI} except at low L_{AI} (cf. Fig. 3.4). Animal intake (Fig. 9.10d) is the same at 700 vpm CO_2 as in the standard treatment because intake is saturated with respect to L_{AI} (Fig. 4.2). The four treatments of Fig. 9.10 underline the great extent to which the rainfall response depends on other factors.

9.7 Responses to Climate Change in Two Different Environments

In this section a comparison is made between selected steady-state variables for southern and northern Britain, when a given component of the environment or management is altered. The conditions at these two sites are given in Chapter 7 (Tables 7.2 and 7.3). Relative to southern Britain, the northern Britain environment has lower temperatures, higher rainfall and humidity, and lower radiation inputs. The atmospheric N input is much lower. A standard grazing regime with 12 sheep ha^{-1} for 7 months is assumed for southern Britain [Eqn (7.6f)]; for northern Britain the stocking density is decreased to 8 sheep ha^{-1} for the same 7-month period. Both

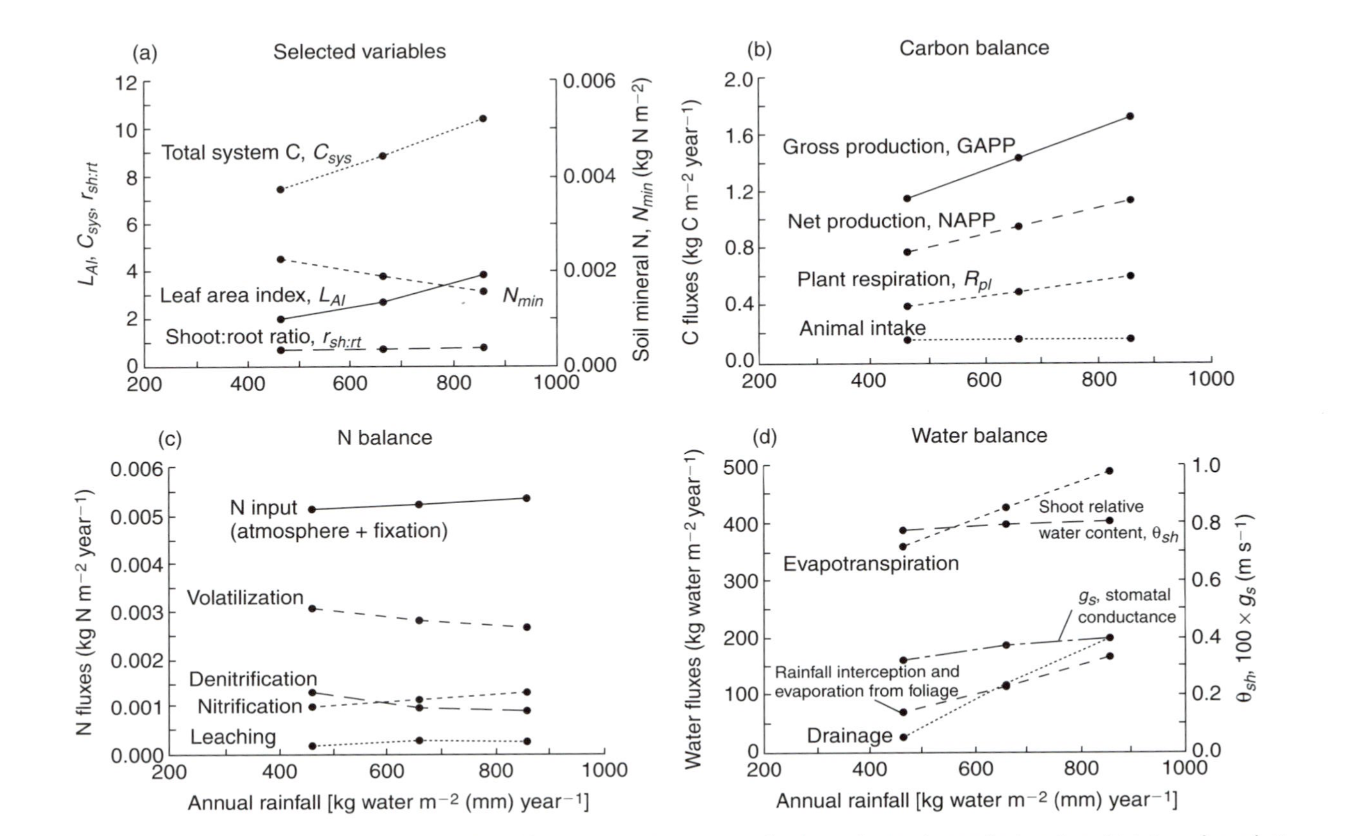

Fig. 9.9. Steady-state responses of rainfall (Section 7.5, Table 7.3, Fig. 7.1c). A multiplying factor is applied to Eqn (7.4a) so the relative seasonal variation in rainfall is unchanged. The standard southern Britain climate is otherwise used (Chapter 7, Tables 7.2 and 7.3). Quantities shown are annual averages (e.g. of L_{AI}), or for fluxes, annual sums. (a) Leaf area index [Eqn (3.1b)], total system C [kg C m^{-2}, Eqn (8.1b)], shoot:root ratio [Eqn (3.1l)], soil mineral N [Eqn (5.5g)]. (b) Carbon balance fluxes. (c) Nitrogen balance fluxes. (d) Water balance fluxes, shoot relative water content [Eqn (6.5d), $i = sh$] and stomatal conductance [Eqn (3.2u)]. The last two quantities are for 1 July at 15.00 h when water stress is relatively high.

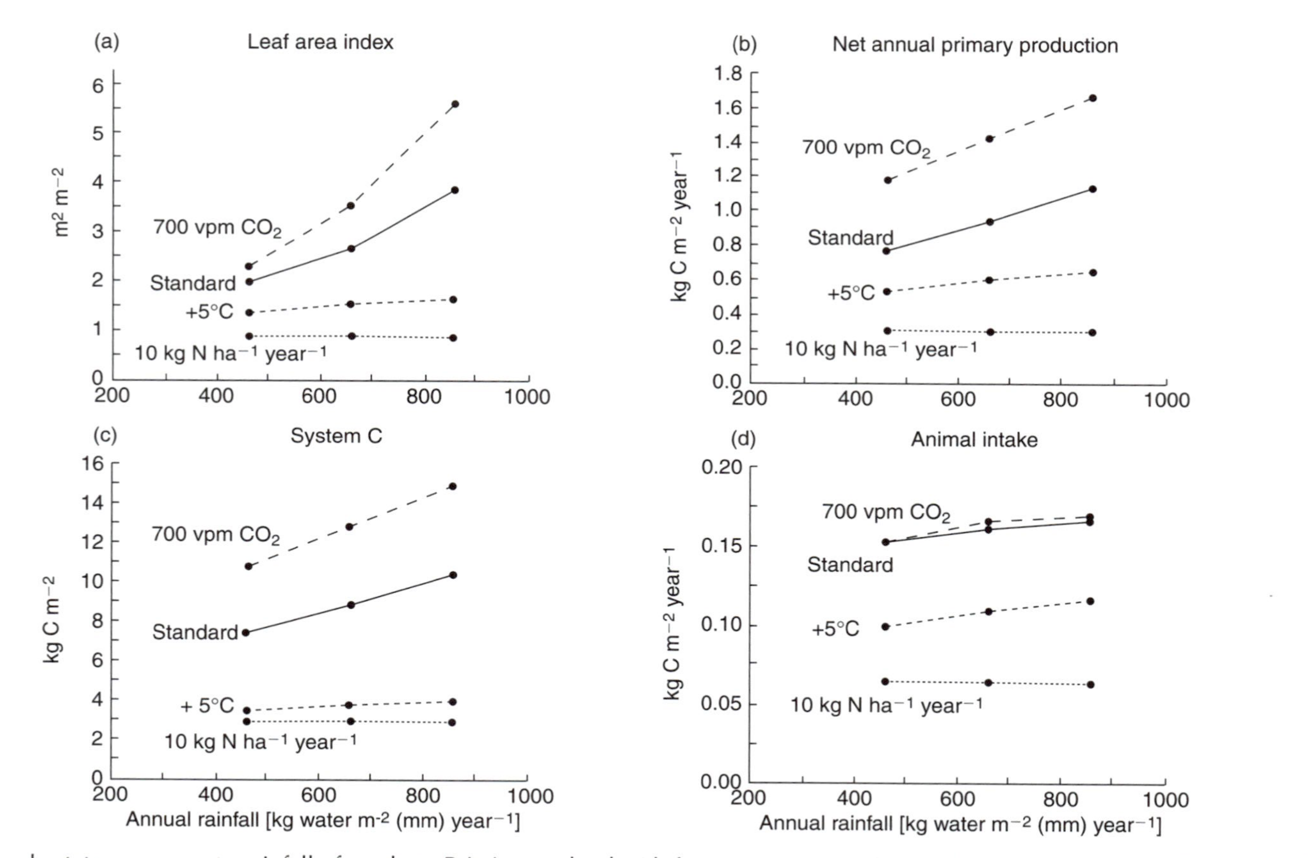

Fig. 9.10. Steady-state responses to rainfall of southern Britain grassland with four treatments: (i) the standard environment; (ii) 700 vpm CO_2 (instead of 350); (iii) 10 kg N ha^{-1} year^{-1} (instead of 50); (iv) temperature increased by 5°C. A multiplying factor is applied to Eqn (7.4a) so the relative seasonal variation in rainfall is unchanged. Otherwise the standard southern Britain climate is used (Chapter 7, Tables 7.2 and 7.3). Quantities shown are annual averages (e.g. L_{Al}), or for fluxes, annual sums. (a) Leaf area index [Eqn (3.1b)]. (b) Net annual primary production [from Eqn (8.1a)]. (c) Total system C [kg C m^{-2}, Eqn (8.1b)]. (d) Animal intake [$I_{C,pl \to an}$, Eqn (4.2c)].

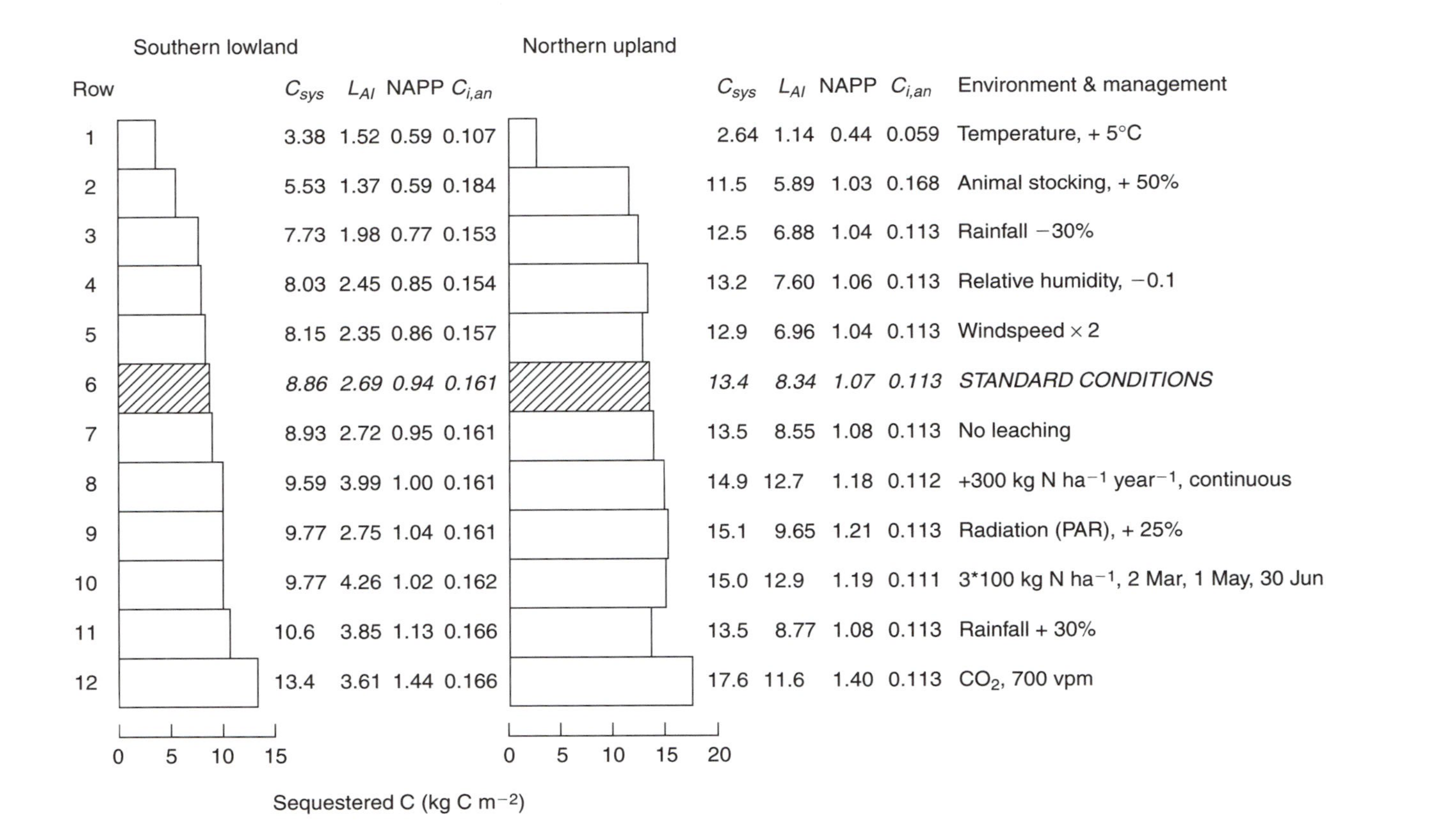

Fig. 9.11. Effects of changing elements of the environment on a southern Britain (lowland) and a northern Britain (upland) grassland. See Chapter 7 (Tables 7.2 and 7.3) for details of the sites. Steady-state annual means and fluxes are given: C_{sys} [Eqn (8.1b)] is total sequestered C in the system (kg C m^{-2}); L_{AI} is the leaf area index [Eqn (3.1b)] [m^2 (m^2 ground)$^{-1}$]; NAPP is the net annual primary production [from Eqn (8.1a)] (kg C m^{-2} $year^{-1}$); $C_{i,an}$ denotes the annual C input to the animals [kg C m^{-2} $year^{-1}$, from $I_{C,pl \rightarrow an}$, Eqn (4.2c)]. The decreased rainfall and relative humidity treatments at the lowland site give rise to biennial oscillations (Section 8.14). The notation 3 * 100 indicates that this item is repeated three times.

sites are lightly grazed, the northern site very much so, as shown by the leaf area indices in Fig. 9.11.

In Fig. 9.11, the treatments are ordered according to the steady-state sequestered C (C_{sys}) in the lowland (southern Britain) environment. Note again that sequestered C is usually but not always increased by increases in primary production (cf. rows 9 and 10, lowland; Figs 9.1a, b, 9.2b, c, 9.8b, c). At the lowland site, decreasing rainfall decreases all variables greatly whereas at the upland sites the variables change much less (cf. rows 3 and 6). However, increasing rainfall brings large benefits to the lowland grass, but little benefit to upland grass (cf. rows 11 and 6). Increasing photosynthetically active radiation (not net radiation) gives similar benefits at both sites (cf. rows 10 and 6). High CO_2 (row 12) gives the highest values of sequestered C (C_{sys}) and primary production (NAPP) at both sites, although leaf area index (L_{AI}) is higher in the fertilized treatments at both sites (rows 8 and 10), and higher in the high rainfall treatment at the lowland site (row 11). The differences between the lowland and upland sites underlines the subtlety in the responses of a grassland ecosystem, and the impossibility of giving simple answers about the consequences of changing the environment even within a small area such as the British Isles.

The ACSL Program 10

10.1 Introduction

ACSL is a software tool designed for modelling dynamical systems described by non-linear differential equations, as outlined in Chapter 1. ACSL denotes **A**dvanced **C**ontinuous **S**imulation **L**anguage; it was developed by Mitchell and Gauthier many years ago, and is continually being extended, improved, and adapted to new platforms and compilers (Mitchell and Gauthier, 1993). ACSL provides a friendly front-end for writing Fortran-like programs for simulating continuous systems. An early standard for these languages was established by a Technical Committee on Continuous System Simulation Languages (CSSL, 1967). A Continuous System Simulation Language (CSSL) provides many standard mathematical, housekeeping, input/output routines which would be tedious and time-consuming to program, and easy to get wrong. As discussed in Section 2.5, the grassland ecosystem model has a standard structure, and is therefore well-suited to ACSL. Because ACSL is non-procedural, statements can be placed in an order that makes biological sense. The program for the grassland simulator, pasture.csl, is kept as close to the biology and mathematics as possible. Furthermore, the ACSL runtime environment (Section 10.2) makes it possible to interact closely with the program in a way which is helpful for debugging, evaluation, development and application.

It is puzzling that, among biologists and ecologists, the use of a CSSL language like ACSL is relatively uncommon, although there are many agricultural, ecological and environmental problems that are of exactly the right type. The use of a package, whether statistical, mathematical or graphical, can distance the user from the actual operations being carried out. A degree of confidence that the correct numerical outcome is obtained is required. Perhaps many CSSLs fail to evoke the confidence needed. On the other hand, a good package can enable the scientist to grapple more closely with the realities of the problem, rather than the mechanics of obtaining solutions. While ACSL is not without difficulties, it is much more efficient than programming the problem in Fortran, Pascal, C, etc. A CSSL provides a generic front-end which remains essentially the same as the model, and the program, are changed.

An ecosystem model rarely remains unchanged for more than a month or two, and it is then inefficient to develop a problem-specific front-end, which may be needed if the problem is programmed in Fortran.

10.2 Example of an ACSL Program*

There are two parts to the inputs to ACSL. First, the *source* PROGRAM is placed in a .csl file (e.g. the file fig1x1.csl shown in Fig. 10.1; the grassland simulator is in file pasture.csl), and defines the system being modelled. Second the *runtime* commands define how the model is to be exercised: e.g. the length of the run, the frequency for outputting results, the plots that are needed, and parameters and switches that are to be set/changed etc. The runtime commands can be input from the keyboard, but it is usually more convenient to place them in a *command* file with extension .cmd, e.g. fig1x1.cmd (Fig. 10.2). The source file filename.csl is translated by an ACSL translator into a Fortran file filename.for; the Fortran code is compiled and linked with

```
PROGRAM to generate Fig. 1.1. Water pouring into a leaky bucket.
  !File: Fig1x1.csl. Listing is given in Fig. 10.1.

!ACSL PARAMETERS
  Algorithm ialg = 3          !Euler's method of integration.
  Cinterval cint = 0.1        !Communication interval for results.
  Maxterval maxt = 0.01       !Integration interval.
  Constant tstop = 2          !Stop simulation after 2 days.

INITIAL   !Calculations to be performed at time t = 0 are placed here.
          !There are none in this model.
END       !Initial.

DERIVATIVE
  termt   ( t .GE. tstop )    !Stop simulation when time t .GE. tstop.

  !Differential eqn for state variable V, volume of water in bucket (m3):
    DV = IW - ( kL / A ) * V    !DV denotes dV/dt (m3 day-1).
  !Integration statement (V0 is the value of V at t = 0 day):
     V = integ ( DV, V0 )
  !Parameters are:
     Constant V0 = 0   !m3.                        Initially bucket is empty.
     Constant IW = 1   !m3 water day-1.            Volume input rate to bucket.
     Constant kL = 2   !m2 day-1.                  Leakage rate of bucket.
     Constant A = 1    !m2.                        Area of cylindrical bucket.
END  !Derivative.
END  !Program.
```

Fig. 10.1. Listing of file fig1x1.csl, the ACSL program for the leaky bucket model of Fig. 1.1. The DOS command, acsl fig1x1, translates the file fig1x1.csl, and then compiles and links it to give a file fig1x1.exe. All characters following '!' on the same line are comments.

* This account of ACSL is based on my experience with ACSL version 11.2.2 (13 November 1995), PC/DOS, with Watcom Fortran 10.5. ACSL for Windows is widely used, but does not execute as quickly as DOS/ACSL. For working with the pasture model, execution speed can be an important constraint. Double precision is essential.

```
!Command file Fig1x1.cmd for ACSL program Fig1x1.csl.

prepar/clear        !Clear the prepar list.
prepar t, V, DV     !Save these variables in a .rrr file for printing/plotting.

output/clear        !Clear the output list.
output t, V, DV     !When the program is run (with <start>), these variables
                    !are sent to the screen.
set nciout = 5, nciprn = 2  !Items are output to the screen every 5 cint
                    !intervals. Items can be printed every 2 cint intervals.

start; pause  !This runs the simulation, and then waits for a <return>.

!Plotting stuff:
  set ftsplt = .t., grdcpl = .f.  !Suppress flyback trace, grid lines.
  d xticpl, xincpl, yticpl, yincpl
  set xticpl = 1.2, xincpl = 6    !Set x tick interval, length of x axis.
  set yticpl = 0.8, yincpl = 4    !Set y tick interval, length of y axis.

  plot /xhi = 2, V                !Plot V. Plots are to screen by default.
  set devplt = 5, plt = 15        !Send graphs in postscript to file plot15.
  plot V                          !Generates a postscript file plot15.

quit  !This quits ACSL.  If omitted, control returns to the keyboard.
```

Fig. 10.2. Listing of file fig1x1.cmd, the command file of runtime commands for use with the ACSL program fig1x1.csl of Fig. 10.1. All characters following '!' on the same line are comments. Running the program fig1x1.exe with the DOS command acsl -r fig1x1 automatically looks for runtime commands in fig1x1.cmd. These runtime commands generate screen text output, which is recorded in the log file fig1x1.log of Fig. 10.3, a screen graph similar to Fig. 1.1, and a postscript file of the screen graph.

an ACSL library giving a filename.exe file. Execution of the filename.exe file normally reads commands from the command file filename.cmd if this file is present. That is, execution of fig1x1.exe reads and executes the runtime commands in fig1x1.cmd (Fig. 10.2). If the filename.cmd file is not present, then the ACSL prompt ACSL> appears on the screen, and runtime commands can be entered at the keyboard. In work with the pasture model and other large models, the use of command files which define how the model is exercised has been found to be highly efficient. If the file of runtime commands in filename.cmd is read and executed, and if at the end of the filename.cmd file ACSL is not exitted with quit, then an ACSL runtime prompt, ACSL>, appears on the screen and ACSL waits for further runtime commands from the keyboard. Further commands can be entered at the keyboard, or placed in another command file (e.g. myfile.cmd) which can be referenced by typing set cmd = 10, in which case the runtime executive asks for a filename, and, on entering myfile.cmd at the keyboard, the contents of myfile.cmd are then executed.

In Fig. 10.1, the ACSL source program fig1x1.csl for the leaky bucket model of Fig. 1.1 is given. At the DOS command acsl fig1x1, the file fig1x1.csl is translated, compiled and linked to give the file fig1x1.exe; this executable file is then run and looks for the command file fig1x1.cmd (Fig. 10.2) for runtime statements. These are executed giving some screen output, a screen graph, and a postscript file of the

```
ACSL Runtime Exec Version 11.2.2  97/06/12 10:53:19

Switching CMD unit to 4 to read fig1x1.cmd

!Command file Fig1x1.cmd for ACSL program Fig1x1.csl.
prepar/clear          !Clear the prepar list.
prepar t, V, DV       !Save these variables in a .rrr file for printing/plotting.

output/clear          !Clear the output list.
output t, V, DV       !When the program is run (with <start>), these variables
                      !are sent to the screen.
set nciout = 5, nciprn = 2  !Items are output to the screen every 5 cint
                      !intervals. Items can be printed every 2 cint intervals.

start; pause  !This runs the simulation, and then waits for a <return>.
          T 0.                    V 0.                    DV 1.00000000
          T 0.50000000            V 0.31791500            DV 0.36417000
          T 0.99999900            V 0.43369000            DV 0.13262000
          T 1.50000000            V 0.47485200            DV 0.04829600
          T 2.00000000            V 0.49120600            DV 0.01758800
          T 2.02000000            V 0.49155400            DV 0.01723620

!Plotting stuff:
  set ftsplt = .t., grdcpl = .f.  !Suppress flyback trace, grid lines.
  d xticpl, xincpl, yticpl, yincpl
    XTICPL   1.00000000       XINCPL 5.00000000         YTICPL 1.00000000
    YINCPL   5.00000000
  set xticpl = 1.2, Xincpl = 6   !Set x tick interval, length of x axis.
  set yticpl = 0.8, yincpl = 4   !Set y tick interval, length of y axis.

  plot /xhi = 2, V               !Plot V. Plots are to screen by default.
Drawing plot number 1
  set devplt = 5, plt = 15       !Send graphs in postscript to file plot15.
  plot V                         !Generates a postscript file plot15.
Drawing plot number 2.
Opening new plot unit number 15 as C:\JT\BOOK\GRAPHICS\plot15

quit  !This quits ACSL.  If omitted, control returns to the keyboard.
```

Fig. 10.3. The file fig1x1.log shown here is the output file from running fig1x1.exe with the commands in file fig1x1.cmd of Fig. 10.2.

screen graph. The last statement in Fig. 10.2 quits ACSL. All screen output (not the graph) is logged in the file fig1x1.log which is shown in Fig. 10.3. The screen graph is similar to Fig. 1.1.

10.3 About the ACSL Pasture Program

The grassland simulator, pasture.csl, the other ACSL programs (.csl) and command files (.cmd) used to generate the data and graphs in this book are available *gratis* by ftp (budbase.nbu.ac.uk/pub/tree/Book). The most recent version of the Hurley Pasture Model is at budbase.nbu.ac.uk/pub/tree/Pasture. Alternatively, the author can be contacted by e-mail (john.thornley@unixmail.nbu.ac.uk). If items are required on floppy disk, then please send a floppy, and if possible the return postage. The structure of the grassland simulator program is similar to that of the simple program of Fig. 10.1. As far as possible, it follows the mathematical presentation used in the book, although the chapter order is not followed. Each pool (state variable) forms a subsection: first the differential equation is given, and then the inputs and outputs are specified.

Some of the ACSL runtime facilities that have been particularly useful with the grassland ecosystem simulator include the following:

1. Run the model with **start** until time tstop (the time when the simulation stops); change parameters or the environment; change (increase) tstop; **cont**inue the run until the new tstop.
2. Look at all the nitrogen inputs and outputs to, say, the soil ammonium pool, with **d**isplay IN*Namm, ONamm*.
3. save; restore. The state of the system at the end of a simulation run can be **save**d. On another occasion you can **restore** that system state, and then run the system starting from that saved state (see some of the command files used in generating the graphs in this book).
4. Print variables to a file using print/f = 'myfile.dat' *list* for use in a graphing package.

Appendix
Useful Constants and Conversion Factors

The numerical values of some constants are shown below.

Absolute zero of temperature	−273.15°C
Atmospheric pressure	101,325 Pa
Avogadro constant	6.022×10^{23} mol^{-1}
Gas constant	8.314 J K^{-1} mol^{-1}
Gravitational acceleration	9.81 m s^{-2}
Planck constant	6.6262×10^{-34} J s
Speed of light (vacuum)	2.998×10^{8} m s^{-1}
Volume of 1 mol of ideal gas at 0°C and atmospheric pressure	0.02241 m^3

Carbon dioxide

The conversion of CO_2 'concentration' in vpm (volumes per million volumes) to a true CO_2 concentration or density is achieved by

$$CO_2 \text{ concentration} \left(\text{kg } CO_2 \text{ m}^{-3}\right) = \frac{CO_2\text{concentration(vpm)}}{10^6 \text{ vpm}} \times \frac{273.15 \text{ K}}{T(\text{K})} \frac{P\,(\text{Pa})}{101{,}325 \text{ Pa}} \frac{44.01 \text{ kg } CO_2}{22.4136 \text{ m}^3}, \quad \text{(A1)}$$

where T (K) is the absolute temperature, P (Pa) is the pressure, standard temperature is 273.15 K (freezing point of water), standard pressure is 101,325 Pa (atmospheric pressure), 44.01 is the relative molecular mass of CO_2 and 22.4136 m^3 is the volume occupied by 44.01 kg of CO_2, assuming it is an ideal gas.

At 20°C (293.15 K) and atmospheric pressure (101,325 Pa), therefore

$$350 \text{ vpm } CO_2 = 0.6404 \times 10^{-3} \text{ kg } CO_2 \text{ m}^{-3}. \tag{A2}$$

Radiation

The conversion of radiation in energy units (J) into quanta is given by

$$1 \text{ J of wavelength } \lambda \text{ (m)} = \frac{1 \text{ (J)} \lambda \text{ (m)}}{6.6262 \times 10^{-34} \text{ J s} \times 2.998 \times 10^{8} \text{ m s}^{-1}} \text{ quanta,} \tag{A3}$$

where 6.6262×10^{-34} J s is the Planck constant and 2.998×10^8 m s^{-1} is the speed of light. Therefore, using a wavelength of $\lambda = 0.55 \times 10^{-6}$ m in the yellow region of the visible spectrum:

$$1 \text{ J of wavelength } 0.55 \times 10^{-6} \text{ m} = 2.77 \times 10^{18} \text{ quanta} = \frac{2.77 \times 10^{18} \text{ quanta}}{6.022 \times 10^{23} \text{ entities mol}^{-1}} = 4.6 \times 10^{-6} \text{ mol.} \tag{A4}$$

Here, 6.022×10^{23} entities mol^{-1} is the Avogadro constant. Recognizing that there is no exact conversion between photosynthetically active radiation (PAR) in energy units and in quanta, we use this conversion of 1 J PAR = 4.6 μmol PAR in calculations.

Photosynthetic efficiency

This may be expressed as mass of CO_2 (kg) fixed per energy unit of photosynthetically active radiation (J), or as the number of molecules of CO_2 fixed per quantum of photosynthetically active radiation. The conversion makes use of Eqn (A4):

$$1 \times 10^{-8} \text{ kg } CO_2 \text{ (J PAR)}^{-1} = 1 \times 10^{-8} \text{ kg } CO_2 \text{ (J PAR)}^{-1} \times \frac{1000 \text{ g kg}^{-1}}{44.01 \text{ g (mol } CO_2)^{-1}} \frac{1}{4.6 \times 10^{-6} \text{ mol PAR (J PAR)}^{-1}} = 0.0494 \text{ mol } CO_2 \text{ (mol PAR)}^{-1}. \tag{A5}$$

About 20 quanta are required for the reduction of one CO_2 molecule. In this equation, the unsatisfactory definition of the SI mole is apparent.

Water potential

First note that the SI unit of pressure, the pascal (Pa), has dimensions of force per unit area (N m^{-2}) or energy per unit volume (J m^{-3}). Taking the density of water ρ_W = 1000 kg m^{-3}, we can write down a relationship between our preferred energy units of water potential, ψ_E [J (kg water)$^{-1}$], and the frequently used pressure units of water potential, ψ_P (Pa), as

$$\psi_E\left[J\,(\text{kg water})^{-1}\right] = \frac{\psi_P\left[\text{J}\,(\text{m}^3\ \text{water})^{-1}\right]}{\rho_W\left[\text{kg water}\,(\text{m}^3\ \text{water})^{-1}\right]}$$

$$= \frac{\psi_P\ (\text{Pa})}{1000\ \text{kg m}^{-3}}. \tag{A6}$$

Thus

$$1\ \text{J kg}^{-1} \equiv 10^3\ \text{Pa} = 1\ \text{kPa}. \tag{A7}$$

References

Abbès, C., Robert, J.L. and Parent, L.E. (1996) Mechanistic modeling of coupled ammonium and nitrate uptake by onions using the finite element method. *Soil Society of America Journal* 60, 1160–1167.

Acock, B. and Grange, R.I. (1981) Equilibrium models of leaf water relations. In: Rose, D.A. and Charles-Edwards, D.A. (eds) *Mathematics and Plant Physiology*. Academic Press, London, pp. 29–47.

Addiscott, T.M. and Whitmore, A.P. (1991) Simulation of solute leaching in soils of differing permeabilities. *Soil Use Management* 7, 94–102.

Adger, W.N., Brown, K., Shiel, R.S. and Whitby, M.C. (1992) Carbon dynamics of land use in Great Britain. *Journal of Environmental Management* 36, 117–133.

Almeida, J.S., Reis, M.A.M. and Carrondo, M.J.T. (1997) A unifying kinetic model of denitrification. *Journal of Theoretical Biology* 189, 241–249.

Amthor, J.S. (1984) The role of maintenance respiration in plant growth. *Plant, Cell and Environment* 7, 561–569.

ARC (Agricultural Research Council) (1980) *The Nutrient Requirements of Ruminant Livestock*. Commonwealth Agricultural Bureaux, Farnham Royal.

Arah, J.R.M. and Smith, K.A. (1989) Steady-state denitrification in aggregated soils: a mathematical model. *Journal of Soil Science* 40, 139–149.

Aulakh, M.S., Doran, J.W. and Mosier, A.R. (1992) Soil denitrification – significance, measurement and effects of management. *Advances in Soil Science* 18, 1–57.

Baker, G.L. amd Gollub, J.P. (1996) *Chaotic Dynamics*. Cambridge University Press, Cambridge, UK, 272 pp.

Barber, D.A. and Martin, J.K. (1976) The release of organic substances by cereal roots into soil. *New Phytologist* 76, 69–80.

Battaglia, M., Beadle, C. and Loughhead, S. (1996) Photosynthetic temperature responses of *Eucalyptus globulus* and *Eucalyptus nitens*. *Tree Physiology* 16, 81–89.

Baxter, R., Ashenden, T.W., Sparks, T.H and Farrar, J.F. (1994) Effects of elevated carbon dioxide on three montane grass species. I. Growth and dry matter partitioning. *Journal of Experimental Botany* 45, 305–315.

Berry, J. and Björkman, O. (1980) Photosynthetic response and adaptation to temperature in higher plants. *Annual Reviews in Plant Physiology* 31, 491–543.

Berryman, A.A. and Milstein, J.A. (1989) Are ecological systems chaotic – and if not, why not? *Trends in Ecology and Evolution* 4, 26–28.

Bloom, A.J., Sukrapanna, S.S. and Warner, R.I. (1992) Root respiration associated with ammonium and nitrate absorption and assimilation by barley. *Plant Physiology* 99, 1294–1301.

Boote, K.J. and Loomis, R.S. (1991) The prediction of canopy photosynthesis. In: Boote, K.J. and Loomis, R.S. (eds) *Modelling Photosynthesis – from Biochemistry to Canopy*. CSSA Special Publication Number 19. Crop Science Society of America, Inc., American Society of Agronomy, Inc. Madison, Wisconsin, USA, pp. 109–140.

Bouma, T.J. (1995) *Utilization of Respiratory Energy in Higher Plants. Requirements for 'Maintenance' and Transport Processes*. Department of Theoretical Production Ecology, Wageningen Agricultural University, P.O. Box 430, 6700 AK Wageningen, The Netherlands. ISBN 90–5485–350–6.

Bouma, T.J. and De Visser, R. (1993) Energy requirements for maintenance of ion concentrations in roots. *Physiologia Plantarum* 89, 133–142.

Bouma, T.J., Spitters, C.J.T. and De Visser, R. (1992) Variation in leaf respiration rate between potato cultivars: effect of development stage. In: Lambers, H. and van der Plas, L.H.W. (eds.) *Molecular, Biochemical and Physiological Aspects of Plant Respiration*. SPB Academic Publishing, The Hague.

Bouma, T.J., De Visser, R., Van Leeuwen, P.H., De Kock, M.J. and Lambers, H. (1994) Respiratory energy requirements and rate of protein turnover *in vivo* determined by the use of an inhibitor of protein synthesis and a probe to assess its effect. *Physiologia Plantarum* 92, 585–594.

Boyer, J.S. (1968) Relationship of water potential to growth of leaves. *Plant Physiology* 43, 1056–1062.

Bunce, J.A. (1977) Leaf elongation in relation to water potential in soybean. *Journal of Experimental Botany* 28, 156–161.

Campbell, G.S. (1974) A simple method for determining unsaturated conductivity from moisture retention data. *Soil Science* 117, 311–315.

Campbell, G.S. (1977) *An Introduction to Environmental Biophysics*. Springer-Verlag, New York.

Cannell, M.G.R. and Thornley, J.H.M. (1998) N-poor systems may respond more to elevated CO_2 than N-rich ones in the long term. A model analysis of grassland. *Global Change Biology* (in press).

Carpenter, S.R. (1981) Decay of heterogenous detritus: a general model. *Journal of Theoretical Biology* 89, 539–547.

Casella, E., Soussana, J.F. and Loiseau, P. (1996) Long-term effects of CO_2 enrichment and temperature increase on a temperate grass sward. I. Productivity and water use. *Plant and Soil* 182, 83–99.

Castellví, F., Perez, P.J., Villar, J.M. and Rosell, J.I. (1996) Analysis of methods for estimating vapor pressure deficits and relative humidity. *Agricultural and Forest Meteorology* 82, 29–45.

Cazelles, B. and Ferriere, R.H. (1992) How predictable is chaos? *Nature* 355, 25–26.

Charles-Edwards, D.A. (1981) *The Mathematics of Photosynthesis and Productivity*. Academic Press, London, 127 pp.

Charles-Edwards, D.A. and Acock, B. (1977) Growth response of a *Chrysanthemum* crop to the environment. II. A mathematical analysis relating photosynthesis and growth. *Annals of Botany* 41, 49–58.

Charles-Edwards, D.A., Doley, D. and Rimington, G.M. (1986) *Modelling Plant Growth and Development*. Academic Press, North Ryde, N.S.W., Australia.

Chartier, P. (1966) Étude théorique de l'assimilation brute de la feuille. *Annales de Physiologie végétale* 6, 167–195.

Chen, D-X. and Coughenour, M.B. (1994) GEMTM – A general-model for energy and mass-transfer of land surfaces and its application at the FIFE sites. *Agricultural and Forest Meteorology* 68, 145–171.

Childs, E.C. and Collis-George, N. (1950) The permeability of porous materials. *Proceedings of the Royal Society, London, A* 201, 392–405.

Christensen, B.T. (1996) Matching measurable soil organic matter fractions with conceptual pools in simulation models of carbon turnover: revision of model structure. In: Powlson, D.S., Smith, P. and Smith, J.P. (eds) *Evaluation of Soil Organic Matter Models using Long-Term Datasets*. NATO ASI Series I, volume 38. Springer-Verlag, Heidelberg, pp. 225–230.

Cienciala, C., Eckersten, H., Lindroth, A. and Hällgren, J-E. (1994) Simulated and measured water uptake by *Picea abies* under non-limiting soil water conditions. *Agricultural and Forest Meteorology* 71, 147–164.

Clarkson, D.T., Hopper, M.J. and Jones, L.H.P. (1986) The effect of root temperature on the uptake of nitrogen and the relative size of the root system in *Lolium perenne*. I. Solutions containing both NH_4^+ and NO_3^-. *Plant, Cell and Environment* 9, 535–545.

Clarkson, D.T., Jones, L.H.P. and Purves, J.V. (1992) Absorption of nitrate and ammonium ions by *Lolium perenne* from flowing solution cultures at low root temperatures. *Plant, Cell and Environment* 15, 99–106.

Corrall, A.J. and Fenlon, J.S. (1978) A comparative method for describing the seasonal distribution of production from grasses. *Journal of Agricultural Science, Cambridge* 91, 61–67.

Coughenour, M.B. (1984) A mechanistic simulation analysis of water use, leaf angles, and grazing in East African graminoids. *Ecological Modelling* 26, 203–230.

CSSL (1967) The SCi Continuous system simulation language. *Simulation* 9, 281–303.

Dahlman, R.C. and Kucera, C.L. (1965) Root productivity and turnover in native grassland. *Ecology* 46, 84–89.

Davidson, E.A., Nepstad, D.C., Klink, C. and Trumbore, S.E. (1995) Pasture soils as carbon sink. *Nature* 376, 472–473.

Davies, A. (1977) Structure of the grass sward. In: Gilsenan, B. (ed.) *Proceedings of an International Meeting on Animal Production from Temperate Grassland.* An Foras Taluntais, Dublin, pp. 36–44.

de Wit, C.T. (1970) Dynamic concepts in biology. In: Setlik, I. (ed.) *Prediction and Measurement of Photosynthetic Productivity*. Pudoc, Wageningen, pp. 17–23.

de Wit, C.T. *et al.* (1978) *Simulation of Assimilation, Respiration and Transpiration in Crops.* Pudoc, Wageningen.

Dewar, R.C. (1995) Interpretation of an empirical model for stomatal conductance in terms of guard cell function. *Plant, Cell and Environment* 18, 365–372.

Dewar, R.C. (1996) The correlation between plant growth and intercepted radiation: an interpretation in terms of optimal plant nitrogen content. *Annals of Botany* 78, 125–136.

Dixon, M. and Webb, E.C. (1964) *Enzymes*. Longmans, London, 950 pp.

Doyle, C.J. (1981) Economics of irrigating grassland in the United Kingdom. *Grass and Forage Science* 36, 297–306.

Ehleringer, J. and Björkman, O. (1977) Quantum yields for CO_2 uptake in C_3 and C_4 plants. *Plant Physiology* 59, 86–90.

Ephrath, J.E., Goudriaan, J. and Marani, A. (1996) Modelling diurnal patterns of air temperature, radiation, wind speed and relative humidity by equations from daily characteristics. *Agricultural Systems* 51, 377–393.

Evans, J.R. (1989) Photosynthesis and nitrogen relationships in leaves of C_3 plants. *Oecologia* 78, 9–19.

Evans, J.R. and Farqhar, G.D. (1991) Modeling canopy photosynthesis from the biochemistry of the C_3 chloroplast. In: Boote, K.J. and Loomis, R.S. (eds) *Modelling Photosynthesis – from Biochemistry to Canopy*. CSSA Special Publication Number 19. Crop Science Society of America, Inc., American Society of Agronomy, Inc., Madison, Wisconsin, USA, pp. 109–140.

Evans, J.R. and Terashima, I. (1988) Photosynthetic characteristics of spinach leaves grown with different nitrogen treatments. *Plant and Cell Physiology* 29, 157–165.

Evans, J.R., Jakobsen, I. and Ögren, E. (1993) Photosynthetic light response curves. 2. Gradients of light absorption and photosynthetic capacity. *Planta* 189, 191–200.

Farquhar, G.D. (1988) Models relating subcellular effects of temperature to whole plant responses. In: Long, S.P. and Woodward, F.I. (eds) *Plants and Temperature*. Symposia of the Society for Experimental Botany, No. 42. Cambridge University Press, Cambridge, pp. 395–409.

Farquhar, G.D. and von Caemmerer, S. (1982) Modelling of photosynthetic response to environment. In: Lange, O., Nobel, P.S., Osmond, C.B. and Ziegler, H. (eds) *Encyclopedia of Plant Physiology*. New Series. Vol 12B. Physiological Plant Ecology II. Springer-Verlag, Berlin, pp. 549–587.

Feng, Y. and Boersma, L. (1995) Kinematics of axial root growth. *Journal of Theoretical Biology* 174, 109–117.

Fisher, M.E., Rao I.M., Ayarza, M.A., Lascano, C.E., Sanz, J.I., Thomas, R.J. and Vera, R.R. (1994) Carbon storage by introduced deep-rooted grasses in the South American savannas. *Nature* 371, 236–238.

Fisher, M.E., Rao I.M., Lascano, C.E., Sanz, J.I., Thomas, R.J., Vera, R.R. and Ayarza, M.A. (1995) Pasture soils as carbon sink. Reply. *Nature* 376, 473.

Forbes, J.M. and France, J. (1993) *Quantitative Aspects of Ruminant Digestion and Metabolism*. CAB International, Wallingford, Oxon.

France, J. and Thornley, J.H.M. (1984) *Mathematical Modelling in Agriculture*. Butterworths, London, 335 pp.

Fruton, J.S. and Simmonds, S. (1958) *General Biochemistry*. Wiley, New York, 1077 pp.

Gabrielle, B. and Kengni, L. (1996) Analysis and field-evaluation of the CERES models' soil components: nitrogen transfer and transformations. *Soil Society of America Journal* 60, 142–149.

Gardner, W.R. (1991) Modelling water uptake by roots. *Irrigation Science* 12, 109–114.

Garwood, E.A. (1967a) Seasonal variation in appearance and growth of grass roots. *Journal of the British Grassland Society* 22, 121–130.

Garwood, E.A. (1967b) Studies on the roots of grasses. *Annual Report of the Grassland Research Institute, Hurley* 1966, 72–79.

Gastal, F. and Bélanger, G. (1993) The effects of nitrogen fertilization and the growing season on photosynthesis of field-grown tall fescue (*Festuca arundinacea* Schreb.) canopies. *Annals of Botany* 72, 401–408.

Gastal, F. Bélanger, G. and Lemaire, G. (1992) A model of the leaf extension rate of tall fescue in response to nitrogen and temperature. *Annals of Botany* 70, 437–442.

Gill, M., Beever, D.E. and Osbourn, D.F. (1989) The feeding value of grass and grass products. In: Holmes, W. (ed.) *Grass: its Production and Utilization*. Blackwell, Oxford, pp. 89–129.

Gordon, A.J., Mitchell, D.F., Ryle, G.J.A. and Powell, C.E. (1987) Diurnal production and utilization of photosynthate in nodulated white clover. *Journal of Experimental Botany* 38, 84–89.

Gradshteyn, I.S. and Ryzhik, J.W. (ed. A. Jeffrey, 5th edn) (1994) *Tables of Integrals, Series and Products*. Academic Press, Boston, 1204 pp.

Grange, R.I. (1985) Carbon partitioning in mature leaves of pepper: effects of daylength. *Journal of Experimental Botany* 36, 1749–1759.

Greenfield, S.B. and Smith, D. (1974) Diurnal variations of nonstructural carbohydrates in the individual parts of switchgrass shoots at anthesis. *Journal of Range Management* 27, 466–469.

Gregson, K., Hector, D.J. and McGowan, M. (1987) A one-parameter model for the soil water characteristic. *Journal of Soil Science* 38, 483–486.

Hahn, B.D. (1991) Photosynthesis and photorespiration: modelling the essentials. *Journal of Theoretical Biology*, 151, 123–139.

Hassink, J. and Whitmore, A.P. (1997) A model of the physical protection of the organic matter in soils. *Soil Science Society of America Journal* 61, 131–139.

Hatch, D.J. and Macduff, J.H. (1991) Concurrent rates of N_2 fixation, nitrate and ammonium uptake by white clover in response to growth at different root temperatures. *Annals of Botany* 67, 265–274.

Hearon, J.Z. (1952) Rate behaviour of metabolic systems. *Physiological Reviews* 32, 499–523.

Hillel, D. and van Bavel, C.H.M. (1976) Simulation of profile water storage as related to soil hydraulic properties. *Soil Science Society of America Journal* 40, 807–815.

Hirose, T. and Werger, M.J.A. (1987) Nitrogen use efficiency in instananeous and daily photosynthesis of leaves in the canopy of a *Solidago altissima* stand. *Physiologia Plantarum* 70, 215–222.

Hodgson, J., Bircham, J.S., Grant, S.A. and King, J. (1981) The influxence of cutting and grazing management of herbage growth and utilization. In: Wright, C.E. (ed.) *Plant Physiology and Herbage Production. Occasional Symposium of the British Grassland Society*, No. 13, pp. 51–62.

Holmes, W. (ed.) (1989) *Grass: its Production and Utilization*. Blackwell, Oxford, 306 pp.

Holt, D.A. and Hilst, A.R. (1969) Daily variation in carbohydrate content of selected forage crops. *Agronomy Journal* 61, 239–242.

Hopkins, J.C. and Leipold, R.J. (1996) On the dangers of adjusting the parameter values of mechanism-based mathematical models. *Journal of Theoretical Biology* 183, 417–427.

Hörmann, G., Branding, A., Clemen, T., Herbst, M., Hinrichs, A. and Thamm, T. (1996) Calculation and simulation of wind controlled canopy interception of a beech forest in Northern Germany. *Agricultural and Forest Meteorology* 79, 131–148.

Hsiao, T.C. and Bradford, K.J. (1983) Physiological consequences of cellular water deficits. In: Taylor, H.M., Jordan, W.R. and Sinclair, T.M. (eds) *Limitations to Efficient Water Use in Crop Production*. American Society of Agronomy, Crop Science Society of America, Soil Science Society of America, Madison, WI, USA, pp. 227–265.

Hughes, R. (1974) Grassland Agronomy. In: *Annual Report for Welsh Plant Breeding Station* 1973. Welsh Plant Breeding Station, Aberystwyth, Wales, pp. 29–32.

Hunt, W.F. (1983) Nitrogen cycling through senescent leaves and litter in swards of Ruanui and Nui ryegrass with high and low nitrogen inputs. *New Zealand Journal of Agricultural Research* 26, 461–471.

Hunt, H.W., Trlica, M.J., Redente, E.F., Moore, J.C., Detling, J.K., Kittel, T.G.F., Walter, D.E., Fowler, M.C., Klein, D.A. and Elliott, E.T. (1991) Simulation model for the effects of climate change on temperate grassland ecosystems. *Ecological Modelling* 53, 205–246.

Jackson, R.D. (1972) On the calculation of hydraulic conductivity. *Soil Science Society of America Proceedings* 36, 380–382.

Janzen, H.H. (1990) Deposition of nitrogen into the rhizosphere by wheat roots. *Soil Biology and Biochemistry* 22, 1155–1160.

Jeffrey, D.W. (1988) Mineral nutrients and the soil environment. In: Jones, M.B. and Lazenby, A. (eds) *The Grass Crop*. Chapman and Hall, London, pp. 179–204.

Jenkinson, D.S. (1990) The turnover of organic carbon and nitrogen in soil. *Philosophical Transactions of the Royal Society B* 329, 361–368.

Jenkinson, D.S. and Powlson, D.S. (1980) Measurement of microbial biomass in intact soil cores and in sieved soil. *Soil Biology and Biochemistry* 12, 579–581.

Jensen, C.R., Svendsen, H., Andersen, M.N. and Lösch, R. (1993) Use of the root contact concept, an empirical leaf conductance model and pressure–volume curves in simulating water relations. *Plant and Soil* 149, 1–26.

Johnson, I.R. and Parsons, A.J. (1985a) A theoretical analysis of grass growth under grazing. *Journal of Theoretical Biology* 112, 345–367.

Johnson, I.R. and Parsons, A.J. (1985b) Use of a model to analyse the effects of continuous grazing managements on seasonal patterns of grass production. *Grass and Forage Science* 40, 449–458.

Johnson, I.R., Melkonian, J.J., Thornley, J.H.M. and Riha, S.J. (1991) A model of water flow through plants incorporating shoot/root 'message' control of stomatal conductance. *Plant, Cell and Environment* 14, 531–544.

Jones, H.G. (1978) Modelling diurnal trends of leaf water potential in transpiring wheat. *Journal of Applied Ecology* 15, 613–626.

Jones, M.B. (1988) Water relations. In: Jones, M.B. and Lazenby, A. (eds) *The Grass Crop*. Chapman and Hall, London, pp. 205–242.

Jones, H.G. (1992) *Plants and Microclimate*. Cambridge University Press, Cambridge, 428 pp.

Jones, M.B. and Jongen, M. (1996) Sensitivity of temperate grassland species to elevated CO_2 and the interaction with temperature and water stress. *Agricultural and Food Science in Finland* 5, 271–283.

Jones, M.B. and Lazenby, A. (eds) (1988) *The Grass Crop*. Chapman and Hall, London, 369 pp.

Jones, M.B., Leafe, E.L. and Styles, W. (1980) Water stress in field-grown perennial ryegrass. I. Its effect on leaf water status, stomatal resistance and leaf morphology. *Annals of Applied Biology* 96, 87–101.

Jones, M.B., Ashenden, T.W., Raschi, A., Parsons, A.J., Payer, H.D. and Thornley, J.H.M. (1994) *An Investigation of the Effect of Increasing Concentrations of Atmospheric CO_2 and Changing Climate on Natural and Managed Grassland Communities in Europe Using Open-top and Closed Chambers*. Final report of contract no. EPOC-CT90–0022.

Jones, M.B., Jongen, M. and Doyle, T. (1996) Effects of elevated carbon dioxide concentrations on agricultural grassland production. *Agricultural and Forest Meteorology* 79, 243–252.

Kameli, A. and Lösel, D.M. (1993) Carbohydrates and water status in wheat plants under water stress. *New Phytologist* 125, 609–614.

Keith, H., Oades, J.M. and Martin, J.K. (1986) Input of carbon to soil from wheat plants. *Soil Biology and Biochemistry* 18, 445–449.

Kelliher, F.M., Leuning, R., Raupach, M.R. and Schulze, E.-D. (1995) Maximum conductances for evaporation from global vegetation types. *Agricultural and Forest Meteorology* 73, 1–16.

Kirchmann, H. (1991) Carbon and nitrogen mineralization of fresh, aerobic and anaerobic farm animal manures during incubation with soil. *Swedish Journal of Agricultural Research* 21, 165–173.

Kirschbaum, M.U.M. and Farquhar, G.D. (1984) Temperature dependence of whole-leaf photosynthesis in *Eucalyptus pauciflora* Sieb. ex Spreng. *Australian Journal of Plant Physiology* 11, 519–538.

Kramer, P.J. (1983) *Water Relations of Plants*. Academic Press, New York, 489 pp.

Lawlor, D.W. (1987) *Photosynthesis: Metabolism, Control and Physiology*. Longman, Harlow, Essex, 262 pp.

Leaver, J.D. (1985) Milk-production from grazed temperate grassland. *Journal of Dairy Research* 52, 313–344.

Leuning, R. (1995) A critical appraisal of a combined stomatal-photosynthesis model for C_3 plants. *Plant, Cell and Environment* 18, 339–355.

Leuning, R., Condon, A.G., Dunin, F.X., Zegelin, S. and Denmead, O.T. (1994) Rainfall interception and evaporation from soil below a wheat canopy. *Agricultural and Forest Meteorology* 67, 221–238.

Long, S.P. (1991) Modification of the response of photosynthetic productivity to rising temperature by atmospheric CO_2 concentrations: has its importance been underestimated? *Plant, Cell and Environment* 14, 729–739.

Loomis, R.S. and Connor, D.J. (1992) *Crop Ecology*. Cambridge University Press, Cambridge, 538 pp.

Ludlow, M.M. (1980) Adaptive significance of stomatal response to water stress. In: Turner, M.C., Kramer, P.J. (eds) *Adaption of Plants to Water and High Temperature*. Wiley, New York, pp. 123–138.

Ludlow, M.M. (1983) External factors influencing photosynthesis and respiration. In: Dale, J.E., Milthorpe, F.L. (eds) *The Growth and Function of Leaves*. Cambridge University Press, Cambridge, pp. 347–380.

Ludlow, M.M. (1987) Defining root water status in the most meaningful way to relate to physiological processes. In: Hanks, R.J., Brown, R.W. (eds) *Measurement of Soil and Plant Water Status*. Utah Agricultural Experiment Station, Logan, UT, USA, pp. 47–53.

Lundgren, B. and Söderström, B. (1983) Bacterial numbers in a pine forest soil in relation to environmental factors. *Soil Biology and Biochemistry* 15, 625–630.

Lutze, J.L and Gifford, R.M. (1995) Carbon storage and productivity of a carbon dioxide enriched nitrogen limited grass sward after one year's growth. *Journal of Biogeography* 22, 227–233.

Mäkelä, A.A. and Sievänen, R.P. (1987) Comparison of two shoot–root partitioning models with respect to substrate utilization and functional balance. *Annals of Botany* 59, 129–140.

MAFF (Ministry of Agriculture, Fisheries and Food) (1976) *The Agricultural Climate of England and Wales*. Technical Bulletin No. 35. HMSO, London, 147 pp.

Maron, S.H. and Prutton, C.F. (1965) *Principles of Physical Chemistry*. Macmillan, New York, 886 pp.

Martin, J.K. (1977) Factors affecting the influence of loss of organic carbon from wheat roots. *Soil Biology and Biochemistry* 9, 1–7.

Martínez-Lozano, J.A., Tena, F., Onrubia, J.E. and De la Rubia, J. (1984) The historical evolution of the Ångström formula and its modifications: review and bibliography. *Agricultural and Forest Meteorology* 33, 109–128.

May, R.M. (1976) Simple mathematical models with very complicated dynamics. *Nature* 261, 459–467.

McCree, K.J. (1970) An equation for the respiration of white clover plants grown under controlled conditions. In: Setlik, I. (ed.) *Prediction and Measurement of Photosynthetic Productivity*. Pudoc, Wageningen, pp. 221–229.

McDonald, P. (1981) *The Biochemistry of Silage*. Wiley, Chichester.

McMurtrie, R.E. and Wang, Y-P. (1993) Mathematical models of the photosynthetic response of tree stands to rising CO_2 concentrations and temperatures. *Plant, Cell and Environment* 16, 1–13.

Medlyn, B.E. (1996) Interactive effects of atmospheric carbon dioxide and leaf nitrogen concentration on canopy light use efficiency: a modeling analysis. *Tree Physiology* 16, 201–209.

Meteorological Office (1982) *Tables of Temperature, Relative Humidity, Precipitation and Sunshine for the World. Part III Europe and the Azores*. Met.O.856c. UDC 551.582.2(4). Meteorological Office, Bracknell, Berkshire, UK.

Mitchell and Gauthier Associates (MGA), Inc. (1993) *Advanced Continuous Simulation Language (ACSL) Reference Manual*. 200 Baker Avenue, Concord, MA 01742–2100, USA. Tel: (800) 647–2275. Fax: (508) 369–0013. E-mail: software@mga.com WWW: www.mga.com

Monsi, M. and Saeki, T. (1953) Uber den Lichtfaktor in den Pflanzengesellschaften und seine Bedeutung für die Stoffproduktion. *Japanese Journal of Botany* 14, 22–52.

Monteith, J.L. (1965) Evaporation and the environment. *Symposium of the Society of Experimental Biology* 19, 205–234.

Morison, J.I.L. (1985) Intercellular CO_2 concentration and stomatal responses to CO_2. In: Zeiger, E., Cowan, I., Farquhar, G.D. (eds) *Stomatal Function*. University Press, Stanford, pp. 229–251.

Morris, R.M. (1969) The pattern of grazing in 'continuously' grazed swards. *Journal of the British Grassland Society* 24, 65–71.

Murray, J.D. (1977) *Lectures on Nonlinear-Differential-Equation Models in Biology*. Oxford University Press, Oxford, 370 pp.

Newman, J.A., Parsons, A.J., Thornley, J.H.M. Penning, P.D. and Krebs, J.R. (1995) Optimal diet selection by a generalist grazing herbivore. *Functional Ecology* 9, 255–268.

Newton, P.C.D. (1991) Direct effects of increasing carbon dioxide on pasture plants and communities. *New Zealand Journal of Agricultural Science* 34, 1–24.

Newton, P.C.D., Clark, H., Bell, C.C., Glasgow, E.M., Tate, K.R., Ross, D.J. and Yeates, G.W. (1995) Plant growth and soil processes in temperate grassland communities at elevated CO_2. *Journal of Biogeography* 22, 235–240.

Newton, P.C.D., Clark, H., Bell, C.C. and Glasgow, E.M. (1996) Interaction of soil moisture and elevated CO_2 on the above-ground growth rate, root length density and gas exchange of turves from temperate pasture. *Journal of Experimental Botany* 47, 771–779.

Nobel, P.S. (1991) *Physicochemical and Environmental Plant Physiology*. Academic Press, San Diego, 635 pp.

Nonhebel, S. (1994) The effects of use of average instead of daily weather data in crop growth simulation models. *Agricultural Systems* 44, 377–396.

Noy-Meir, I. (1975) Stability of grazing systems: an application of predator-prey graphs. *Journal of Ecology* 63, 459–481.

Ogink-Hendricks, M.J. (1995) Modelling surface conduction and transpiration of an oak forest in The Netherlands. *Agricultural and Forest Meteorology* 74, 99–118.

Olsen, L.F. and Degn, H. (1985) Chaos in biological systems. *Quarterly Reviews of Biophysics* 18, 161–199.

Orr, R.J., Parsons, A.J., Treacher, T.T. and Penning, P.D. (1988) Seasonal patterns of grass production under cutting or continuous stocking managements. *Grass and Forage Science* 43, 199–207.

Orr, R.J., Penning, P.D., Parsons, A.J. and Champion, R.A. (1995) Herbage intake and N excretion by sheep grazing monocultures or a mixture of grass and white clover. *Grass and Forage Science* 50, 31–40.

Pachepsky, L.B., Haskett, J.D. and Acock, B. (1996) An adequate model of photosynthesis – I parameterization, validation and comparison of models. *Agricultural Systems* 50, 209–225.

Parnas, H. (1975) Model for decomposition of organic matter by microorganisms. *Soil Biology and Biochemistry* 7, 161–169.

Parsons. A.J. (1988) The effects of season and management on the growth of grass swards. In: Jones, M.B. and Lazenby, A. (eds) *The Grass Crop*. Chapman and Hall, London, pp. 129–177.

Parsons. A.J. (1994) Exploiting resource capture – grassland. In: Monteith, J.L., Scott, R.K.

and Unsworth, M.H. (eds) *Resource Capture by Crops*. Nottingham University Press, Nottingham, pp. 315–349.

Parsons, A.J., Leafe, E.L., Collett, B., Penning, P.D. and Lewis, J. (1983) The physiology of grass production under grazing. II. Photosynthesis, crop growth and animal intake of continuously grazed swards. *Journal of Applied Ecology* 20, 127–139.

Parsons, A.J., Johnson, I.R. and Harvey, A. (1988a) Use of a model to optimize the interaction between frequency and severity of intermittent defoliation and to provide a fundamental comparison of the continuous and intermittent defoliation of grass. *Grass and Forage Science* 43, 49–59.

Parsons, A.J., Johnson, I.R. and Williams, J.H.H. (1988b) Leaf age structure and canopy photosynthesis in rotationally and continuously grazed swards. *Grass and Forage Science* 43, 1–14.

Parsons, A.J., Harvey, A. and Woledge, J. (1991a) Plant–animal interactions in a continuously grazed mixture. 1. Difference in the physiology of leaf expansion and the fate of leaves of grass and clover. *Journal of Applied Ecology* 28, 619–634.

Parsons. A.J., Orr, R.J., Penning, P.D., Lockyer, D.R. and Ryden, J.C. (1991b) Uptake, cycling and fate of nitrogen in grass–clover swards continuously grazed by sheep. *Journal of Agricultural Science, Cambridge* 116, 47–61.

Parsons, A.J., Thornley, J.H.M., Newman, J.A. and Penning, P.D. (1994) A mechanistic model of foraging and diet selection in a two species sward. *Functional Ecology* 8, 187–204.

Parton, W.J., Schimel, D.S., Cole, C.V. and Ojima, D.S., (1987) Analysis of factors controlling soil organic matter levels in Great Plain grasslands. *Soil Society of America Journal* 51, 1173–1179.

Parton, W.J., Scurlock, J.M.O., Ojima, D.S., Gilmanov, T.G., Scoles, R.J., Schimel, D.S., Kirchner, T., Menaut, J-C., Seastedt, T., Garcia Moya, E., Apinan Kamnalrut and Kinyamario, J.I. (1993) Observations and modeling of biomass and soil organic matter dynamics for the grassland biome worldwide. *Global Biogeochemical Cycles* 7, 785–809.

Parton, W.J., Mosier, A.R., Ojima, D.S., Valentine, D.W., Schimel, D.S., Weir, K. and Kulmala, A.E. (1996) Generalized model for N_2 and N_2O production from nitrification and denitrification. *Global Biogeochemical Cycles* 10, 401–412.

Passioura, J.B. (1984) Hydraulic resistance of plants I. constant or variable? *Australian Journal of Plant Physiology* 11, 333–339.

Passioura, J.B. (1994) The physical chemistry of the primary cell wall: implications for the control of expansion rate. *Journal of Experimental Botany* 45, 1675–1682.

Penman, H.L. (1948) Natural evaporation from open water, bare soil and grass. *Proceedings of the Royal Society of London, Series A* 193, 120–145.

Penning, P.D., Parsons, A.J., Orr, R.J. and Treacher, T.T. (1991) Intake and behaviour responses by sheep to changes in sward characteristics under continuous stocking. *Grass and Forage Science* 46, 15–28.

Penning, P.D., Parsons, A.J., Orr, R.J., Harvey, A. and Champion, R.A (1995) Intake and behaviour responses by sheep, in different physiological states, when grazing monocultures of grass or white clover. *Applied Animal Behaviour Science* 45, 63–78.

Penning de Vries, F.W.T. (1975) The cost of maintenance processes in plant cells. *Annals of Botany* 39, 77–82.

Penning de Vries, F.W.T., Brunsting, A.H.M. and Van Laar, H.H. (1974) Products, requirements and efficiency of biosynthesis: a quantitative approach. *Journal of Theoretical Biology* 45, 339–377.

Pettersson, R. and McDonald, J.S. (1994) Effects of nitrogen supply on the acclimation of photosynthesis to elevated CO_2. *Photosynthesis Research* 39, 389–400.

Pirt, S.J. (1975) *Principles of Microbe and Cell Cultivation*. Blackwell, Oxford, 274 pp.

Polley, H.W., Johnson, H.B., Marino, B.D. and Mayeux, H.S. (1993) Increase in C3 plant water-use efficiency and biomass over glacial to present CO_2 concentrations. *Nature* 361, 61–64.

Pollock, C.J. and Jones, T. (1979) Seasonal patterns of fructosan metabolism in forage grasses. *New Phytologist* 83, 9–15.

Pool, R. (1989) Is it chaos, or is it just noise? *Science* 243, 25–28.

Prioul, J-L., Brangeon, J. and Reyss, A. (1980) Interaction between external and internal conditions in the development of photosynthetic features in a grass leaf. I. Regional responses along a leaf during and after low-light or high-light acclimation. *Plant Physiology* 86, 762–769.

Rabinowitch, E.I. (1951) *Photosynthesis and Related Processes* 2 (1). Interscience, New York, pp. 603–1208.

Robson, M.J. (1973) The growth and development of simulated swards of perennial ryegrass. I. Leaf growth and dry weight changes as related to ceiling yield of a seedling sward. *Annals of Botany* 37, 487–500.

Robson, M.J. and Sheehy, J.E. (1981) Leaf area and light interception. In: Hodgson, J., Baker, R.D., Davies, A., Laidlaw, A.S. and Leaver, J.D. (eds) *Sward Measurement Handbook*. British Grassland Society, Hurley, pp. 115–139.

Robson, M.J., Ryle, G.J.A. and Woledge, J. (1988) The grass plant – its form and function. In: Jones, M.B. and Lazenby, A. (eds) *The Grass Crop*. Chapman and Hall, London, pp. 25–83.

Robson, M.J., Parsons, A.J. and Williams, T.E. (1989) Herbage production: grasses and legumes. In: Holmes, W. (ed.) *Grass: its Production and Utilization*. Blackwell, Oxford, pp. 7–88.

Rogers, H.H., Runion, G.B. and Krupa, S.V. (1994) Plant responses to atmospheric CO_2 enrichment with emphasis on roots and the rhizosphere. *Environmental Pollution* 83, 155–189.

Rowell, D.L. (1994) *Soil Science: Methods and Applications*. Longman, Harlow, Essex, 350 pp.

Royal Society (1975) *Quantities, Units and Symbols*. Symbols Committee, Royal Society of London, London, UK, 54 pp.

Russo, D., Jury, W.A. and Butters, G.L. (1989) Numerical analysis of solute transport during transient irrigation. 2. The effect of immobile water. *Water Resources Research* 10, 2119–2127.

Ryan, M.G. (1995) Foliar maintenance respiration of subalpine and boreal trees and shrubs in relation to nitrogen content. *Plant, Cell and Environment* 18, 765–772.

Ryden, J.C. (1983) Denitrification loss from a grassland soil in the field receiving different rates of nitrogen as ammonium nitrate. *Journal of Soil Science* 34, 355–365.

Sands, P.J. (1995) Modelling canopy production. II. From single-leaf photosynthetic parameters to daily canopy photosynthesis. *Australian Journal of Plant Physiology* 22, 603–614.

Schimel, D.S., Braswell, B.H., Holland, E.A., McKeown, R., Ojima, D.S., Painter, T.H., Parton, W.J. and Townsend, A.R. (1994) Climatic, edaphic, and biotic controls over storage and turnover of carbon in soils. *Global Biogeochemical Cycles* 8, 279–293.

Schuster, W.S. and Monson, R.K. (1990) An examination of the advantages of C_3-C_4 intermediate photosynthesis in warm environments. *Plant, Cell and Environment* 13, 903–912.

Schwinning, S. and Parsons, A.J. (1996a) Analysis of the coexistence mechanisms for grasses and legumes in grazing systems. *Journal of Ecology* 84, 799–813.

Schwinning, S. and Parsons, A.J. (1996b) A spatially explicit population model of stoloniferous N-fixing legumes in mixed pasture with grass. *Journal of Ecology* 84, 815–826.

Sheehy, J.E. (1989) How much dinitrogen fixation is required in grazed grassland? *Annals of Botany* 64, 159–161.

Silk, W.K. and Erickson, R.O. (1979) Kinematics of plant growth. *Journal of Theoretical Biology* 76, 481–501.

Sims, P.L. and Singh, J.S. (1978) The structure and function of ten western North American grasslands. III. Net primary production, turnover and efficiencies of energy capture and water use. *Journal of Ecology* 66, 573–597.

Sinell, H.J. (1980) Interacting factors affecting mixed populations. In: Silliker, J.H., Elliott, R.P., Baird-Parker, A.C., Bryan, F.L., Christian, J.H.B., Clark, D., Olson, J.C. and Roberts, T.A. (eds) *Microbiology of Foods*, Vol. I. Academic Press, New York, pp. 215–231.

Smith, L.P. (1984) *The Agricultural Climate of England and Wales*. MAFF Reference Book 435. HMSO, London.

Smucker, A.J.M. (1984) Carbon utilization and losses by plant root systems. In: *Roots, Nutrient and Water Influx*. American Society of Agronomy, Madison.

Soussana, J.F., Casella, E. and Loiseau, P. (1996) Long-term effects of CO_2 enrichment and temperature increase on a temperate grass sward. II. Plant nitrogen budgets and root fraction. *Plant and Soil* 182, 101–114.

Stevens, P.A., Hornung, M. and Hughes, S. (1989) Solute concentrations, fluxes and major nutrient cycles in a mature Sitka-spruce plantation in Beddgelert Forest, North Wales. *Forest Ecology and Management* 27, 1–20.

Stitt, M. (1991) Rising CO_2 levels and their potential significance for carbon flow in photosynthetic cells. *Plant, Cell and Environment* 14, 741–762.

Stuff, R.G. and Dale, R.F. (1973) A simple method of calendar conversion in computer applications. *Agricultural Meteorology* 12, 441–442.

Svenning, M.M., Junttila, O. and Macduff, J.H. (1996) Differential rates of inhibition of N_2 fixation by sustained low concentrations of NH_4^+ and NO_3^- in northern ecotypes of white clover (*Trifolium repens* L.) *Journal of Experimental Botany* 47, 729–738.

Talsma, T. (1985) Prediction of hydraulic conductivity from soil water retention data. *Soil Science* 140, 184–188.

Tang, P.S. and Wang, J.S. (1941) A thermodynamic formulation of the water relations in an isolated plant cell. *Journal of Physical Chemistry* 45, 443–453.

Tate, R.L. (1987) *Soil Organic Matter: Biological and Ecological Effects*. Wiley, New York, 291 pp.

Taylor, H.M. and Klepper, B. (1978) The role of rooting characteristics in the supply of water to plants. *Advances in Agronomy* 30, 99–128.

Thompson, D.W. (1942) *On Growth and Form*. Cambridge University Press, Cambridge.

Thornley, J.H.M. (1970) Respiration, maintenance and growth in plants. *Nature (London)* 227, 304–305.

Thornley, J.H.M. (1972) A balanced quantitative model for root:shoot ratios in vegetative plants. *Annals of Botany* 36, 431–441.

Thornley, J.H.M. (1991a) A model of leaf tissue growth, acclimation and senescence. *Annals of Botany* 67, 219–228.

Thornley, J.H.M. (1991b) A transport-resistance model of forest growth and partitioning. *Annals of Botany* 68, 211–226.

Thornley, J.H.M. (1995) Shoot:root allocation with respect to C, N and P: an investigation and comparison of resistance and teleonomic models. *Annals of Botany* 75, 391–405.

Thornley, J.H.M. (1996) Modelling water in crops and plant ecosystems. *Annals of Botany* 77, 261–275.

Thornley, J.H.M. (1997) Modelling allocation with transport/conversion processes. *Silva Fennica* 31, 341–355.

Thornley, J.H.M. and Cannell, M.G.R. (1992) Nitrogen relations in a forest plantation-soil organic matter ecosystem model. *Annals of Botany* 70, 137–151.

Thornley, J.H.M. and Cannell, M.G.R. (1996) Temperate forest responses to carbon dioxide, temperature and nitrogen: a model analysis. *Plant, Cell and Environment* 19, 1331–1348.

Thornley, J.H.M. and Cannell, M.G.R. (1997) Temperate grassland responses to climate change: an analysis using the Hurley pasture model. *Annals of Botany* 80, 205–221.

Thornley, J.H.M. and Johnson, I.R. (1990) *Plant and Crop Modelling*. Oxford University Press, Oxford, 669 pp.

Thornley, J.H.M. and Johnson, I.R. (1991) Making mountains out of moles? *Chemistry in Britain* 27, 427.

Thornley, J.H.M. and Verberne, E.L.J. (1989) A model of nitrogen flows in grassland. *Plant, Cell and Environment* 12, 863–886.

Thornley, J.H.M., Bergelson, J. and Parsons, A.J. (1995) Complex dynamics in a carbon–nitrogen model of a grass–legume pasture. *Annals of Botany* 75, 79–94.

Tilman, D. and Wedlin, D. (1991) Oscillations and chaos in the dynamics of a perennial grass. *Nature* 353, 643–655.

Troughton, A. (1981) Length of life of grass roots. *Grass and Forage Science* 36, 117–120.

Usher, M.B. (1970) An algorithm for estimating the length and direction of shadows with reference to the shadows of shelter belts. *Journal of Applied Ecology* 7, 141–145.

Vaadia, Y. (1960) Autonomic diurnal fluctuations in rate of exudation and root pressure in decapitated sunflower plants. *Physiologia Plantarum* 13, 701–717.

van Bavel, C.H.M. (1974) Soil water potential and plant behaviour: a case modeling study with sunflower. *Oecologia Plantarum* 9, 89–109.

van Genuchten, M. Th. (1980) A closed-form equation for predicting the hydraulic conductivity of unsaturated soils. *Soil Science Society of America Journal* 44, 892–98.

van Veen, J.A. and Paul, E.A. (1981) Organic carbon dynamics in grassland soils. I. Background information and computer simulation. *Canadian Journal of Soil Science* 61, 185–201.

van Volkenburgh, E., Hunt, S. and Davies, W.J. (1983) A simple instrument for measuring cell-wall extensibility. *Annals of Botany* 51, 669–672.

Vaughn, C.E., Center, D.M. and Jones, M.B. (1986) Seasonal fluctuations in nutrient availability in some northern Californian annual range soils. *Soil Science* 141, 43–51.

Verberne, E. (1992) Simulation of the nitrogen and water balance in a system of grassland and soil. *Nota 258*. DLO-Institut voor Bodemvruchtbaarheid, Oosterweg 92, Postbus 30003, 9750 RA HAREN, The Netherlands.

Verberne, E.L.J., Hassink, J., De Willigen, P., Groot, J.J.R. and Van Veen, J.A. (1990) Modelling organic matter dynamics in different soils. *Netherland Journal of Agricultural Science* 38, 221–238.

Volenec, J.J. and Nelson, C.J. (1982) Diurnal leaf elongation of contrasting fescue genotypes. *Crop Science* 22, 531–535.

Whitehead, D.C. (1995) *Grassland Nitrogen*. CAB International, Wallingford, Oxon, 397 pp.

Wild, A., Jones, L.H.P. and Macduff, J.H. (1987) Uptake of mineral nutrients and crop growth: the use of flowing nutrient solutions. *Advances in Agronomy* 41, 171–219.

Wilson, J.B. (1988) A review of evidence of the control of shoot:root ratio, in relation to models. *Annals of Botany* 61, 433–449.

Wilson, G.V., Jardine, P.M. and Gwo, J.P. (1992) Modelling the hydraulic-properties of a multiregion soil. *Soil Science Society of America Journal* 56, 1731–1737.

Woledge, J. and Pearse, J.P. (1985) The effect of nitrogenous fertilizer on the photosynthesis of leaves of a ryegrass sward. *Grass and Forage Science* 40, 305–309.

Wolfenden, J. and Diggle, P.J. (1995) Canopy gas exchange and growth of upland pasture swards in elevated CO_2. *New Phytologist* 130, 369–380.

Zur, B. and Jones, J.W. (1981) A model for the water relations, photosynthesis, and expansive growth of crops. *Water Resources Research* 17, 311–320.

Index